# MIMO Communications - Fundamental Theory, Propagation Channels, and Antenna Systems

*Edited by Ahmed A. Kishk and Xiaoming Chen*

Published in London, United Kingdom

MIMO Communications - Fundamental Theory, Propagation Channels, and Antenna Systems
http://dx.doi.org/10.5772/intechopen.110927
Edited by Ahmed A. Kishk and Xiaoming Chen

Contributors
Yu Luo, Xiaoxuan Guo, Kasturi Vasudevan, Surendra Kota, Lov Kumar, Himanshu Bhusan Mishra, Jiguang He, Chongwen Huang, Li Wei, Yuan Xu, Merouane Debbah, Ahmed AlHammadi, Ali Araghi, Mohsen Khalily, Kun Chen-Hu, Manuel José López Morale, Ana García Armada, Yiying Wang, Kun Yang, Xin Li, Motoyuki Sato, Yunlong Cai, Qiyu Hu, Guangyi Zhang, Kai Kang, Mohammad Reza Soleymani, Sareh Majidi Ivari, Yousef R. Shayan, Faisal Al-Kamali, Francois Chan, Mohamed Alouzi, Claude D'Amours, Yan Wang, Xiaoxue Fan, Cecil Bruce Boye, Andreas Fechner, Moritz Krebbel, Robert Luyken, Telse David, Thomas Eisenbarth, Katarina Vuckovic, Nazanin Rahnavard, Ang Li, Haonan Wang



First published in London, United Kingdom, 2023 by IntechOpen
IntechOpen is the global imprint of INTECHOPEN LIMITED, registered in England and Wales, registration number: 11086078, 5 Princes Gate Court, London, SW7 2QJ, United Kingdom

British Library Cataloguing-in-Publication Data
A catalogue record for this book is available from the British Library

Additional hard and PDF copies can be obtained from orders@intechopen.com

MIMO Communications - Fundamental Theory, Propagation Channels, and Antenna Systems
Edited by Ahmed A. Kishk and Xiaoming Chen
p. cm.
Print ISBN 978-1-83768-999-6
Online ISBN 978-1-83769-000-8
eBook (PDF) ISBN 978-1-83769-001-5

For EU product safety concerns:
IN TECH d.o.o., Prolaz Marije Krucifikse Kozulić 3, 51000 Rijeka, Croatia,
info@intechopen.com or visit our website at intechopen.com.

# Meet the editors

Prof. Ahmed A. Kishk received a BSc in Electronics and Communication Engineering from Cairo University, Egypt, in 1977. He obtained an MEng and Ph.D. from the University of Manitoba, Canada, in 1983 and 1986, respectively. In 1986, he joined the Department of Electrical Engineering, University of Mississippi, USA as an assistant professor, becoming a full professor in 1995. He is currently a professor at Concordia University, Canada, and Canada Research Chair in Advanced Antenna Systems. He was a distinguished lecturer for the Antennas and Propagation Society in 2013–2015. He was previously an editor of *IEEE Antennas and Propagation Magazine*, editor-in-chief of the ACES Journal, and an editorial board member of several other journals. He was also a technical program committee member for several international conferences, a member of AP-S AdCom, and president of AP-S.

Xiaoming Chen received a BSc in Electrical Engineering from Northwestern Polytechnical University, Xi'an, China, in 2006 and an MSc and Ph.D. in Electrical Engineering from Chalmers University of Technology, Gothenburg, Sweden, in 2007 and 2012, respectively. From 2013 to 2014, he was a postdoctoral researcher at the same university. From 2014 to 2017, he was with Qamcom Research & Technology AB, Gothenburg, Sweden. Since 2017, he has been a professor at Xi'an Jiaotong University, Xi'an, China. His research areas include MIMO antennas, over-the-air testing, and reverberation chambers. He has published more than 150 journal articles on these topics. Prof. Chen is a senior associate editor for *IEEE Antennas and Wireless Propagation Letters* and an associate editor for *IEEE Transactions on Antennas and Propagation*. He was the general chair of the IEEE International Conference on Electronic Information and Communication Technology (ICEICT) in 2021. He won the first prize in universities' scientific research results in Shaanxi province, China, in 2022. He received the IEEE Outstanding Associate Editor awards six times from 2018 to 2023 and the URSI (International Union of Radio Science) Young Scientist Award in 2017 and 2018.

# Contents

# Preface

We are pleased to present this edited volume, *MIMO Communications – Fundamental Theory, Propagation Channels, and Antenna Systems*. This book introduces and discusses the latest research and developments in MIMO communications, serving as a comprehensive resource for scholars, researchers, industry professionals, and engineers.

The volume is divided into four parts: "Wireless Communications", "Antenna Techniques", "Channel Modeling", and "Autonomous Driving and Radars". Each part features contributions from esteemed experts in the field, covering a wide range of topics, including capacity analysis of MIMO channels, beamforming and antenna array design, channel modeling and estimation, and the applications of autonomous driving and radars.

Section 1, "Wireless Communications", lays out the fundamental theory of MIMO communications, providing a detailed introduction to the key concepts and techniques involved in designing and analyzing MIMO systems.

Section 2, "Antenna Techniques", discusses various types of antenna arrays, radiation patterns, and design techniques, and covers critical aspects of MIMO communication systems such as antenna selection and beamforming.

Section 3, "Channel Modeling", examines the various aspects of MIMO channels, including channel models, multipath fading, spatial correlation, and more advanced topics such as channel estimation and feedback.

Section 4, "Autonomous Driving and Radars", provides a comprehensive overview of the field of autonomous driving, including the history, challenges, and applications of this technology, and covers the basics of radar, including its principle, design, and implementation.

We want to take this opportunity to acknowledge the authors' hard work and contributions and the support of the editorial team. We hope this volume will serve as a valuable resource for scholars, researchers, and industry professionals and that the insights and perspectives presented in this volume will help advance the MIMO communication fields.

**Ahmed A. Kishk**
Concordia University,
Montreal, Canada

**Xiaoming Chen**
Xi'an Jiaotong University,
Xi'an, China

Section 1

# Wireless Communications

Chapter 1

# Architectures for Hybrid Precoding and Combining Techniques in Massive MIMO Systems Operating in the mmWave Band

*Faisal Al-Kamali, Mohamed Alouzi, Claude D'Amours and Francois Chan*

## Abstract

Hybrid precoding and combining techniques in millimeter-wave (mmWave) multiple-input multiple-output (MIMO) systems with various array architectures have attracted significant interest as a promising technology for the development of 6G wireless communication systems. This approach presents numerous advantages, including reduced complexity, cost, and power consumption, when compared to traditional analog precoding methods. In this chapter, we investigate hybrid precoding and combining techniques for massive MIMO systems operating in the millimeter-wave (mmWave) band, with a focus on different architectures, such as full array (FA), subarray (SA), and hybrid array (HA) architectures. We discuss the system model of each architecture. Additionally, we solve the hybrid precoding and combining optimization problem to maximize the spectral efficiency of each architecture. We then propose iterative hybrid precoding and combining algorithms for all architectures, as well as compare their performance to that of traditional hybrid design methods to demonstrate that the proposed algorithms achieve superior performance with lower complexity and hardware requirements.

**Keywords:** full array architecture, subarray architecture, overlapped architecture, hybrid precoding and combining, massive MIMO systems

## 1. Introduction

The use of millimeter-wave (mmWave) communications has become increasingly popular as a potential solution for current and upcoming cellular systems, mainly due to the extensive yet underutilized mmWave frequency range [1]. To ensure an adequate link margin and achieve array gain, massive multiple-input multiple-output (MIMO) antenna arrays are required for mmWave systems [2]. The use of traditional analog precoding and combining schemes in mmWave MIMO systems is not practical due to the high hardware cost and power consumption of the radio frequency (RF) chains. In light of this, hybrid precoding and combining schemes are viewed as a promising technology that can strike a balance between system performance and hardware complexity. There are two primary hybrid precoding and combining architectures used in millimeter-wave systems: full array (FA) [2–15] and subarray (SA) [16–23] architectures. The FA architecture is commonly employed in hybrid precoding and combining systems. With this architecture, phase shifters (PSs) connect each RF chain to each antenna, leading to a linear increase in the number of PSs with the number of antennas. In contrast, the SA architecture connects each RF chain to a subset of antennas, requiring fewer PSs than the FA architecture.

In the literature, FA hybrid architecture for mmWave systems has received significant attention. The authors of [2] proposed a hybrid precoding/combining algorithm based on simultaneous orthogonal matching pursuit, achieving performance comparable to that of optimal digital beamforming with high complexity. In [4], we introduced an iterative low-complexity hybrid design algorithm based on gradient descent. The work in [5] proposed hybrid designs for Mini-Mental State Examination (MMSE)-based rate balancing in mmWave multiuser MIMO systems and the work in [7] proposed joint hybrid precoding and combining for massive MIMO systems. In [8], a greedy approach is introduced without assumptions about channel structure or array geometry. The work in [9] presented the hybrid design by alternating minimization (HD-AM) algorithm, which achieves high spectral efficiency but is limited to equal numbers of data streams and RF chains. Manifold optimization-based hybrid precoding algorithm in [10] achieves high spectral efficiency but with high computation complexity. Sohrabi and Yu in [11] proposed a heuristic hybrid beamforming algorithm, while the authors of [12, 13] developed gradient projection algorithms for hybrid beamforming design. In multiuser scenarios [14, 15], digital beamforming removes interuser interference, and the analog precoders and combiners maximize user signal power.

Although FA hybrid architecture led to the lower complexity of hybrid precoding and combining algorithms compared to the analog one, the high cost, power consumption, and hardware complexity of this architecture persist due to the need for a phase shifter (PS) to connect each RF chain to every antenna [16, 17]. To address these challenges, the SA architecture has gained popularity as a practical solution for hybrid precoding and combining designs that offer a balance between performance, complexity, and cost. SA architectures for hybrid precoding can be classified as fixed SA [16–18], adaptive SA [19], and dynamic SA [20, 21]. In fixed SA, each RF chain is connected to a subarray of antennas, while switches are used in dynamic SA. Dynamic SA achieves similar performance as FA with high complexity as compared to the fixed SA. Overlapped SA architecture with hybrid precoding can improve the spectral efficiency of the SA architecture and still lower the complexity compared to the FA architecture [18]. A study in [16] presented an energy-efficient hybrid precoding technique for the fixed SA architecture. The technique utilized successive interference cancelation and assumed a diagonal digital precoder with real elements. Two low-complexity hybrid precoding algorithms for mmWave MIMO systems with fixed SA architecture were proposed and studied in [17]. In [18], we proposed and highlighted the use of overlapped SA architecture for improved spectral efficiency. An adaptive hybrid precoding approach for SA architecture was studied in [19]. Dynamic SA architectures in [20, 21] provided higher spectral efficiency but with increased complexity. In [20, 21], it is found that the dynamic SA architectures perform better than fixed SA architectures, but with higher hardware complexity and power consumption due to the linear increase in the switches with the number of transmit antennas. To reduce the complexity of the dynamic SA, the authors of [22, 23] proposed partially SA structures. However, the partially dynamic precoders in [22, 23] still result in greater computational and hardware complexities, as well as higher power consumption, compared to fixed SA precoders. Recently, deep learning-based hybrid designs have been explored in [24–26]. A new hybrid design approach for SA was studied in [27]. In [27], an iterative algorithm that begins by designing a hybrid precoding and combining matrix for the FA structure and then converts it into a SA matrix by setting certain entries to zero while achieving better performance was proposed and studied.

While the cost and hardware complexity of hybrid precoding and combining for SA architecture are lower than those for the FA architecture, the spectral efficiency achieved through SA architectures is still inferior to that of optimal digital precoding and combining [17]. Therefore, proposing a new hybrid array architecture that balances spectral efficiency, cost, and power consumption is an essential topic. In this chapter, we introduce a new HA architecture for mmWave MIMO systems that aims to achieve a balance between spectral efficiency, cost, and power consumption. Initially, the antennas at the transmitter/receiver are partitioned into subarrays, each containing the same number of antennas as the number of RF chains at the transmitter/receiver, and then divided into nonoverlapping subsets called groups. Finally, the antennas in each group are connected to a group of RF chains in a way similar to the connections in the FA architecture.

The main contributions of this chapter are summarized as follows:

- The FA, SA, and HA architectures' system models for mmWave MIMO communication systems are derived and explained. The FA architecture employs PSs to connect each RF chain to all antennas. The SA architecture links each RF chain only to a subset of antennas in a subarray. In contrast, the HA architecture divides the antennas at both the transmitter and receiver into a set of subarrays, which is equivalent to the number of RF chains. The resulting subarrays are divided into groups that do not overlap, and each group's antennas are linked to a group of RF chains in the same manner as in the FA architecture.

- The optimization problems for hybrid precoding in the FA, SA, and HA architectures are formulated and solved. In the FA architecture, the hybrid precoding optimization problem for the entire system is solved. For the SA architecture, the hybrid precoding optimization problem for each subarray is independently solved. In the HA architecture, each group's optimization problem is independently solved.

- New iterative algorithms for hybrid precoding and combining are proposed and derived for the FA, SA, and HA architectures. The design derivation takes into account the block structure of the analog precoding/combining matrices in each architecture without relying on any other assumptions or the antenna array geometry. The proposed iterative FA (IFA) algorithm for the FA architecture iteratively determines the hybrid precoding and combining for the entire system. However, for the SA architecture, the proposed iterative SA (ISA) algorithm determines the hybrid precoding and combining for each subarray independently and then for the entire system. In the HA architecture, the proposed iterative HA (IHA) algorithm determines the hybrid precoding and combining for each group independently and then for the entire system.

- The complexities of the proposed algorithms are derived, discussed, and compared to show the simplicity of the proposed algorithms as compared with other existing algorithms.

- Simulations were used to evaluate the proposed algorithms for mmWave MIMO systems with FA, SA, and HA architectures. The results indicate that these algorithms can enhance the mmWave MIMO system's performance and provide a high level of spectral efficiency.

## 2. The mmWave channel model

In this section, the mmWave channel model is discussed. **H** can be written as [2, 4, 17].

$$\mathbf{H} = \sqrt{N_t N_r / N_{cl} N_{ray}} \times \sum_{i}^{N_{cl}} \sum_{l}^{N_{ray}} \left[ \alpha_{il} \boldsymbol{\Lambda}_r \left( \phi_{il}^r, \theta_{il}^r \right) \times \boldsymbol{\Lambda}_t \left( \phi_{il}^t, \theta_{il}^t \right) \mathbf{a}_r \left( \phi_{il}^r, \theta_{il}^r \right) \mathbf{a}_t \left( \phi_{il}^t, \theta_{il}^t \right)^* \right] \tag{1}$$

where $N_t$ is the number of antennas at the transmitter and $N_r$ is the number of antennas at the receiver. The numbers of clusters and paths are denoted by $N_{cl}$ and $N_{ray}$, respectively. $\alpha_{il}$ is the complex gain of the *lth* path in the *ith* cluster. $\phi_{il}^t \left( \phi_{il}^r \right)$ and $\theta_{il}^t \left( \theta_{il}^r \right)$ are the azimuth (elevation) angles of departure and arrival of the *lth* path in the *ith* cluster, respectively. The transmitter and receiver antenna element gains at their departure and arrival angles are denoted by $\boldsymbol{\Lambda}_t \left( \phi_{il}^t, \theta_{il}^t \right)$ and $\boldsymbol{\Lambda}_r \left( \phi_{il}^r, \theta_{il}^r \right)$, respectively. $\mathbf{a}_t \left( \phi_{il}^t, \theta_{il}^t \right)$ and $\mathbf{a}_r \left( \phi_{il}^r, \theta_{il}^r \right)$ are the antenna array responses at the transmitter and receiver, respectively. The array response vector in a uniform planar array can be defined as [2, 4, 17].

$$\mathbf{a}_{UPA(\phi,\theta)} = \frac{1}{\sqrt{\mathrm{N_t}}} \left[ 1, \ldots, e^{jkd(x sin(\phi) \sin(\theta) + y cos(\theta))}, \ldots, e^{jkd((w-1) \sin(\phi) \sin(\theta) + (h-1) \cos(\theta))} \right]^T \tag{2}$$

where $k = \frac{2\pi}{\lambda}$, $1 \leq x \leq (w-1)$, and $1 \leq y \leq (h-1)$. $d = \frac{\lambda}{2}$, $w$, and $h$ are the interantenna spacing, width, and height of the antenna array, respectively. The transmitter's array size is $N_t = wh$. In this chapter, we assume perfect channel estimation at the transmitter and receiver.

## 3. Full array architecture system model

### 3.1 FA architecture

This subsection presents a discussion on the system model of the FA architecture. First, the baseband digital precoder $\mathbf{P_D}$ is applied to the signal at the transmitter, after which it is precoded by the FA analog precoder $\mathbf{P_{AFA}}$. At the receiver's end, the FA analog combiner $\mathbf{W_{AFA}}$ and the digital combiner $\mathbf{W_D}$ are both applied. The structure

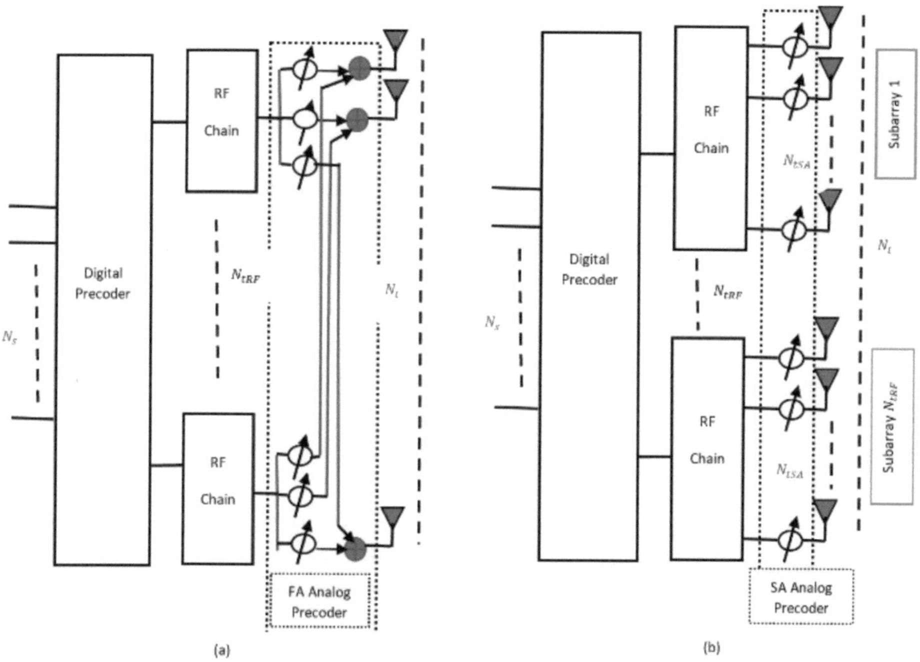


**Figure 1.**
*Hybrid precoding at the base station (BS). (a) Full array (FA) architecture. (b) Subarray (SA) architecture.*

of the hybrid precoding for FA architecture is depicted in **Figure 1(a)**. The received signal in the FA architecture can be expressed as [4].

$$\mathbf{y} = \sqrt{\rho}\mathbf{W_D}^{\mathbf{H}}\mathbf{W_{AFA}}^{\mathbf{H}}\mathbf{H}\mathbf{P_{AFA}}\mathbf{P_D}\mathbf{s} + \mathbf{n} = \sqrt{\rho}\mathbf{W_{FA}}^{\mathbf{H}}\mathbf{H}\mathbf{P_{FA}}\mathbf{s} + \mathbf{n} \tag{3}$$

The channel matrix is represented by $\mathbf{H} \in \mathbb{C}^{N_r \times N_t}$, and $\rho$ denotes the average power of the received signal. $\mathbf{P_{AFA}}$ and $\mathbf{P_D}$ are $N_t \times N_{tRF}$ and the $N_{tRF} \times N_s$ matrices of analog and digital precoding matrices, respectively. Similarly, $\mathbf{W_{AFA}}$ and $\mathbf{W_D}$ are $N_r \times N_{rRF}$ and $N_{rRF} \times N_s$ matrices, respectively, and represent the FA analog and digital combiner matrices. The transmitted signal is represented by the $N_s \times 1$ vector $\mathbf{s}$, with $\mathbb{E}[\mathbf{ss}^*] = \frac{1}{N_s}\mathbf{I}_{N_s}$. The additive white Gaussian noise $\mathbf{n}$ is represented by the $N_s \times 1$ vector of independent and identical distribution. The matrices $\mathbf{P_{FA}} = \mathbf{P_{AFA}}\mathbf{P_D}$ and $\mathbf{W_{FA}} = \mathbf{W_{AFA}}\mathbf{W_D}$. All the elements in $\mathbf{P_{AFA}}$ and $\mathbf{W_{AFA}}$ have a constant amplitude, which is equal to $1/\sqrt{N_t}$ and $1/\sqrt{N_r}$, respectively [4]. The digital precoder and combiner satisfy the total power constraint and are normalized as $\|\mathbf{P_{AFA}}\mathbf{P_D}\|_F^2 = N_s$ and $\|\mathbf{W_{AFA}}\mathbf{W_D}\|_F^2 = N_s$. The spectral efficiency of the FA architecture can be written as [4]

$$R = \log_2\left(\left|\mathbf{I}_{\mathbf{N_r}} + \frac{\rho}{N_s}\mathbf{Q}_{\mathbf{k}}^{-1}\mathbf{W}_{\mathbf{B}}^{\mathbf{k}\,\mathbf{H}}\mathbf{W}_{\mathbf{AFA}}^{\mathbf{k}\;\;\mathbf{H}}\mathbf{H}\mathbf{F_{AFA}}\mathbf{F_B}\mathbf{P}_{\mathbf{B}}^{\mathbf{H}}\mathbf{P}_{\mathbf{AFA}}^{\mathbf{H}}\mathbf{H}_{\mathbf{k}}^{\mathbf{H}}\mathbf{W}_{\mathbf{AFA}}^{\mathbf{k}}\mathbf{W}_{\mathbf{B}}^{\mathbf{k}}\right|\right) \tag{4}$$

where $\mathbf{Q}_k = \sigma_n^2\mathbf{W}_{\mathbf{B}}^{\mathbf{k}\,\mathbf{H}}\mathbf{W}_{\mathbf{AFA}}^{\mathbf{k}\;\;\mathbf{H}}\mathbf{W}_{\mathbf{AFA}}^{\mathbf{k}}\mathbf{W}_{\mathbf{B}}^{\mathbf{k}}$. To optimize the spectral efficiency in (4), it is important to take into account both the total transmitted power constraint and the constraints on $\mathbf{F_{AFA}}$ and $\mathbf{W_{AFA}}$ during the hybrid precoder/combiner design process.

$$\begin{array}{c} \max \quad R \\ \mathbf{P_{AFA}}, \mathbf{P_D}, \mathbf{W_{AFA}}, \mathbf{W_D} \\ \text{st.} \mathbf{P_{AFA}} \in \mathcal{F}_{\mathbf{AFA}} \text{ and } \mathbf{W_{AFA}} \in \mathfrak{J}_{\mathbf{AFA}} \\ \|\mathbf{P_{AFA}} \mathbf{P_D}\|_F^2 = N_s \end{array} \tag{5}$$

where $\mathcal{F}_{\text{AFA}}$ and $\mathfrak{J}_{\text{AFA}}$ contain all possible precoding and combining matrices, respectively, that fulfill the amplitude constraint. The maximization of the objective function of the precoding in Eq. (5) can be expressed in a more concise manner as [4].

$$\begin{array}{c} \left(\mathbf{P}_{\mathbf{AFA}}^{\mathbf{opt}}, \mathbf{P}_{\mathbf{D}}^{\mathbf{opt}}\right) = \quad \arg \quad \min \left\|\mathbf{P}_{\mathbf{FA}}^{\mathbf{opt}} - \mathbf{P_{AFA}} \mathbf{P_D}\right\|_F^2 \\ \mathbf{P_{AFA}}, \quad \mathbf{P_D} \\ \text{st.} \mathbf{P_{AFA}} \in \mathcal{F}_{\mathbf{AFA}}, \\ \|\mathbf{P_{AFA}} \mathbf{P_D}\|_F^2 = N_s \end{array} \tag{6}$$

Clearly, the optimization problem in Eq. (6) is non-convex and finding its optimal solution is challenging. Nonetheless, the optimal unconstrained hybrid precoding can be determined by setting $\mathbf{P}_{\mathbf{FA}}^{\mathbf{opt}}$ equal to $\mathbf{V_1}$, which represents the first $N_s$ column of the matrix $\mathbf{V}$ obtained through singular value decomposition (SVD) of $\mathbf{H}$, i.e., $\mathbf{H} = \mathbf{U\Sigma V^H}$. Similarly, the optimal unconstrained hybrid combiner can be obtained by setting optimal precoding equal to $\mathbf{U_1}$, which represents the first $N_s$ column of the matrix $\mathbf{U}$ [2, 4].

### 3.2 Subarray architecture system model

This subsection presents a system model of the SA architecture. The structure of the hybrid precoding for SA architecture is depicted in **Figure 1(b)**. The received signal of the SA is given by

$$\mathbf{y} = \sqrt{\rho}\mathbf{W_D}^{\mathbf{H}}\mathbf{W_{ASA}}^{\mathbf{H}}\mathbf{H}\mathbf{P_{ASA}}\mathbf{P_D}\mathbf{s} + \mathbf{n} = \sqrt{\rho}\mathbf{W_{SA}}^{\mathbf{H}}\mathbf{H}\mathbf{P_{SA}}\mathbf{s} + \mathbf{n} \tag{7}$$

where $\mathbf{P_{ASA}}$ and $\mathbf{P_D}$ are the analog and the digital precoding matrices of the SA architecture, respectively, and $\mathbf{W_{ASA}}$ and $\mathbf{W_D}$ are the analog and the digital combining matrices, respectively. $\mathbf{P_{ASA}}$ and $\mathbf{W_{ASA}}$ can be expressed as

$$\mathbf{P_{ASA}} = \begin{bmatrix} \mathbf{p}_{A1} & \mathbf{0}_{N_{tSA}\times 1} & \cdots & \mathbf{0}_{N_{tSA}\times 1} \\ \mathbf{0}_{N_{tSA}\times 1} & \mathbf{p}_{A2} & \cdots & \vdots \\ \vdots & \vdots & \ddots & \mathbf{0}_{N_{tSA}\times 1} \\ \mathbf{0}_{N_{tSA}\times 1} & \cdots & \mathbf{0}_{N_{tSA}\times 1} & \mathbf{p}_{AN_{tSA}} \end{bmatrix} \tag{8}$$

and

$$\mathbf{W_{ASA}} = \begin{bmatrix} \mathbf{w}_{A1} & \mathbf{0}_{N_{rSA}\times 1} & \cdots & \mathbf{0}_{N_{rSA}\times 1} \\ \mathbf{0}_{N_{rSA}\times 1} & \mathbf{w}_{A2} & \cdots & \vdots \\ \vdots & \vdots & \ddots & \mathbf{0}_{N_{rSA}\times 1} \\ \mathbf{0}_{N_{rSA}\times 1} & \cdots & \mathbf{0}_{N_{rSA}\times 1} & \mathbf{w}_{AN_{rSA}} \end{bmatrix} \tag{9}$$

In Eq. (8), the $N_{tSA} \times 1$ analog precoding vector for the *lth* subarray $(l = 1, 2, \ldots\ldots, N_{tRF})$ is denoted as $\mathbf{p}_{\mathrm{Al}}$. Its elements have equal amplitude of $1/\sqrt{N_{tSA}}$, but varying phases. In Eq. (9), the $N_{rSA} \times 1$ analog combining vector for the lth subarray $(l = 1, 2, \ldots\ldots, N_{rRF})$ is represented by $\mathbf{w}_{\mathrm{Al}}$. Its elements have equal amplitude of $1/\sqrt{N_{rSA}}$, but varying phases. Here, $N_{tSA} = N_t/N_{tRF}$ and $N_{rSA} = N_r/N_{rRF}$ denote the number of elements in each subarray at the transmitter and receiver, respectively. The optimization problem of the *lth* subarray can be written as [17].

$$\begin{aligned}\left(\mathbf{p}_{\mathrm{Al}}^{\mathrm{opt}}, \mathbf{p}_{\mathrm{Dl}}^{\mathrm{opt}}\right) = \underset{\mathbf{p}_{\mathrm{Al}},\ \mathbf{p}_{\mathrm{Dl}}}{\arg\ \min} \left\|\mathbf{P}_{\mathrm{l}}^{\mathrm{opt}} - \mathbf{p}_{\mathrm{Al}}\mathbf{p}_{\mathrm{Dl}}\right\|_{\mathrm{F}}^{2} \\ \mathrm{st.}\mathbf{p}_{\mathrm{Al}} \in \overline{\mathcal{F}}_{\mathrm{A}}, \\ \left\|\mathbf{P}_{\mathrm{ASA}}\mathbf{P}_{\mathrm{D}}\right\|_{\mathrm{F}}^{2} = \mathrm{N_s}\end{aligned} \tag{10}$$

where $\mathbf{P}_{l}^{opt} = \mathbf{V_1}(((l-1)N_{tSA} + 1 : lN_{tSA}),\ :)$ denotes the optimum unconstrained hybrid precoding solution of the *lth* subarray. $\mathbf{p}_{\mathrm{Dl}}$, the *lth* row of the $\mathbf{P}_{\mathrm{D}}$. $\overline{\mathcal{F}}_{\mathrm{A}}$, and includes all possible $N_{tSA} \times 1$ vectors satisfying the amplitude constraint.

### 3.3 Hybrid architecture system model

In this subsection, we discuss the system model for hybrid precoding and combining in mmWave MIMO systems with HA architecture. The structure of the hybrid precoding for HA is depicted in **Figure 2**. Antennas are divided into subarrays and grouped with RF chains. $N_{tg}$ and $N_{rg}$ groups are assumed for transmitter and receiver, respectively, with $N_{tSAg} = N_{tSA}/N_{tg}$ and $N_{rSAg} = N_{rSA}/N_{rg}$ being the number of

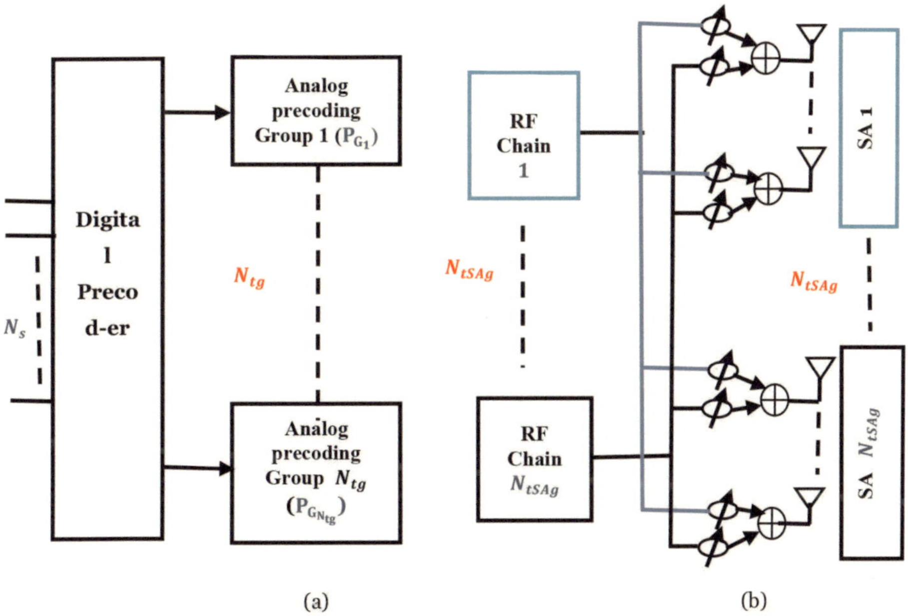


**Figure 2.**
*The proposed hybrid array (HA) architecture at the base station (BS). (a) Block diagram of the hybrid precoding in the HA architecture. (b) The structure of analog precoding in the **ngth** group.*

subarrays in each group. The chapter assumes the same number of RF chains and subarrays in all groups and the number of groups must not exceed the number of RF chains but acknowledges that future work may explore cases with different numbers. In HA, the received signal can be expressed as

$$\mathbf{y} = \sqrt{\rho}\mathbf{W}_{\mathrm{D}}{}^{\mathrm{H}}\mathbf{W}_{\mathrm{AHA}}{}^{\mathrm{H}}\mathbf{H}\mathbf{P}_{\mathrm{AHA}}\mathbf{P}_{\mathrm{D}}\mathbf{s} + \mathbf{n} = \sqrt{\rho}\mathbf{W}_{\mathrm{HA}}{}^{\mathrm{H}}\mathbf{H}\mathbf{P}_{\mathbf{HA}}\mathbf{s} + \mathbf{n} \tag{11}$$

where $\mathbf{P}_{\mathrm{AHA}}$ is the matrix of the HA analog precoding, with dimension $N_t \times N_{tRF}$. $\mathbf{W}_{\mathrm{AHA}}$ is the matrix of the HA analog combining and has dimensions $N_r \times N_{rRF}$. The amplitudes of all elements in $\mathbf{P}_{\mathrm{AHA}}$ and $\mathbf{W}_{\mathrm{AHA}}$ are $1/\sqrt{N_t/N_{tg}}$ and $1/\sqrt{N_r/N_{rg}}$, respectively. Note that $\mathbf{P}_{\mathrm{HA}}$ and $\mathbf{W}_{\mathrm{HA}}$ must satisfy $\|\mathbf{P}_{\mathrm{HA}}\|_F^2 = N_s$ and $\|\mathbf{W}_{\mathrm{HA}}\|_F^2 = N_s$, where $\mathbf{P}_{\mathrm{HA}} = \mathbf{P}_{\mathrm{AHA}}\mathbf{P}_{\mathrm{D}}$ and $\mathbf{W}_{\mathrm{HA}} = \mathbf{W}_{\mathrm{AHA}}\mathbf{W}_{\mathrm{D}}$. The general structure of $\mathbf{P}_{\mathrm{AHA}}$ in the HA architecture can be expressed as

$$\mathbf{P}_{\mathrm{AHA}} = \begin{bmatrix} \mathbf{P}_{\mathrm{G}_1} & 0 & \dots & 0 \\ 0 & \mathbf{P}_{\mathrm{G}_2} & \dots & 0 \\ \vdots & \vdots & \vdots & \vdots \\ 0 & 0 & \dots & \mathbf{P}_{\mathrm{G}_{\mathrm{N_{tg}}}} \end{bmatrix} \tag{12}$$

where $\mathbf{P}_{\mathrm{G_{ng}}}$ is an $(N_t/N_{tg}) \times (N_{tRF}/N_{tg})$ matrix representing the analog precoding matrix of the *ngth* group, where $1 \le ng \le N_{tg}$. $N_{tg} = 2^n$, where $n = 0, 1, \dots, \log_2 N_{tSA}$. Note that, $N_{tg} = 1$ when $n = 0$, resulting in an FA structure [4], and $N_{tg} = N_{tSA}$ when $n = \log_2 N_{tSA}$, resulting in a conventional SA structure [17]. For the HA architecture, $1 \le n \le (\log_2 N_{tSA}) - 1$. Similarly, $\mathbf{W}_{\mathrm{AHA}}$ can be expressed as

$$\mathbf{W}_{\mathrm{AHA}} = \begin{bmatrix} \mathbf{W}_{\mathrm{G}_1} & \mathbf{0} & \dots & \mathbf{0} \\ \mathbf{0} & \mathbf{W}_{\mathrm{G}_2} & \dots & \mathbf{0} \\ \vdots & \vdots & \vdots & \vdots \\ \mathbf{0} & \mathbf{0} & \dots & \mathbf{W}_{\mathrm{G}_{\mathrm{N_{rg}}}} \end{bmatrix} \tag{13}$$

where $\mathbf{W}_{\mathrm{G_{ng}}}$ is an $(N_r/N_{rg}) \times (N_{rRF}/N_{rg})$ matrix representing the analog combining matrix of the *ngth* group, $1 \le ng \le N_{rg}$. $N_{rg}$ can be computed by the same method as $N_{tg}$, by only replacing $N_{tSA}$ by $N_{rSA}$. The hybrid precoding optimization problem of the HA architecture can be written as

$$\begin{aligned} \left(\mathbf{P}_{\mathrm{AHA}}^{\mathrm{opt}}, \mathbf{P}_{\mathrm{D}}^{\mathrm{opt}}\right) = \quad & \arg \min_{\mathbf{P}_{\mathrm{AHA}}, \; \mathbf{P}_{\mathrm{D}}} \|\mathbf{P}^{\mathrm{opt}} - \mathbf{P}_{\mathrm{AHA}}\mathbf{P}_{\mathrm{D}}\|_{\mathrm{F}}^2 \\ & \mathrm{st.}\mathbf{P}_{\mathrm{AHA}} \in \mathcal{F}_{\mathrm{AHA}}, \\ & \|\mathbf{P}_{\mathrm{AHA}}\mathbf{P}_{\mathrm{D}}\|_{\mathrm{F}}^2 = \mathrm{N_s} \end{aligned} \tag{14}$$

where $\mathcal{F}_{\mathrm{AHA}}$ includes all possible precoding matrices that satisfy the amplitude constraint of the HA structure. $\mathbf{P}^{\mathrm{opt}} = \mathbf{V}_1$ is the optimal solution of the unconstrained hybrid precoding. Similarly, the hybrid combining optimization problem of the HA architecture can be expressed as that given in Eq. (14). However, the problem in Eq. (14) is non-convex with a difficult optimal solution. Due to the block nature of the hybrid precoding matrix in the HA architecture, $\mathbf{P}_{\mathrm{HA}}$ can be written as

$$\begin{aligned}\mathbf{P}_{\mathrm{HA}} = \mathbf{P}_{\mathrm{AHA}}\mathbf{P}_{\mathrm{D}} &= \begin{bmatrix} \mathbf{P}_{\mathrm{AG}_1} & 0 & \cdots & 0 \\ 0 & \mathbf{P}_{\mathrm{AG}_2} & \cdots & 0 \\ \vdots & \vdots & \vdots & \vdots \\ 0 & 0 & \cdots & \mathbf{P}_{\mathrm{AG}_{\mathrm{N_{tg}}}} \end{bmatrix} \begin{bmatrix} \mathbf{P}_{\mathrm{DG}_1} \\ \mathbf{P}_{\mathrm{DG}_2} \\ \vdots \\ \mathbf{P}_{\mathrm{DG}_{\mathrm{N_{tg}}}} \end{bmatrix} \\ &= \begin{bmatrix} \mathbf{P}_{\mathrm{AG}_1}\mathbf{P}_{\mathrm{DG}_1} \\ \mathbf{P}_{\mathrm{AG}_2}\mathbf{P}_{\mathrm{DG}_2} \\ \vdots \\ \mathbf{P}_{\mathrm{AG}_{\mathrm{N_{tg}}}}\mathbf{P}_{\mathrm{DG}_{\mathrm{N_{tg}}}} \end{bmatrix} = \begin{bmatrix} \mathbf{P}_{\mathrm{HA}_1} \\ \mathbf{P}_{\mathrm{HA}_2} \\ \vdots \\ \mathbf{P}_{\mathrm{HA}_{\mathrm{N_{tg}}}} \end{bmatrix}\end{aligned} \tag{15}$$

where $\mathbf{P}_{\mathrm{DG_{ng}}}$ is an $(N_{tRF}/N_{tg}) \times N_s$ matrix representing the digital precoder of the *ngth* group and $\mathbf{P}_{\mathrm{HA_{ng}}}$ is an $(N_t/N_{tg}) \times N_s$ matrix denoting the hybrid precoding of the *ngth* group. Furthermore, the optimal hybrid precoding can be decomposed according to the HA architecture as

$$\mathbf{P}^{\mathrm{opt}} = \begin{bmatrix} \mathbf{P}^{\mathrm{opt}}_{\mathrm{G}_1} \\ \mathbf{P}^{\mathrm{opt}}_{\mathrm{G}_2} \\ \vdots \\ \mathbf{P}^{\mathrm{opt}}_{\mathrm{G}_{\mathrm{N_{tg}}}} \end{bmatrix} \tag{16}$$

where $\mathbf{P}^{\mathrm{opt}}_{\mathrm{G_{ng}}}$ is the optimal digital precoding of the *ngth* group in the HA architecture. Based on Eqs. (15) and (16), the problem in Eq. (8) can be decomposed into a series of $N_{tg}$ independent subproblems as

$$\|\mathbf{P}^{\mathrm{opt}} - \mathbf{P}_{\mathrm{AHA}}\mathbf{P}_{\mathrm{D}}\|_{\mathrm{F}}^2 = \sum_{ng=1}^{N_{tg}} \left\|\mathbf{P}^{\mathrm{opt}}_{\mathrm{G_{ng}}} - \mathbf{P}_{\mathrm{AG_{ng}}}\mathbf{P}_{\mathrm{DG_{ng}}}\right\|_{\mathrm{F}}^2 \tag{17}$$

Now, minimizing the objective function in Eq. (8) can be performed by optimizing the $N_{tg}$ subproblems as

$$\begin{aligned}\left(\mathbf{P}^{\mathrm{opt}}_{\mathrm{AG_{ng}}}, \mathbf{P}^{\mathrm{opt}}_{\mathrm{DG_{ng}}}\right) = \underset{\mathbf{P}_{\mathrm{AG_{ng}}},\ \mathbf{P}_{\mathrm{DG_{ng}}}}{\arg\min} & \left\|\mathbf{P}^{\mathrm{opt}}_{\mathrm{G_{ng}}} - \mathbf{P}_{\mathrm{AG_{ng}}}\mathbf{P}_{\mathrm{DG_{ng}}}\right\|_{\mathrm{F}}^2 \\ \mathrm{st.}\mathbf{P}_{\mathrm{AGng}} & \in \mathcal{F}_{\mathrm{AHA}}, \\ \left\|\mathbf{P}_{\mathrm{AG_{ng}}}\mathbf{P}_{\mathrm{DG_{ng}}}\right\|_{\mathrm{F}}^2 & = N_s/N_{tg}\end{aligned} \tag{18}$$

The optimal combining matrices can be achieved by optimizing the $N_{rg}$ subproblems in a similar fashion.

## 4. Proposed hybrid precoding and combining algorithms

In this section, the proposed IFA, ISA, and IHA hybrid precoding and combining algorithms for mmWave MIMO system will be derived and discussed. We only derive the equations that relate to the precoder since the derivation of the combiner is similar.

### 4.1 Iterative full array (IFA) algorithm

In this subsection, we propose the low-complexity IFA hybrid precoding algorithm with equal power allocation per stream. In addition, we do not assume any constraint on the optimization problem, which is related to Eq. (6). The derivation of the combiner is similar. The optimization problem in Eq. (6) is not convex and its solution is NP-hard. Therefore, we propose an iterative solution to solve the problem in Eq. (6). Specifically, we solve the following optimization problem iteratively, which is related to Eq. (6):

$$\left(\mathbf{P}_{\mathbf{D}}^{\mathbf{opt}}\right) = \arg \min_{\mathbf{P}_{\mathbf{D}}} \left\| \mathbf{P}_{\mathbf{FA}}^{\mathbf{opt}} - \mathbf{P}_{\mathbf{AIFA}}\mathbf{P}_{\mathbf{D}} \right\|_F^2 \tag{19}$$

where $\mathbf{P}_{\mathbf{AIFA}}$ is the proposed IFA analog precoder. The objective function can be expanded as

$$\begin{aligned} \left\| \mathbf{P}_{\mathbf{FA}}^{\mathbf{opt}} - \mathbf{P}_{\mathbf{AIFA}}\mathbf{P}_{\mathbf{D}} \right\|_F^2 &= \mathrm{tr}\left(\mathbf{P}_{\mathbf{FA}}^{\mathbf{opt}^{\mathbf{H}}}\mathbf{P}_{\mathbf{FA}}^{\mathbf{opt}}\right) - 2\mathrm{tr}\left(\mathbf{P}_{\mathbf{FA}}^{\mathbf{opt}^{\mathbf{H}}}\mathbf{P}_{\mathbf{AIFA}}\mathbf{P}_{\mathbf{D}}\right) \\ &\quad + \mathrm{tr}\left(\mathbf{P}_{\mathbf{D}}^{\mathbf{H}}\mathbf{P}_{\mathbf{AIFA}}^{\mathbf{H}}\mathbf{P}_{\mathbf{AIFA}}\mathbf{P}_{\mathbf{D}}\right) \\ &= N_S - 2\mathrm{tr}\left(\mathbf{P}_{\mathbf{FA}}^{\mathbf{opt}^{\mathbf{H}}}\mathbf{P}_{\mathbf{AIFA}}\mathbf{P}_{\mathbf{D}}\right) + \mathrm{tr}\left(\mathbf{P}_{\mathbf{D}}^{\mathbf{H}}\mathbf{P}_{\mathbf{AIFA}}^{\mathbf{H}}\mathbf{P}_{\mathbf{AIFA}}\mathbf{P}_{\mathbf{D}}\right) \end{aligned} \tag{20}$$

To minimize over $\mathbf{P}_{\mathbf{D}}$, we set the derivative of Eq. (20) with respect to $\mathbf{P}_{\mathbf{D}}$ equal to zero, which yields the following minimized proposed baseband precoder $\mathbf{P}_{\mathbf{D}}$ (least-squares solution)

$$\mathbf{P}_{\mathbf{D}} = \left(\mathbf{P}_{\mathbf{AIFA}}^{\mathbf{H}}\mathbf{P}_{\mathbf{AIFA}}\right)^{-1}\mathbf{P}_{\mathbf{AIFA}}^{\mathbf{H}}\mathbf{P}_{\mathbf{FA}}^{\mathbf{opt}} \tag{21}$$

Then, we keep $\mathbf{P}_{\mathbf{D}}$ fixed and solve the same optimization problem but now minimizing over $\mathbf{P}_{\mathbf{AIFA}}$

$$\left(\mathbf{P}_{\mathbf{AIFA}}^{\mathbf{opt}}\right) = \arg \min_{\mathbf{P}_{\mathbf{AIFA}}} \left\| \mathbf{P}_{\mathbf{FA}}^{\mathbf{opt}} - \mathbf{P}_{\mathbf{AIFA}}\mathbf{P}_{\mathbf{D}} \right\|_F^2 \tag{22}$$

Similar to Eq. (20), expanding the objective function yields:

$$\left\| \mathbf{P}_{\mathbf{FA}}^{\mathbf{opt}} - \mathbf{P}_{\mathbf{AIFA}}\mathbf{P}_{\mathbf{D}} \right\|_F^2 = N_S - 2\mathrm{tr}\left(\mathbf{P}_{\mathbf{FA}}^{\mathbf{opt}^{\mathbf{H}}}\mathbf{P}_{\mathbf{AIFA}}\mathbf{P}_{\mathbf{D}}\right) + \mathrm{tr}\left(\mathbf{P}_{\mathbf{D}}^{\mathbf{H}}\mathbf{P}_{\mathbf{AIFA}}^{\mathbf{H}}\mathbf{P}_{\mathbf{AIFA}}\mathbf{P}_{\mathbf{D}}\right) \tag{23}$$

We again set the derivative of (23) with respect to $\mathbf{P}_{\mathbf{AIFA}}$ as equal to zero, which yields the following equation:

$$\nabla_f(\mathbf{P}_{\mathbf{AIFA}}) = -\mathbf{P}_{\mathbf{FA}}^{\mathbf{opt}}\mathbf{P}_{\mathbf{D}}^{\mathbf{H}} + \mathbf{P}_{\mathbf{AIFA}}\mathbf{P}_{\mathbf{D}}\mathbf{P}_{\mathbf{D}}^{\mathbf{H}} = 0 \tag{24}$$

Since $\mathbf{P}_{\mathbf{D}}\mathbf{P}_{\mathbf{D}}^{\mathbf{H}}$ cannot be inverted when $N_S < N_{tRF}$, we used the gradient descent method to obtain:

$$\mathbf{P_{AIFA}}^{k+1} = \mathbf{P_{AIFA}}^{k} - \alpha\nabla_f\left(\mathbf{P_{AIFA}}^{k}\right)$$
$$\mathbf{P_{AIFA}}^{k+1} = \mathbf{P_{AIFA}}^{k} + \alpha(\mathbf{P_{FA}^{opt}} - \mathbf{P_{AIFA}}^{k}\mathbf{P_D})\mathbf{P_D^H}$$
$$\mathbf{P_{AIFA}}^{k+1} = \mathbf{P_{AIFA}}^{k} + \alpha\mathbf{P_{res}}\mathbf{P_D^H} \tag{25}$$

where the residual precoding matrix, denoted as $\mathbf{P_{res}}$, is obtained by subtracting the product of $\mathbf{P_{AIFA}}$ and $\mathbf{P_D}$ from the optimized $\mathbf{P_{FA}^{opt}}$, and $\alpha$ is the step size. This approach is valid even when $N_S$ equals $N_{tRF}$. However, if $N_S$ is less than $N_{tRF}$, the $\mathbf{P_{AIFA}}$ matrix with dimensions of $N_t$ x $N_{tRF}$ needs to be completed after initialization. In each iteration, the column that results in the largest reduction of the residual is added to $\mathbf{P_{AIFA}}$. This column is chosen from the basis of the range of the residual, which is obtained by normalizing the first singular vector of the residual element-wise. Algorithm 1 provides the pseudo-code for the proposed IFA hybrid precoder, denoted as $\mathbf{P_{IFA}}$. In a mmWave system that employs hybrid precoding, the base station (BS) or mobile station (MS) can support up to $\min(N_{tRF}, N_{rRF})$ [2]. The inputs to the algorithm 1 are $\mathbf{P_{FA}^{opt}} \in C^{N_t \text{x} N_S}$ and the maximum number of iterations $K$. When $N_S < N_{tRF}$, $K$ should be greater than or equal to $N_{tRF}$ - $N_S$ to compute the $N_t$ x $N_{tRF}\mathbf{P_{AIFA}}$ matrix. When $N_{tRF} = N_S$, $K$ should be greater than or equal to 1.

In the general case where $N_S \geq 1$ and $N_S \leq N_{tRF}$, the algorithm initializes $\mathbf{P_{AIFA}}$ by normalizing the first $N_s$ columns of $\mathbf{P_{FA}^{opt}}$ element-wise, i.e., $\mathbf{P_{AIFA}} = \mathbf{P_{FA}^{opt}} \oslash \left(\left|\mathbf{P_{FA}^{opt}}\right|\sqrt{N_t}\right)$. Next, step 2 calculates $\mathbf{P_D}$ using least squares. Steps 3 and 4 compute the residual precoding matrix $\mathbf{P_{res}}$ and the proposed IFA analog precoder $\mathbf{P_{AIFA}}$, respectively. Step 5 ensures that the constant-magnitude entries in $\mathbf{P_{AIFA}}$ can be applied at radio frequency (RF) using analog phase shifters.

In steps 7 and 8, when $N_S < N_{tRF}$, the $N_t$ x $N_{tRF}\mathbf{P_{AIFA}}$ needs to be completed by adding the element-wise normalization of the first singular vector of $\mathbf{P_{res}}$ to $\mathbf{P_{AIFA}}$. After $K$ iterations, the algorithm finds the $N_t$x $N_{tRF}$ proposed IFA analog precoding matrix $\mathbf{P_{AIFA}}$ and the $N_{tRF}$ x $N_S$ baseband precoder $\mathbf{P_D}$, such that $\left\|\mathbf{P_{FA}^{opt}} - \mathbf{P_{AIFA}}\mathbf{P_D}\right\|_F$ is minimized. In steps 12 and 13, the algorithm ensures that the transmit power constraint is satisfied and returns the proposed IFA hybrid precoder $\mathbf{P_{IFA}} = \mathbf{P_{AIFA}}\mathbf{P_D}$. The proposed hybrid combiner $\mathbf{W_{IFA}}$ can be calculated in the same manner.

Remark 1 - Convergence of the proposed IFA hybrid precoder to local minimum points: Note that when $N_S = N_{tRF}$ or $N_S < N_{tRF}$, $\mathbf{P_D}$ is a square matrix that is approximately unitary $\mathbf{P_D^H}\mathbf{P_D} \approx \mathbf{P_D}\mathbf{P_D^H} \approx \mathbf{I_{N_s}}$ or a non-square matrix that is approximately semi-unitary $\mathbf{P_D^H}\mathbf{P_D} \approx \mathbf{I_{N_s}}$, respectively [2].

Algorithm 1. Proposed IFA Hybrid Precoding.

**Input**: The optimum unconstrained solution $\mathbf{P_{FA}^{opt}} \in C^{N_t \text{x} N_S}$. and the maximum number of iterations $K$

**Output**: Analog $\mathbf{P_{AIFA}} \in C^{N_t \text{x} N_{tRF}}$. with the element-wise normalization and baseband $\mathbf{P_D} \in C^{N_{tRF} \text{x} N_S}$ such that $\left\|\mathbf{P_{FA}^{opt}} - \mathbf{P_{IFA}}\right\|_F$ is reduced and $\left\|\mathbf{P_{IFA}}\right\|_F^2 = N_S$, where $\mathbf{P_{IFA}} = \mathbf{P_{AIFA}}\mathbf{P_D}$

Initialization: analog precoder $\mathbf{P_{AIFA}}^1 = \mathbf{P_{FA}^{opt}} \oslash \left(\left|\mathbf{P_{FA}^{opt}}\right|\sqrt{N_t}\right)$.

1: for $k = 1 : K$ do

2: Update: $\mathbf{P_D} = \left(\mathbf{P_{AIFA}^H}^{k}\mathbf{P_{AIFA}}^{k}\right)^{-1}\mathbf{P_{AIFA}^H}^{k}\mathbf{P_{FA}^{opt}}.$

3: Update the residual: $\mathbf{P_{res}} = \mathbf{P_{FA}^{opt}}$ - $\mathbf{P_{AIFA}}^{k}\mathbf{P_D}$

4: Update: $\mathbf{P_{AIFA}}^{k+1} = \mathbf{P_{AIFA}}^{k} + \alpha \mathbf{P_{res}} \mathbf{P_D^H}$
5: Element-Wise Normalization:

$$\mathbf{P_{AIFA}}^{k+1} = \mathbf{P_{AIFA}}^{k+1} \oslash \left( \left| \mathbf{P_{AIFA}}^{k+1} \right| \sqrt{N_t} \right)$$

6: If $i \leq N_{tRF}$ - $N_S$
7: $\mathbf{F_{res}} = \mathbf{U \Sigma V^H}$
8: Append the element-wise normalization of the first vector of **U** as a new column to $\mathrm{P_{AIFA}}$:

$$\mathbf{P_{AIFA}}^{k+1} = \left[ \mathbf{P_{AIFA}}^{k+1} (\mathbf{U})_1 \oslash \left( \left| (\mathbf{U})_1 \right| \sqrt{N_t} \right) \right]$$

9: end if
10: end for
11: $\mathbf{P_D} = \left( \mathbf{P^H_{AIFA}} \mathbf{P_{AIFA}} \right)^{-1} \mathbf{P^H_{AIFA}} \mathbf{P^{opt}_{FA}}$
12: $\mathbf{P_D} = \sqrt{N_S} \frac{\mathbf{P_D}}{\|\mathbf{P_{AIFA}} \mathbf{P_D}\|_F}$
13: return $\mathbf{P_{IFA}} = \mathbf{P_{AIFA}} \mathbf{P_D}$

Thus, each iteration in Algorithm 1 minimizes the objective function $\|\mathbf{P^{opt}_{FA}} - \mathbf{P_{AIFA}} \mathbf{P_D}\|_F$ and the error term decreases monotonically with each iteration. Since the objective function has a lower bound, the proposed method must converge to local optimum points.

### 4.2 Iterative subarray (ISA) algorithm

In this subsection, the proposed ISA hybrid precoding is derived and discussed. The optimization problem of the *lth* subarray in Eq. (10) can be solved for each subarray in a similar way as in Eq. (22), and the following iterative solution can be obtained:

$$\mathbf{p}_{Al}^{k+1} = \mathbf{p}_{Al}^{k} + \mathbf{P}_{res} \mathbf{p}_{Dl}^{H} \tag{26}$$

The residual precoding matrix for the *lth* subarray, $\mathbf{P}_{res}$, is calculated as $\mathbf{P}_{res} = \mathbf{P}_l^{opt} - \mathbf{p}_{Al} \mathbf{p}_{Dl}$. Eq. (26) shows that the updated value of $\mathbf{p}_{Al}^{k+1}$ for the lth subarray can be obtained by adding $\mathbf{P}_{res} \mathbf{p}_{Dl}^{H}$ to the value of $\mathbf{p}_{Al}^{k}$ from the previous iteration. Once initial values of $\mathbf{p}_{Al}$ and $\mathbf{p}_{Dl}$ have been obtained, these can be used to iteratively solve the optimization problem given in Eq. (26). After convergence, the resulting $\mathbf{p}_{Al}$ of the lth subarray must be normalized to satisfy the constraint in Eq. (10).

Algorithm 2. Proposed ISA Hybrid Precoding scheme

1. Input $\mathbf{V}_1, K$

2. Decompose $\mathbf{V_1}$ as $\mathbf{V}_1 = \left[ \tilde{\mathbf{V}}_1 \tilde{\mathbf{V}}_2 \ldots \tilde{\mathbf{V}}_{L_t} \right]^T$

---

3. For $1 \leq l \leq L_t$

4. Find $\mathbf{p}_{Al}^{initial} = \frac{1}{\sqrt{N_{tSA}}} e^{jangle(\tilde{\mathbf{V}}_l)}$

5. Find $\mathbf{p}_{Dl}^{initial} = (\mathbf{p}_{Al})^H * \tilde{\mathbf{V}}_l$

6. Initial $\mathbf{p}_{Al}{}^k = \mathbf{p}_{Al}^{initial}$

7. Initial $\mathbf{p}_{Dl} = \mathbf{p}_{Dl}^{initial}$

8. For $1 \leq k \leq K$

9. Compute the residual $\mathbf{P}_{res} = \tilde{\mathbf{V}}_l - \mathbf{p}_{Al}{}^k \mathbf{p}_{Dl}$

10. Update $\mathbf{p}_{Al}{}^{k+1} = \mathbf{p}_{Al}{}^k + \mathbf{P}_{res}\mathbf{p}_{Dl}^H$

11. Normalize $\mathbf{p}_{Al}{}^{k+1} = \mathbf{p}_{Al}{}^{k+1} / \left( \left| \mathbf{p}_{Al}{}^{k+1} \right| * \sqrt{N_{tSA}} \right)$

12. $\mathbf{p}_{Dl} = \left(\mathbf{p}_{Al}{}^{k+1}\right)^H * \tilde{\mathbf{V}}_l$

13. end for

14. end for

15. Construct $\mathbf{P}_D$ and $\mathbf{P}_{AISA}$

16. Normalize $\mathbf{P}_D$ as $\mathbf{P}_D = \frac{\sqrt{N_s}}{\|\mathbf{P}_{AISA}\mathbf{P}_D\|_F} \mathbf{P}_D$

17. Return $\mathbf{P}_{ISA} = \mathbf{P}_{AISA}\mathbf{P}_D$

---

The normalized $\mathbf{p}_{Al}{}^{k+1}$ can be expressed as

$$\mathbf{p}_{Al}{}^{k+1} = \mathbf{p}_{Al}{}^{k+1} / \left( \left| \mathbf{p}_{Al}{}^{k+1} \right| * \sqrt{N_{tSA}} \right) \tag{27}$$

In summary, the pseudo-code of the proposed ISA hybrid precoding can be summarized in Algorithm 2, which can be explained as follows. First, the initial values of $\mathbf{p}_{Al}$ and $\mathbf{p}_{Dl}$ of the $lth$ subarray must be obtained. Then, the iterative solution for the $lth$ subarray is applied to obtain the optimal $\mathbf{p}_{Al}$ and $\mathbf{p}_{Dl}$ and this operation will be repeated for all subarrays. Finally, $\mathbf{P}_D$ and $\mathbf{P}_A$ are constructed. Note that the initial solution of $\mathbf{p}_{Al}$ and $\mathbf{p}_{Dl}$ for each subarray can be obtained as follows:

$$\mathbf{p}_{Al}^{initial} = \frac{1}{\sqrt{N_{tSA}}} e^{jangle(\tilde{\mathbf{V}}_l)} \tag{28}$$

and

$$\mathbf{p}_{Dl}^{initial} = \left(\mathbf{p}_{Al}^{initial}\right)^{H} * \tilde{\mathbf{V}}_1 \tag{29}$$

where $\tilde{\mathbf{V}}_1$ is an $N_{tSA} \times N_s$ matrix that represents the optimal precoder of the *lth* subarray.

The algorithm can be summarized as follows:

1. Obtain the initial values $\mathbf{p}_{Al}^{initial}$ and $\mathbf{p}_{Dl}^{initial}$ for the *lth* subarray.

2. Apply an iterative solution for the *lth* subarray to acquire the optimal $\mathbf{p}_{Al}$ and $\mathbf{p}_{Dl}$.

3. Repeat the above operation for all subarrays.

4. Construct the proposed ISA analog precoder, $\mathbf{P}_{AISA}$, and digital precoder, $\mathbf{P}_D$.

5. Construct the hybrid precoder of the proposed ISA as $\mathbf{P}_{ISA} = \mathbf{P}_{AISA}\mathbf{P}_D$.

Eq. (26) satisfies the property of the gradient descent method as it minimizes the objective function $\left\|\mathbf{P}_l^{opt} - \mathbf{p}_{Al}\mathbf{p}_{Dl}\right\|_F^2$ in each iteration from step 8 to 13 of Algorithm 2. This guarantees the convergence of $\mathbf{p}_{Al}{}^{k+1}$ to a local optimal point.

Note that the proposed ISA algorithm in this subsection differs from the proposed IFA algorithm in the previous subsection in its approach. The proposed ISA algorithm independently obtains hybrid precoding for each subarray and then combines them to find the hybrid precoding for the entire system. This makes the proposed ISA algorithm simpler than the IFA algorithm, which computes the hybrid precoding directly for the entire system.

### 4.3 Iterative hybrid array (IHA) algorithm

In this subsection, we introduce a low-complexity IHA hybrid precoding algorithm. The combiner derivation follows a similar approach. We know that the structure of optimal precoder of the *ngth* group $\mathbf{V}_{\mathbf{G}_{ng}}$ is non-square semi-unitary, meaning $\mathbf{V}_{\mathbf{Gng}}^{\mathbf{H}}\mathbf{V}_{\mathbf{G}_{ng}} = \mathbf{I}_{\mathbf{N}_s}$ when $N_{tg} = 1$ and $\mathbf{V}_{\mathbf{Gng}}^{\mathbf{H}}\mathbf{V}_{\mathbf{G}_{ng}} \approx \mathbf{I}_{\mathbf{N}_s}$ when $N_{tg} > 1$. Therefore, the HA precoder design must also be non-square semi-unitary, i.e., $\mathbf{P}_{\mathbf{DGng}}^{\mathbf{H}}\mathbf{P}_{\mathbf{AGng}}^{\mathbf{H}}\mathbf{P}_{\mathbf{AG}_{ng}}\mathbf{P}_{\mathbf{DG}_{ng}} = \mathbf{I}_{\mathbf{N}_s}$. This structure will be used to solve the optimization problem and make the HA precoder $\mathbf{P}_{\mathbf{AG}_{ng}}\mathbf{P}_{\mathbf{DG}_{ng}}$ approach the optimal precoder $\mathbf{V}_{\mathbf{G}_{ng}}$ as closely as possible. Thus, by assuming the structure of the HA hybrid precoder $\mathbf{P}_{\mathbf{AG}_{ng}}\mathbf{P}_{\mathbf{DG}_{ng}}$ as a semi-unitary matrix, we need to solve the following optimization problem:

$$\begin{aligned}\left(\mathbf{P}_{\mathbf{AG}_{ng}}^{\mathbf{opt}}, \mathbf{P}_{\mathbf{DG}_{ng}}^{\mathbf{opt}}\right) = \; & \arg \min_{\mathbf{P}_{\mathbf{AG}_{ng}},\, \mathbf{P}_{\mathbf{DG}_{ng}}} \left\|\mathbf{P}_{\mathbf{G}_{ng}}^{\mathbf{opt}} - \mathbf{P}_{\mathbf{AG}_{ng}}\mathbf{P}_{\mathbf{DG}_{ng}}\right\|_F^2 \\ & st.\mathbf{P}_{\mathbf{AG}_{ng}} \in \mathcal{F}_{AHA}, \\ \mathbf{P}_{\mathbf{AG}_{ng}}^{\mathbf{H}}\mathbf{P}_{\mathbf{AG}_{ng}} = \mathbf{I}_{\mathbf{N}_{\mathbf{tRF}}} & \text{ and } \mathbf{P}_{\mathbf{DG}_{ng}}^{\mathbf{H}}\mathbf{P}_{\mathbf{DG}_{ng}} = \mathbf{I}_{\mathbf{N}_s} \\ & \left\|\mathbf{P}_{\mathbf{AG}_{ng}}\mathbf{P}_{\mathbf{DG}_{ng}}\right\|_F^2 = N_s/N_{tg}\end{aligned} \tag{30}$$

The problem in Eq. (30) is a non-convex optimization problem whose solution is mathematically intractable. However, in this subsection, we will use iterative algorithms to solve (30):

$$\left(\mathbf{P}_{\mathbf{AG}_{\mathbf{ng}}}^{\mathbf{opt}}, \mathbf{P}_{\mathbf{DG}_{\mathbf{ng}}}^{\mathbf{opt}}\right) = \underset{\mathbf{P}_{\mathbf{AG}_{\mathbf{ng}}},\ \mathbf{P}_{\mathbf{DG}_{\mathbf{ng}}}}{\arg\min} \left\|\mathbf{P}_{\mathbf{G}_{\mathbf{ng}}}^{\mathbf{opt}} - \mathbf{P}_{\mathbf{AG}_{\mathbf{ng}}}\mathbf{P}_{\mathbf{DG}_{\mathbf{ng}}}\right\|_{\mathrm{F}}^{2} \tag{31}$$

We first need to find the baseband precoder $\mathbf{P}_{\mathbf{DG}_{\mathbf{ng}}}$ of the *ngth* group in the HA architecture that minimizes the Euclidean distance using the initialization of the proposed HA precoder $\mathbf{P}_{\mathbf{AG}_{\mathbf{ng}}}$ of the *ngth* group in the HA architecture, which is calculated by taking the first $N_{tRF}/N_{tg}$ columns from $\mathbf{P}_{\mathbf{G}_{\mathbf{ng}}}^{\mathbf{opt}}$ and then normalizing them such that each entry has constant magnitude, i.e., $\mathbf{P}_{\mathbf{AG}_{\mathbf{ng}}} = \left(\mathbf{P}_{\mathbf{G}_{\mathbf{ng}}}^{\mathbf{opt}} \oslash \left(\left|\mathbf{P}_{\mathbf{G}_{\mathbf{ng}}}^{\mathbf{opt}}\right| \sqrt{(N_t/N_{tg})}\right)\right)$. We then find the RF precoder $\mathbf{P}_{\mathbf{AG}_{\mathbf{ng}}}$ such that the IHA hybrid precoder $\mathbf{P}_{\mathbf{AG}_{\mathbf{ng}}}\mathbf{P}_{\mathbf{DG}_{\mathbf{ng}}}$ of the *ngth* group is sufficiently "close" to the optimal unconstrained digital precoder $\mathbf{P}_{\mathbf{G}_{\mathbf{ng}}}^{\mathbf{opt}}$ of the *ngth* group in the HA architecture. Specifically, we would like to solve the following optimization problem first, which is related to (31):

$$\left(\mathbf{P}_{\mathbf{DGng}}^{\mathbf{opt}}\right) = \arg\min_{\mathbf{P}_{\mathbf{DGng}}} \left\|\mathbf{P}_{\mathbf{G}_{\mathbf{ng}}}^{\mathbf{opt}} - \mathbf{P}_{\mathbf{AG}_{\mathbf{ng}}}\mathbf{P}_{\mathbf{DG}_{\mathbf{ng}}}\right\|_{F}^{2} \tag{32}$$

The objective function can be expanded as

$$\begin{aligned} &\left\|\mathbf{P}_{\mathbf{G}_{\mathbf{ng}}}^{\mathbf{opt}} - \mathbf{P}_{\mathbf{AG}_{\mathbf{ng}}}\mathbf{P}_{\mathbf{DG}_{\mathbf{ng}}}\right\|_{F}^{2} \\ &= \mathrm{tr}\left(\mathbf{P}_{\mathbf{Gng}}^{\mathbf{opt}\ \mathbf{H}}\mathbf{P}_{\mathbf{Gng}}^{\mathbf{opt}}\right) - 2\mathrm{tr}\left(\mathbf{P}_{\mathbf{Gng}}^{\mathbf{opt}\ \mathbf{H}}\mathbf{P}_{\mathbf{AG}_{\mathbf{ng}}}\mathbf{P}_{\mathbf{DG}_{\mathbf{ng}}}\right) + \left\|\mathbf{P}_{\mathbf{AG}_{\mathbf{ng}}}\mathbf{P}_{\mathbf{DG}_{\mathbf{ng}}}\right\|_{F}^{2} \\ &= 2N_S - 2\mathrm{tr}\left(\mathbf{P}_{\mathbf{Gng}}^{\mathbf{opt}\ \mathbf{H}}\mathbf{P}_{\mathbf{AG}_{\mathbf{ng}}}\mathbf{P}_{\mathbf{DG}_{\mathbf{ng}}}\right) \end{aligned} \tag{33}$$

The solution of this problem, which is to find the maximization of $\mathbf{P}_{\mathbf{Gng}}^{\mathbf{opt}\ \mathbf{H}}\mathbf{P}_{\mathbf{AG}_{\mathbf{ng}}}\mathbf{P}_{\mathbf{DG}_{\mathbf{ng}}}$, is solved by what is called the orthonormal Procrustes problem [28] as follows:

$$\mathbf{P}_{\mathbf{DG}_{\mathbf{ng}}} = \mathbf{V}\mathbf{U}^{\mathbf{H}} \tag{34}$$

where $\mathbf{P}_{\mathbf{Gng}}^{\mathbf{opt}\ \mathbf{H}}\mathbf{P}_{\mathbf{AG}_{\mathbf{ng}}} = \mathbf{U}\boldsymbol{\Sigma}\mathbf{V}^{\mathbf{H}}$. Then, we keep $\mathbf{P}_{\mathbf{DG}_{\mathbf{ng}}}$ fixed and solve the same optimization problem but now minimizing over $\mathbf{P}_{\mathbf{AG}_{\mathbf{ng}}}$ as follows:

$$\left(\mathbf{P}_{\mathbf{AGng}}^{\mathbf{opt}}\right) = \arg\min_{\mathbf{P}_{\mathbf{AGng}}} \left\|\mathbf{P}_{\mathbf{G}_{\mathbf{ng}}}^{\mathbf{opt}} - \mathbf{P}_{\mathbf{AG}_{\mathbf{ng}}}\mathbf{P}_{\mathbf{DG}_{\mathbf{ng}}}\right\|_{F}^{2} \tag{35}$$

Similar to (33), expanding the objective function yields:

$$\left\|\mathbf{P}_{\mathbf{G}_{\mathbf{ng}}}^{\mathbf{opt}} - \mathbf{P}_{\mathbf{AG}_{\mathbf{ng}}}\mathbf{P}_{\mathbf{DG}_{\mathbf{ng}}}\right\|_{F}^{2} = 2N_S - 2\mathrm{tr}\left(\mathbf{P}_{\mathbf{Gng}}^{\mathbf{opt}\ \mathbf{H}}\mathbf{P}_{\mathbf{AG}_{\mathbf{ng}}}\mathbf{P}_{\mathbf{DG}_{\mathbf{ng}}}\right) \tag{36}$$

Assuming that the IHA analog precoder $\mathbf{P}_{\mathbf{AG}_{\mathbf{ng}}}$ semi-unitary matrix, the solution that maximizes $\mathbf{P}_{\mathbf{Gng}}^{\mathbf{opt}\ \mathbf{H}}\mathbf{P}_{\mathbf{AG}_{\mathbf{ng}}}\mathbf{P}_{\mathbf{DG}_{\mathbf{ng}}}$ in (35), is also solved by the orthonormal Procrustes problem as follows:

$$\mathbf{P}_{\mathbf{AG}_{\mathbf{ng}}} = \mathbf{V}\mathbf{U}^{\mathbf{H}} \tag{37}$$

where $\mathbf{P}_{\mathbf{DG}_{\mathbf{ng}}}\mathbf{P}_{\mathbf{Gng}}^{\mathbf{opt}\ \mathbf{H}} = \mathbf{U\Sigma V}^{\mathbf{H}}$. Also, there is another way to maximize $\mathbf{P}_{\mathbf{Gng}}^{\mathbf{opt}\ \mathbf{H}}\mathbf{P}_{\mathbf{AG}_{\mathbf{ng}}}\mathbf{P}_{\mathbf{DG}_{\mathbf{ng}}}$ in (35) as follows:

$$\mathbf{P}_{\mathbf{AG}_{\mathbf{ng}}} = \mathbf{P}_{\mathbf{Gng}}^{\mathbf{opt}}\mathbf{P}_{\mathbf{DGng}}^{\mathbf{H}} \tag{38}$$

Both solutions in (37) and (38) are almost the same because we assume that $\mathbf{P}_{\mathbf{DG}_{\mathbf{ng}}}$ is a non-square semi-unitary or square unitary matrix in (36) and the singular values of $\mathbf{P}_{\mathbf{DG}_{\mathbf{ng}}}\mathbf{P}_{\mathbf{Gng}}^{\mathbf{opt}\ \mathbf{H}}$ are almost unity when $N_{tg} = 1$ and close to unity when $N_{tg} > 1$.

The main difference between the hybrid design in this chapter and that in [29] is that our design assumes that $\mathbf{P}_{\mathbf{AG}_{\mathbf{ng}}}$ is non-square semi-unitary matrix, and $\mathbf{P}_{\mathbf{DG}_{\mathbf{ng}}}$ is non-square semi-unitary or square unitary matrix. Our design is more versatile and applicable in various scenarios, including when the number of data streams is equal to or less than the number of RF chains. In contrast, the HD-AM (hybrid design by alternating minimization) technique can only be used when the number of data streams is equal to the number of RF chains [29]. Additionally, our derivation is based on the HA architecture, while HD-AM is only applicable to the FA architecture. Our proposed algorithm is straightforward since it calculates the hybrid precoding for each group in the HA architecture independently before using it to determine that of the entire system. Conversely, the method presented in [29] computes the hybrid precoding directly for the entire system, resulting in higher computational complexity.

---

Algorithm 3: Proposed IHA Hybrid Precoding

---

Input: The optimum unconstrained solution $\mathbf{P}_{\mathbf{Gng}}^{\mathbf{opt}} \in C^{\left(N_t/N_{tg}\right)\ \mathrm{x}\ N_S}$, initialized analog procoder $\mathbf{P}_{\mathbf{AG}_{\mathbf{ng}}} \in C^{\left(N_t/N_{tg}\right)\ \mathrm{x}\ \left(N_{tRF}/N_{tg}\right)}$ with the element-wise normalization, and the maximum number of iterations $K$.

Output: Analog $\mathbf{P}_{\mathbf{HA}_{\mathbf{ng}}} \in C^{N_t \mathrm{x}\ \mathrm{N_{tRF}}}$ such that $\left\|\mathbf{P}_{\mathbf{Gng}}^{\mathbf{opt}} - \mathbf{P}_{\mathbf{HA}_{\mathbf{ng}}}\right\|_F$ is reduced and $\left\|\mathbf{P}_{\mathbf{HA}_{\mathbf{ng}}}\right\|_F^2 = N_S/N_{tg}$, where $\mathbf{P}_{\mathbf{HA}_{\mathbf{ng}}} = \mathbf{P}_{\mathbf{AG}_{\mathbf{ng}}}\mathbf{P}_{\mathbf{DG}_{\mathbf{ng}}}$.

1: for $i = 1 : K$ do

2: Update: $\mathbf{P}_{\mathbf{DG}_{\mathbf{ng}}} = \mathbf{V}\mathbf{U}^{\mathbf{H}}$, where $\mathbf{P}_{\mathbf{Gng}}^{\mathbf{opt}\ \mathbf{H}}\mathbf{P}_{\mathbf{AG}_{\mathbf{ng}}} = \mathbf{U\Sigma V}^{\mathbf{H}}$

3: Update: $\mathbf{P}_{\mathbf{AG}_{\mathbf{ng}}} = \mathbf{P}_{\mathbf{Gng}}^{\mathbf{opt}}\mathbf{P}_{\mathbf{DGng}}^{\mathbf{H}}$

4: Element-Wise Normalization: $\mathbf{P}_{\mathbf{AG}_{\mathbf{ng}}} = \mathbf{P}_{\mathbf{AG}_{\mathbf{ng}}} \oslash \left(\left|\mathbf{P}_{\mathbf{AG}_{\mathbf{ng}}}\right| \sqrt{\left(N_t/N_{tg}\right)}\right)$

5: end for

7: $\mathbf{P}_{\mathbf{DG}_{\mathbf{ng}}} = \mathbf{P}_{\mathbf{AG}_{\mathbf{ng}}}^{\mathbf{H}}\mathbf{P}_{\mathbf{Gng}}^{\mathbf{opt}}$

8: $\mathbf{P}_{\mathbf{D}} = \sqrt{N_S/N_{tg}}\dfrac{\mathbf{P}_{\mathbf{DGng}}}{\left\|\mathbf{P}_{\mathbf{AGng}}\mathbf{P}_{\mathbf{DGng}}\right\|_F}$

9: Return $\mathbf{P}_{\mathbf{HA}_{\mathbf{ng}}} = \mathbf{P}_{\mathbf{AG}_{\mathbf{ng}}}\mathbf{P}_{\mathbf{DG}_{\mathbf{ng}}}$.

---

Algorithm 3 provides the pseudo-code for the proposed IHA precoder. The inputs of the algorithm are $\mathbf{P}^{\mathbf{opt}}_{\mathbf{G_{ng}}} \in C^{(N_t/N_{tg}) \text{ x } N_S}$, initialized analog procoder $\mathbf{P}_{\mathbf{AG_{ng}}} \in C^{(N_t/N_{tg}) \text{ x } (N_{tRF}/N_{tg})}$, i.e., $\mathbf{P}_{\mathbf{AG_{ng}}} = \left(\mathbf{P}^{\mathbf{opt}}_{\mathbf{G_{ng}}} \oslash \left(\left|\mathbf{P}^{\mathbf{opt}}_{\mathbf{G_{ng}}}\right| \sqrt{(N_t/N_{tg})}\,\right)\right)$, and the maximum number of iterations $K$, where $K \geq 1$ for $N_S < (N_{tRF}/N_{tg})$ or $(N_{tRF}/N_{tg}) \leq N_S$. In the general case of $N_S \geq 1$, the algorithm starts by computing the *ngth* group $\mathbf{P}_{\mathbf{DG_{ng}}}$ using the orthonormal Procrustes solution in step 2. After that, the algorithm proceeds to update the *ngth* group RF precoder $\mathbf{P}_{\mathbf{AG_{ng}}}$ in step 3. Step 4 ensures that the proposed RF precoder $\mathbf{P}_{\mathbf{AG_{ng}}}$is satisfied exactly with constant-magnitude entries, which can be applied at RF using analog phase shifters. After the last iteration of the algorithm, $\mathbf{P}_{\mathbf{DG_{ng}}}$ is updated via the maximal ratio combining (MRC), instead of the least solution, which has an impact on the Frobenius norm objective function $\left\|\mathbf{P}^{\mathbf{opt}}_{\mathbf{G_{ng}}} - \mathbf{P}_{\mathbf{AG_{ng}}}\mathbf{P}_{\mathbf{DG_{ng}}}\right\|_F^2$; the least solution becomes MRC after implementing the semi-unitary analog precoder, i.e., $\mathbf{P}^{\mathbf{H}}_{\mathbf{AG_{ng}}}\mathbf{P}_{\mathbf{AG_{ng}}} = \mathbf{I}_{(N_{tRF}/N_{tg})}$. After $K$ iterations, the process is completed and the algorithm finds the $(N_t/N_{tg}) \text{ x } (N_{tRF}/N_{tg})$ proposed RF precoding matrix $\mathbf{P}_{\mathbf{AG_{ng}}}$ and the $(N_{tRF}/N_{tg}) \text{ x } N_S$ baseband precoder $\mathbf{P}_{\mathbf{DG_{ng}}}$ such that $\left\|\mathbf{P}^{\mathbf{opt}}_{\mathbf{G_{ng}}} - \mathbf{P}_{\mathbf{AG_{ng}}}\mathbf{P}_{\mathbf{DG_{ng}}}\right\|_F^2$ is minimized. In steps 8 and 9, we ensure that the transmit power constraint is satisfied for each *ngth* group and return the proposed *ngth* group IHA precoder $\mathbf{P}_{\mathbf{HA_{ng}}} = \mathbf{P}_{\mathbf{AG_{ng}}}\mathbf{P}_{\mathbf{DG_{ng}}}$. The proposed *ngth* group hybrid combiner $\mathbf{W}_{\mathbf{HA_{ng}}}$ can be calculated in the same way.

## 5. Complexity analysis

This section aims to examine the implementation complexities of the proposed hybrid precoding and combining algorithms for various architectures. To simplify the analysis,

we use the following notations: $N = \max\{N_t, N_r\}$ represents the maximum number of antennas, $N_{\text{RF}} = \max\{N_{tRF}, N_{rRF}\}$ represents the maximum number of RF chains, $N_{\text{g}} = \max\{N_{tg}, N_{rg}\}$ represents the maximum number of RF groups, and $K$ denotes the maximum number of iterations for the proposed IFA hybrid design, ISA hybrid design, and IHA hybrid design algorithms. Moreover, we denote the number of antennas for each subarray in the SA design as $N_{\text{SA}}$. Our analysis is based on the total number of floating-point operations (flops) for each hybrid precoding and combining method. **Table 1** shows that the computational complexities of the proposed IFA hybrid design, ISA hybrid design, and IHA hybrid design algorithms are much lower compared to that of the FA sparse hybrid precoding method, which has a complexity of $\text{O}(N^2 N_{RF} N_S)$. Furthermore, the computational complexities of the proposed IHA hybrid design and ISA hybrid design algorithms are lower than that of the IFA Hybrid design, particularly for larger numbers of groups, $N_{\text{g}}$. When $N_{\text{g}} > 1$, the proposed IHA hybrid design and ISA hybrid design algorithms have lower hardware costs than the sparse hybrid design and the proposed IFA hybrid design. To summarize, the proposed IHA hybrid design has lower computational and hardware complexities than the proposed IFA hybrid design and is comparable to that of the proposed ISA hybrid design when $N_{\text{RF}} = N_{\text{g}}$.

| Method | Constraints | Phase Shifters Number | Complexity |
|---|---|---|---|
| Sparse hybrid design [2] | RF precoding/combining codebooks | $NN_{\mathrm{RF}}$ | $O(N^2N_{RF}N_S)$ |
| Proposed IFA hybrid design | None | $NN_{\mathrm{RF}}$ | $O(NN_{RF}^2K)$ |
| Proposed ISA hybrid design | None | $N$ | $O(N_{\mathrm{SA}}N_{\mathrm{RF}}N_SK)$ |
| Proposed IHA hybrid design | None | $(N_{\mathrm{RF}}/N_{\mathrm{g}})N$ | $O(NN_{RF}^2K/N_g^2)$ |

**Table 1.**
*Complexity of the proposed algorithms.*

## 6. Simulation results

This section presents the numerical results to show the performance advantages of the proposed IFA, ISA, and IHA hybrid precoding/combining algorithms. We consider the case where there are only one BS and one MS at a distance of 100 m. The spacing between antenna elements is equal to $\lambda/2$. The system is assumed to operate at a 28 GHz carrier frequency in an outdoor scenario, and with a path loss exponent $n = 3.4$. The channel model is described in (1), with $\overline{P_{\alpha,i}} = 1$ for all clusters. The azimuth and elevation angles of arrival and departure (AoAs/AoDs) of the rays within a cluster are assumed to be randomly Laplacian distributed. The AoAs/AoDs azimuths and elevations of the cluster means are assumed to be uniformly distributed. We use the AoD/AoA beamforming codebooks (the exact array response of the mmWave channel) at the BSs and MSs, respectively, for the sparse hybrid design [2]. The signal-to-noise ratio (*SNR*) in all the plots is defined as $SNR = \rho/\sigma^2$. We assume perfect channel estimation at the BS and MS. For fairness, the same total power constraint is enforced on all precoding/combining solutions. The maximum number of iterations $K$ for the proposed IHA hybrid precoder/combiner, the IFA hybrid precoder/combiner, and the ISA hybrid precoder/combiner is equal to 10 for all data cases.

In this section, we show the spectral efficiencies achieved by the proposed IFA, ISA, and IHA hybrid precoding/combining algorithms, FA sparse hybrid design [2], and the optimal unconstrained digital method at both the BS and the MS.

**Figure 3** shows the spectral efficiencies achieved by the proposed IHA hybrid precoding/combining, the FA sparse hybrid precoding/combining [2], the optimal unconstrained digital design, the proposed IFA hybrid precoding/combining, and the proposed ISA hybrid precoding/combining in a 256 x 64 uniform planar arrays (UPAs) mmWave system for different *SNR* values with $N_S \in \{2, 8\}, N_{tRF} = N_{rRF} \in \{4, 16\}$, and $N_g \in \{1, 2, 4, 8, 16\}$. The spectral efficiency performance of the proposed IFA hybrid precoder/combiner is close to that of the unconstrained digital one and better than those of other methods for all cases. The proposed IHA hybrid precoding/combining method outperforms the ISA hybrid precoder, regardless of the number of data streams $N_S$ and the number of groups $N_g$. Also, the proposed IHA hybrid precoding/combining design outperforms the FA sparse hybrid design when $N_g \in \{1, 2\}$ for $N_S = 2$ and 8. The performance of the proposed IHA hybrid precoding/combining is degraded with the increase of $N_g$, which is equivalent to the decrease of phase shifters, leading to an increase of the interference between data streams. However, when $N_g \in \{4, 16\}$, the proposed IHA hybrid

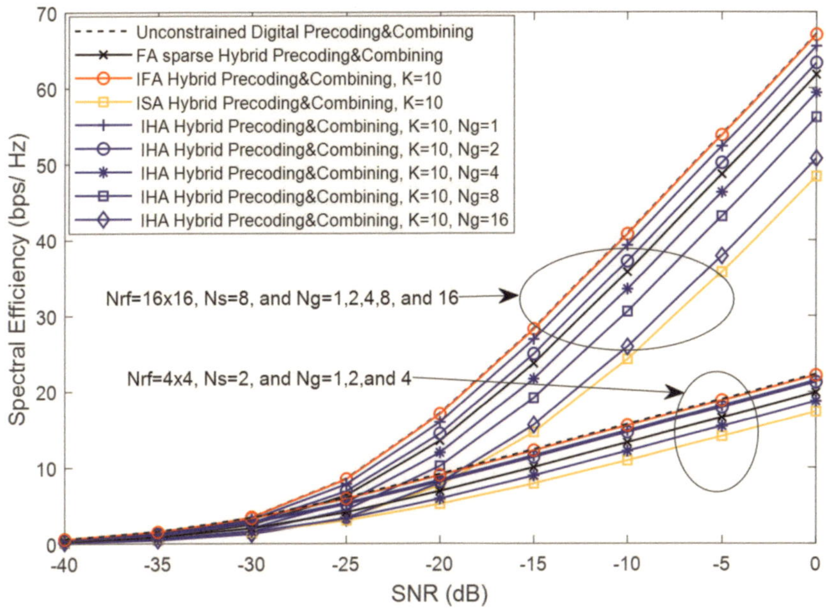


**Figure 3.**
*Average spectral efficiency achieved by the proposed iterative hybrid array (IHA) precoding/combining with $K = 10$, compared to the full array (FA) sparse hybrid precoding/combining design [2], the optimal unconstrained digital precoding/combining, iterative full array (IFA) hybrid precoding/combining design, and the iterative subarray (ISA) hybrid precoding/combining, for a 256 x 64 uniform planar arrays (UPAs) mmWave system for different signal-to-noise ratio (SNR) values with $N_S \in \{2, 8\}$, and $N_{tRF} = N_{rRF} \in \{4, 16\}$.*

precoding/combining becomes similar to the SA architecture with better performance compared to the proposed ISA hybrid precoding/combining. Also, when $N_g = 1$, the performance of IHA hybrid precoding/combining is close to that of the proposed IFA hybrid precoding/combining.

In **Figure 4**, we use the same methods as they were used in **Figure 3** in a 64 x 16 UPAs mmWave system for different SNR values, with $N_S \in \{2, 4\}$, $N_{tRF} = N_{rRF} \in \{4, 8\}$, and $N_g \in \{1, 2, 4, 8\}$. We obtain the same results as in **Figure 3**. However, the proposed IHA hybrid precoding/combining method overlaps with the proposed ISA hybrid precoding/combining when $N_g = 8$ and 4 for $N_S = 4$ and 2, respectively, where the numbers of BS and MS antennas are reduced compared to **Figure 3**.

**Figure 5** shows the performance when the number of RF chains $N_{tRF} = N_{rRF}$ is greater than the number of data streams, where $N_S \in \{2, 4\}$, $N_g \in \{1, 2, 4, 8\}$, and the SNR is fixed to 0 dB over the whole range of RF chains in a 256 x 64 UPAs mmWave system. The spectral efficiency of the proposed IFA hybrid precoding/combining is close to that of the unconstrained digital one with the increase of the RF chains. The performance of the IHA hybrid precoding/combining becomes worse with the increase of $N_g$, where the interference of data streams increases. The performance of the IHA hybrid precoding/combining is much better than that of the proposed ISA hybrid precoding/combining, regardless of the number of $N_g$. Also, the proposed IHA hybrid precoding/combining outperforms the FA sparse hybrid design when $N_g \in \{1, 2\}$ for any data stream $N_S$; however, the FA sparse hybrid design outperforms the proposed IHA hybrid precoding/combining when $N_g \in \{4, 8\}$, but the performance

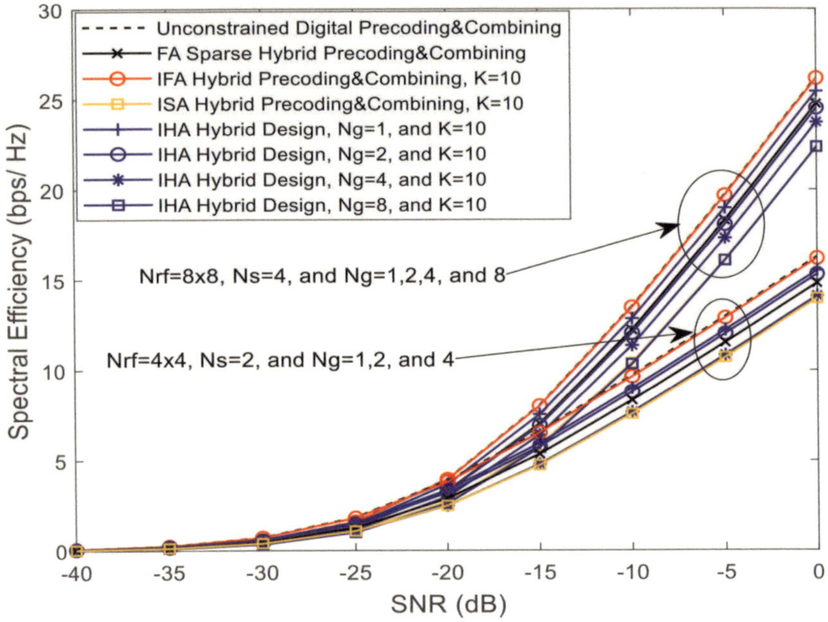


**Figure 4.**
*Average spectral efficiency achieved by the proposed iterative hybrid array (IHA) precoding/combining with $K = 10$, compared to the full array (FA) sparse hybrid precoding/combining design [2], the optimal unconstrained digital precoding/combining, the iterative full array (IFA) hybrid precoding/combining design, and the iterative subarray (ISA) hybrid precoding/combining, for a 64 x 16 UPAs mmWave system for different signal-to-noise ratio (SNR) values with $N_S \in \{2, 4\}$, and $N_{tRF} = N_{rRF} \in \{4, 8\}$.*

gap between them reduces with the increase of the RF chains. Also, when $N_g = 1$, the performance of IHA hybrid precoding/combining is close to that of the proposed IFA hybrid precoding/combining.

**Figure 6** shows the spectral efficiency achieved by the same methods when the number of RF chains equals the number of data streams, varying from 2 to 16, in a 256 x 64 UPAs mmWave system with $N_g \in \{1, 2, 4, 8\}$. The *SNR* is fixed to 0 dB for any number of RF chains. When $N_g = 1$, the performance of IHA hybrid precoder overlaps with that of the proposed IFA hybrid precoding/combining, and both are close to the unconstrained digital one. The performance of the proposed IHA hybrid precoding/combining becomes worse with the increase of $N_g$, where the interference of data streams becomes higher. As seen in **Figure 6**, the proposed IHA hybrid precoding/combining outperforms the FA sparse hybrid precoding/combining and the proposed ISA hybrid precoding/combining, especially for a large number of $N_g$, and $N_S = N_{tRF} = N_{rRF}$.

In conclusion, although we use the proposed IHA hybrid design in both transmitter and receiver, its performance is acceptable, especially for $2 \leq N_g < N_{tRF}$ and $2 \leq N_g < N_{rRF}$, when compared to the higher hardware complexity of FA hybrid designs, such as the FA sparse hybrid design and the proposed IFA hybrid design. All FA hybrid designs require a higher hardware complexity in the BS and MS, with a higher number of phase shifters in the BS and MS, which is equal to $N_t N_{tRF} + N_r N_{rRF}$, whereas the number of phase shifters for the IHA hybrid precoder/combiner is equal to $\left(\frac{N_t N_{tRF}}{N_g}\right) + \left(\frac{N_r N_{rRF}}{N_g}\right)$. The constraint of the analog and baseband precoding/

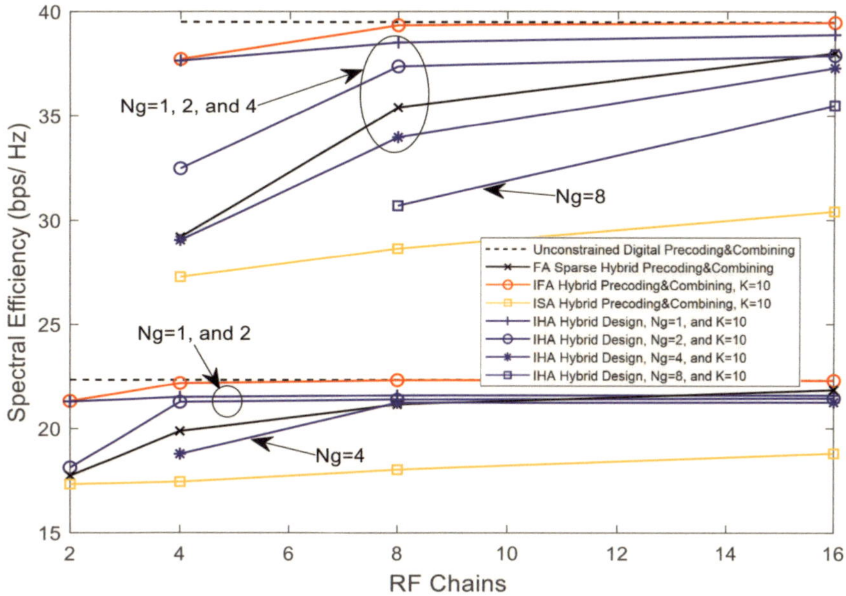


**Figure 5.**
*Average spectral efficiency achieved by the proposed iterative hybrid array (IHA) precoding/combining with $K = 10$ compared to the full array (FA) sparse hybrid precoding/combining [2], the optimal unconstrained digital precoding/combining, the iterative full array (IFA) hybrid precoding/combining with $K = 10$, and the iterative subarray (ISA) hybrid precoding/combining with $K = 10$ for 256 x 64 uniform planar arrays (UPA) mmWave systems for signal-to-noise ratio (SNR) = 0 dB with $N_S \in \{2, 4\}$ and different radio frequency (RF) chains.*

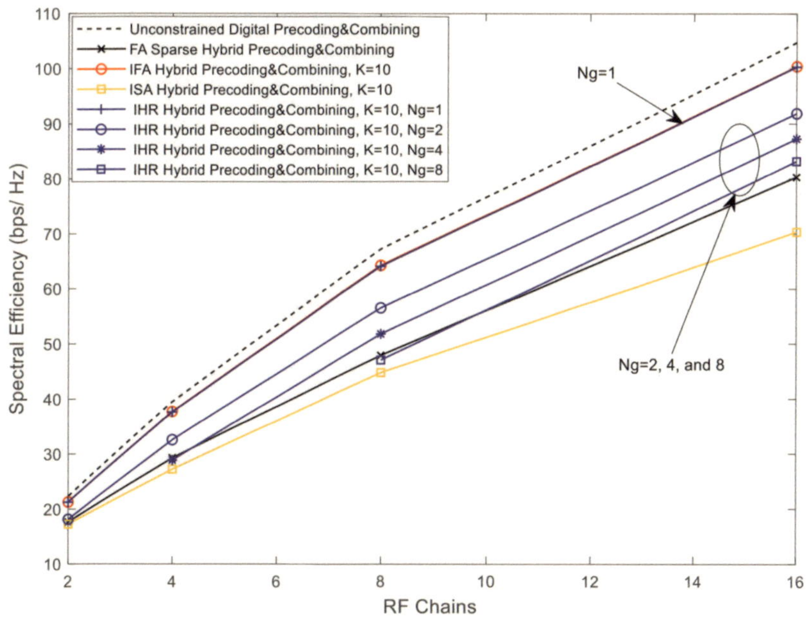


**Figure 6.**
*Average spectral efficiency achieved by the proposed hybrid array (HA) precoding/combining using Algorithm 2 with $K = 10$ compared to the full array (FA) sparse hybrid precoding/combining [2], the optimal unconstrained digital precoding/combining, iterative full array (IFA) hybrid precoding/combining with $K = 10$, and the iterative subarray (ISA) hybrid precoding/combining with $K = 10$ for 256 x 64 uniform planar arrays (UPAs) mmWave systems for signal-to-noise ratio (SNR) = 0 dB with $N_S = N_{tRF} = N_{tRF}$.*

combining matrices helps to build the structure of block diagonal matrices in the proposed IHA hybrid precoder/combiner, yielding higher gains compared to the other methods. When $N_g = N_{tRF}$, the proposed HA structure becomes similar to the SA one; the performance of the proposed IHA hybrid precoding/combining design gives higher gains compared to the proposed ISA hybrid precoding/combining design, especially for a large number of BS antennas.

Also, when $N_g = 1$, the proposed HA structure becomes similar to the FA one, and the performance of the proposed IHA hybrid precoding/combining design is comparable to that of the proposed IFA hybrid precoding/combining design. The number of iterations should be 10 or less because the gain after that will be very small, which is confirmed by our results that we did not include in this chapter.

## 7. Conclusion

In this chapter, we have studied and discussed the issue of hybrid precoding and combining techniques in mmWave MIMO systems for different array architectures. We presented the system models of FA, SA, and HA and solved the optimization problem of hybrid precoding and combining to maximize the spectral efficiency of each architecture. Additionally, we proposed iterative hybrid precoding and combining algorithms for all architectures. The simulation results showed that the proposed algorithms can enhance the spectral efficiency performance of mmWave MIMO systems with lower complexity and hardware requirements than traditional hybrid design methods. The findings of this chapter are expected to be of significant interest to researchers, engineers, and students working in the field of mmWave communications and MIMO systems, as they provide insights into improving the spectral efficiency and performance of wireless communication systems. Overall, this work contributes to the development of efficient and cost-effective solutions for next-generation wireless communication systems.

## Author details

Faisal Al-Kamali[1]*, Mohamed Alouzi[1], Claude D'Amours[1] and Francois Chan[1,2]

1 School of Electrical Engineering and Computer Science, University of Ottawa, Ottawa, Canada

2 Department of Electrical and Computer Engineering, Royal Military College of Canada, Kingston, Canada

*Address all correspondence to: faisalalkamali@gmail.com

## References

[1] Xiao M et al. Millimeter wave communications for future mobile networks. IEEE Journal on Selected Areas in Communications. 2017;**35**(9): 1909-1935. DOI: 10.1109/JSAC.2017.2719924

[2] Ayach O, Rajagopal S, Abu-Surra S, Pi Z, Heath R. Spatially sparse precoding in Millimeter wave MIMO systems. IEEE Transactions on Wireless Communications. 2014;**13**(3):1499-1513. DOI: 10.1109/TWC.2014.011714.130846

[3] Alkhateeb A, Mo J, Gonzalez-Prelcic N, Heath R. MIMO precoding and combining solutions for millimeter-wave systems. IEEE Communications Magazine. 2014;**52**(12):122-131. DOI: 10.1109/MCOM.2014.6979963

[4] Alouzi M, Chan F, D'Amours C. Low complexity hybrid precoding and combining for Millimeter wave systems. IEEE Access. 2021;**9**:95911-95924. DOI: 10.1109/ACCESS.2021.3093880

[5] Park W, Choi J. Hybrid precoding and combining strategy for MMSE-based rate balancing in mmWave multiuser MIMO systems. IEEE Access. 2022;**10**: 88043-88057. DOI: 10.1109/ACCESS.2022.3199875

[6] Wan Q, Fang J, Chen Z, Li H. Hybrid precoding and combining for Millimeter wave/sub-THz MIMO-OFDM systems with beam squint effects. IEEE Transactions on Vehicular Technology. 2021;**70**(8):8314-8319

[7] Wang S et al. A joint hybrid precoding/combining scheme based on Equivalent Channel for massive MIMO systems. IEEE Journal on Selected Areas in Communications. 2022;**40**(10): 2882-2893. DOI: 10.1109/JSAC.2022.3196099

[8] Méndez-Rial R, Rusu C, González-Prelcic N, Heath RW. Dictionary-free hybrid precoders and combiners for mmWave MIMO systems. In: 2015 IEEE 16th International Workshop on Signal Processing Advances in Wireless Communications (SPAWC), Stockholm, Sweden. 2015. pp. 151-155. DOI: 10.1109/SPAWC.2015.7227018

[9] Rusu C, Mèndez-Rial R, González-Prelcic N, Heath RW. Low complexity hybrid precoding strategies for Millimeter wave communication systems. IEEE Transactions on Wireless Communications. Dec 2016;**15**(12):8380-8393. DOI: 10.1109/TWC.2016.2614495

[10] Yu X, Shen J-C, Zhang J, Letaief KB. Alternating minimization algorithms for hybrid precoding in millimeter wave MIMO systems. IEEE Journal on Selected Topics in Signal Processing. 2016;**10**(3): 485-500. DOI: 10.1109/JSTSP.2016.2523903

[11] Sohrabi F, Yu W. Hybrid digital and analog beamforming design for large-scale antenna arrays. IEEE Journal on Selected Topics in Signal Processing. 2016;**10**(3):501-513. DOI: 10.1109/JSTSP.2016.2520912

[12] Chen J-C. Gradient projection-based alternating minimization algorithm for designing hybrid beamforming in millimeter-wave MIMO systems. IEEE Communications Letters. 2019;**23**(1): 112-115. DOI: 10.1109/LCOMM.2018.2878712

[13] Jin J, Zheng YR, Chen W, Xiao C. Hybrid precoding for millimeter wave MIMO systems: A matrix factorization approach. IEEE Transactions on Wireless Communications. 2018;**17**(5):3327-3339. DOI: 10.1109/TWC.2018.2810072

[14] Ni W, Dong X. Hybrid block diagonalization for massive multiuser MIMO systems. IEEE Transactions on Communications. 2016;**64**(1):201-211. DOI: 10.1109/TCOMM.2015.2502954

[15] Nguyen DHN, Le LB, Le-Ngoc T, Heath RW. Hybrid MMSE precoding and combining designs for mmWave multiuser systems. IEEE Access. 2017;**5**: 19167-19181. DOI: 10.1109/access.2017. 2754979

[16] Gao X, Dai L, Han S, Line C, Heath R. Energy-efficient hybrid Analog and digital precoding for mmWave MIMO systems with large antenna arrays. IEEE Journal on Selected Areas in Communications. 2016;**34**(4): 998-1009. DOI: 10.1109/JSAC.2016. 2549418

[17] Al-Kamali F, D'Amours C. Low-complexity hybrid precoding for subarray architecture mmWave MIMO systems. IEEE Access. 2022;**10**: 74921-74930. DOI: 10.1109/ ACCESS.2022.3190511

[18] Al-Kamali F, D'Amours C, Chan F. Hybrid precoding for mmWave MIMO systems with overlapped subarray architecture. IEEE Access. 2022;**10**: 130699-130707. DOI: 10.1109/ ACCESS.2022.3228496

[19] Hu C, Zhang J. Hybrid precoding design for adaptive subconnected structures in millimeter-wave MIMO systems. IEEE System Journal. 2019; **13**(1):137-146. DOI: 10.1109/JSYST. 2018.2816927

[20] Park S, Alkhateeb A, Heath R. Dynamic subarrays for hybrid precoding in wideband mmWave MIMO systems. IEEE Transactions on Wireless Communications. 2017;**16**(5):2907-2920. DOI: 10.1109/TWC.2017.2671869

[21] Yan L, Han C, Yuan J. A dynamic array-of-subarrays architecture and hybrid precoding algorithms for terahertz wireless communications. IEEE Journal on Selected Areas in Communications. 2020;**38**(9): 2041-2056. DOI: 10.1109/JSAC.2020. 3000876

[22] Yang F, Wang J, Cheng M, Wang J, Lin M, Cheng J. A partially dynamic subarrays structure for wideband mmWave MIMO systems. IEEE Transactions on Communications. 2020;**68**(12):7578-7592. DOI: 10.1109/ TCOMM.2020.3020833

[23] Dsouza K, Prasad K, Bhargava V. Hybrid precoding with partially connected structure for millimeter wave massive MIMO OFDM: A parallel framework and feasibility analysis. IEEE Transactions on Wireless Communications. 2018;**17**(12): 8108-8122. DOI: 10.1109/TWC.2018. 2874227

[24] Attiah K, Sohrabi F, Yu W. Deep learning for channel sensing and hybrid precoding in TDD massive MIMO OFDM systems. IEEE Transactions on Wireless Communications. 2022;**21**(12): 10839-10853. DOI: 10.1109/ TWC.2022.3187790

[25] Yuan Q, Liu H, Xu M, Wu Y, Xiao L, Jiang T. Deep learning-based hybrid precoding for terahertz massive MIMO communication with beam squint. IEEE Communications Letters. 2023;**27**(1): 175-179. DOI: 10.1109/LCOMM.2022. 3211514

[26] Elbir AM, Papazafeiropoulos AK. Hybrid precoding for multiuser millimeter wave massive MIMO systems: A deep learning approach. IEEE Transactions on Vehicular Technology. 2020;**69**(1):552-563. DOI: 10.1109/ TVT.2019.2951501

[27] Alouzi M, Al-Kamali F, D'Amours C, Chan F. Direct conversion of hybrid precoding and combining from full Array architecture to subarray architecture for mmWave MIMO systems. IEEE Access. 2023;**11**: 35457-35468. DOI: 10.1109/ACCESS.2023.3264838

[28] Schönemann PH. A generalized solution of the orthogonal Procrustes problem. Psychometrika. 1966;**31**(1): 1-10. DOI: 10.1007/BF02289451

[29] Rusu C, Mèndez-Rial R, González-Prelcic N, Heath RW. Low complexity hybrid precoding strategies for Millimeter wave communication systems. IEEE Transaction on Wireless Communication. 2016;**15**(12):8380-8393. DOI: 10.1109/TWC.2016.2614495

Chapter 2

# Technological Evolution from RIS to Holographic MIMO

*Jiguang He, Chongwen Huang, Li Wei, Yuan Xu, Ahmed Al Hammadi and Merouane Debbah*

## Abstract

Multiple-input multiple-output (MIMO) techniques have been widely applied in current cellular networks. To meet the ever-increasing demands on spectral efficiency and network throughput, more and more antennas are equipped at the base station, forming the well-known concept of massive MIMO. However, traditional design with fully digital precoding architecture brings high power consumption and capital expenditure. Cost- and power-efficient solutions are being intensively investigated to address these issues. Among them, both reconfigurable intelligent surface (RIS) and holographic MIMO (HMIMO) stand out. In this chapter, we will focus on the ongoing paradigm shift from RIS to HMIMO, covering both topics in detail. A wide range of closely related topics, e.g., use cases, hardware architectures, channel modeling and estimation, RIS beamforming, HMIMO beamforming, performance analyses of spectral- and energy-efficiency, and challenges and outlook, will be covered to show their potential to be applied in the next-generation wireless networks as well as the rationales for the technological evolution from RIS to holographic MIMO.

**Keywords:** reconfigurable intelligent surface, holographic MIMO, energy efficiency, channel estimation, hybrid precoding

## 1. Introduction

In order to fully achieve 5G/6G enhanced mobile broadband communications, it is an inevitable trend to move from microwave frequency band, e.g., sub-6 GHz to millimeter wave (mmWave) or even terahertz (THz) frequency bands [1]. However, various challenges are introduced accordingly, such as severe path loss, high power consumption, serious hardware impairment, and frequent blockage. At the early stage, researchers devoted to addressing these aforementioned challenges through hybrid analog-digital precoding along with large-sized antenna arrays at both transmitter and receiver sides, which was verified to offer nearly the same performance in terms of achievable rate compared to its pure digital precoding counterpart under full channel state information (CSI) assumption [2]. However, the energy efficiency (EE) of mmWave or THz system needs to be further enhanced, since the power consumption from digital-to-analog/analog-to-digital converter and up/down converter in the radio frequency (RF) chains is still high. Various novel massive MIMO architectures are analyzed and compared in [3], where reflect/transmit array-based architectures are verified to obtain the best trade-off.

There are two promising candidate technologies, aiming for not only enhancing EE but also spectrum efficiency (SE). One is the reconfigurable intelligent surface (RIS), and the other is holographic multiple-input multiple-output (HMIMO). At the early stage, much more effort were made on RIS-assisted networks compared to HMIMO. Recently, the emphasis of the research community has been shifted toward HMIMO. RIS and HMIMO share some similarities, e.g., made of a large number of cost-efficient low power consumption elements. However, they differ in various aspects. The RIS is usually a passive, tunable, and intelligent metasurface. Unlike the traditional active relays, which can perform either amplify-and-forward (AF), decode-and-forward (DF), or compute-and-forward (CF), RIS does not possess any baseband processing capability. Thus, it can not receive and post-process any incident signals from other network nodes, e.g., base station (BS) or mobile station (MS). Because of this, RIS brings difficulty in efficient yet effective channel estimation for the system. For instance, the BS has to estimate two large dimensional channel coefficient matrices (i.e., MS-RIS and RIS-BS channels) simultaneously via uplink pilot signal transmission. In addition, during the sounding process, coordination and strict synchronization among the MS, RIS, and BS are required. Nevertheless, by deploying an RIS between the BS and the MS, it enables virtual line-of-sight (LoS) transmission especially when the direct LoS between the BS and the MS is temporally blocked. The RIS can also enhance radio localization thanks to the following reasons: (i) The RIS is a natural anchor upon its deployment; (ii) The RIS offers high-resolution angle of departure (AoD) and/or angle of arrival (AoA) estimation; (iii) The RIS can further extend the localization range. Studies show that RIS can also benefit integrated sensing and communication (ISAC) [4].

The full potential of RIS can not be realized unless the CSI acquisition is performed efficiently and effectively. However, there is still a large room for improvement. Meanwhile, an obvious trend has been seen for a paradigm shift from RIS to HMIMO. Under the framework of HMIMO, the tunable metasurface acts as an active transceiver, which is a greener way to implement the massive (mmWave/THz) MIMO systems without the need of a massive number of RF chains. The HMIMO transceiver is made of densely packed meta-atoms, usually with sub-wavelength inter-element spacing (unlike half-wavelength inter-element spacing for RIS), enabling super directivity. However, HMIMO has more severe mutual coupling effect compared to RIS. The large-sized HMIMO aperture pushes the far field further away. Therefore, most studies focus on radiative near-field propagation when modeling the HMIMO channels. An interesting finding that extra degrees of freedom (DoFs) exist even when the MS is located at the LoS path of the BS has been discovered recently [5]. HMIMO surfaces are powerful in transferring the orbital angular momentum (OAM) property, yielding enhance system capacity within a few Rayleigh distances. A comprehensive comparison between RIS and HMIMO can be found in **Table 1**. The foreseeable key performance indicator (KPI) enhancement by the introduction of RIS and/or HMIMO generally includes: (i) Gbps or even Terabit/s level average or peak data rate, (ii) seven 9's reliability, (iii) sub-ms air interface latency, and (iv) cm-level localization accuracy.

From the industrial perspective, ETSI lauched an Industry Specification group on RIS, covering various research aspects. RIS Tech Alliance (RISTA) focuses on bringing together industrial and academic partners, pushing the RIS techniques from theory into practice. The RIS technology white paper was released by RISTA on March 2023.

| Category | RIS | HMIMO |
|---|---|---|
| Passive vs. Active | Passive | Active |
| Role | Passive Reflector/Relay | Transceiver |
| Operation Mode | Full Duplex | Full or Half Duplex |
| Cost | Low | Medium |
| Inter Element Spacing | Half Wavelength | Less than Half Wavelength |
| Mutual Coupling | Mild | Medium |
| Energy Efficiency | High | Medium |
| CSI Acquisition | Very Difficult | Difficult |
| Propagation Environments | Near and Far Fields | Near and Far Fields |
| Degree of Freedom | Low | High |

**Table 1.**
*Comprehensive comparison between RIS and HMIMO.*

Also, 3GPP listed RIS as one of the additional RAN1/2/3 candidate topics. In this sense, it will probably receive a high chance to be studied and included in Release 19. HMIMO techniques have not been studied and included in any global or regional standard development organization (SDO) bodies yet. However, it is highly probable that it will gain tremendous attraction and momentum in the near future as a technique beyond massive MIMO.

## 2. RIS vs. HMIMO hardware architectures

Both RIS and HMIMO are metasurfaces equipped with integrated electronic circuits that can intelligently control the incoming waves, resulting in a programmable electromagnetic field. They are both composed of feeding, substrate, and unit cells. Precisely, the feed can excite the RIS or HMIMO to generate the desired electromagnetic waves, and the substrate support the structure. In addition to the feeding line and substrate, the radiation elements play an important role, which is mounted on the substrate and form uniform/non-uniform radiation patterns, transforming the reference waves into radiated waves. There are some differences in the fabrications or configurations between the RIS and HMIMO.

Specifically, in RIS systems, the feeds are set outside the meta-surface while the feeds are attached to the HMIMO surface in a more flexible behavior [6]. In such way, the electromagnetic waves propagate along the HMIMO surface, and the elements are excited one by one, enabling HMIMO to serve as a transceiver. However, RIS requires additional configuration of external feeding lines to excite unit cells, as shown in **Figure 1**. In addition, since the long feeding line is adopted in RIS, the layout in the implementation is much more complex than the series feeding in HMIMO systems. Thus, HMIMO is more suitable to be implemented in various scenarios compared with RIS. Then we will discuss the details in fabrication from the perspective of feed, substrate, and unit cells for RIS and HMIMO.

In RIS, the spatial feeding techniques are adopted, such as a horn antenna or microstrip antenna array are placed very close to RIS to feed such structures. The feed is adopted to generate a reference wave to excite unit cells in RIS. A single- or a

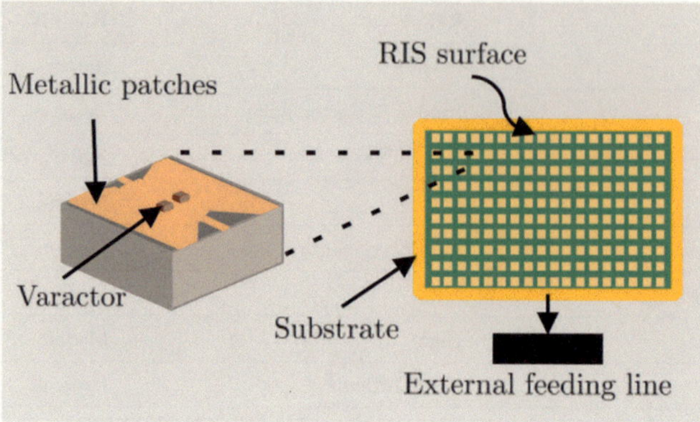


**Figure 1.**
*A schematic view of RIS hardware structures.*

multi-layer stack of planar structure are fabricated using lithography and nano-printing methods [6]. Each RIS element situated on a ground plane adopts varactor diodes or other electrical materials to reflect incoming waves electronically by providing phase modulation. Since no amplifier is employed, RIS consumes less energy and is easily deployed into building facades, room, up to being integrated into human devices. The input voltage to varactor diodes can be controlled to provide variable capacitance, which is an important characteristic of the materials that constitute RIS units. The unit cell in RIS can be fabricated using various varactor types, including metal plate with vias, D-shaped patch, split-ring, and conductive patch separated by the annular slot [7]. In addition to varactor diodes, positive intrinsic-negative (PIN) diodes can also be utilized to tune the impedance in RIS unit cells, i.e., the on−/off-state of PIN diodes exhibits the magnitude and phase difference of the reflection.

Compared with RIS, the fabrication in HMIMO is much more diverse due to development in metamaterials, as shown in **Figure 2**. In HMIMO systems, the feed is integrated into the HMIMO or located externally. The reference wave generated by the feed propagates along the HMIMO surface, then the designed wave is excited from the interference wave of the reference wave and the reflected wave from the object. Clearly, the location of the feed generates a specific propagation mode, for example, a transverse electric propagation mode of the reference wave is supported if the feed is located on the HMIMO surface, while a transverse magnetic propagation mode is supported if the feed is placed on the bottom of HMIMO surface [9]. In addition to the location of feed, the feed material also affects the propagation mode. For example,

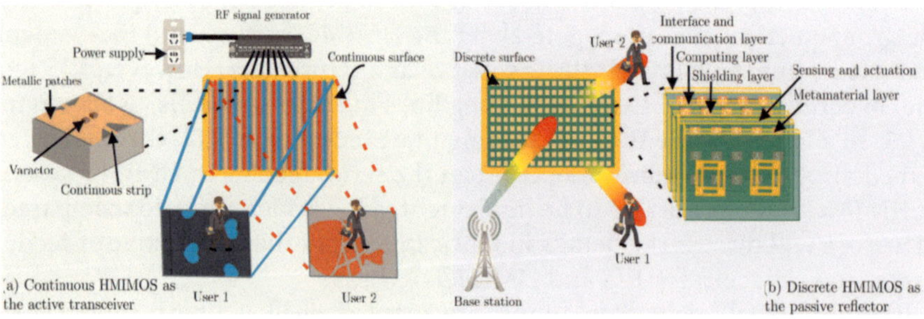


**Figure 2.**
*The two operation modes of HMIMOS systems along with their implementation and hardware structures [8].*

dipoles, planar Vivaldi feed, antipodal Vivaldi feed, and dipole-based Yagi-Uda-feed are capable of exciting transverse electric propagation mode [9].

The substrate enables the reference wave to propagate along the HMIMO surface. The substrate can be in a plate shape that supports the surface wave mode or in a microstrip line shape that supports the waveguide mode [9]. Surface wave mode refers to the propagation of electromagnetic waves along the surface of the HMIMO transceiver. By controlling the states and properties of elements on the surface, HMIMO can guide and manipulate surface waves to achieve specific functions such as wavefront shaping or beamforming. Waveguide mode refers to the guided propagation of electromagnetic waves within internal or integrated waveguides, which can be within the system or attached to it. These waveguides may serve as feeds or interconnections within the HMIMO system. In addition, the substrate materials are also different, i.e., the dielectric substrate and semiconductor substrate [9]. For example, the dielectric substrate is commonly adopted in HMIMO systems, such as printed circuit board, laminates substrate, silicon dioxide substrates, and anisotropic artificial substrates. The silicon dioxide substrate also possesses a low dielectric loss, such as the graphene patches transferred silicon dioxide substrate adopted in tunable THz HMIMO systems. In addition to the above dielectric substrate, the semiconductor substrate is also employed in HMIMO systems for the low cost and excellent conductivity [9].

The radiation elements can be manufactured by metal, dielectric, and graphene materials [9]. Specifically, metal radiation elements exhibit high conductivity and are applicable to low frequencies with insignificant losses. The dielectric radiation elements are more suitable for a wider range of bandwidth. The graphene radiation elements are also the perfect choice for THz HMIMO systems or optical communications.

The HMIMO surface can be divided into contiguous and discrete modes [8]. In contiguous HMIMO systems, a virtually uncountably infinite number of radiation elements are incorporated in a limited area, generating spatially continuous aperture. Such a scheme provides a theoretically infinite number of elements that can approach the inherent capacity and spatial resolution limit. However, the contiguous mode is impractical, thus a discrete HMIMO mode is proposed, which is composed of countable radiation elements. This mode has higher feasibility and lower power consumption while achieving lower spatial resolution and undesired side lobes. Some studies have compared the contiguous mode and discrete mode to investigate the optimal discretization bits. Hu et al. [10] showed that 2 bits quantization is able to approximate the sum rate with contiguous HMIMO systems for multi-user scenario, while 1 bit quantization is enough in the single-user scenario.

In order to achieve both contiguous and discrete apertures, fabrication methods are important. Typically, programmable metamaterials are adopted to approximate the contiguous HMIMO surfaces, where the varactor loading technique is adopted in continuous monolayer metallic structures incorporating a large number of meta-particles [8]. Each meta-particle consists of two metallic trapezoid patches, varactor diodes, and a continuous strip. Specifically, whether the element radiates the energy of the reference wave into free space depends on the state of the diodes [11]. The bias voltage is input to varactor diodes to manipulate the phase and amplitude of each radiation element, resulting in a controlled electromagnetic environment. Different from the contiguous HMIMO systems, the discrete HMIMO systems involve a number of meta-particles, which are composed of a metamaterial layer (graphene material), sensing and actuation layers, shielding layer, computing layer, and interface and communication layer [8].

## 3. RIS vs. HMIMO channel modeling and estimation

Channel modeling plays an essential role in understanding the fundamentals of the channel characteristics, developing cutting-edge signal processing algorithms, and optimizing network resources. The authors in ref. [12] focused on the free-space path loss models for RIS-assisted wireless communications and established a rigorous relationship between them and system parameter setups, e.g., the distances between the transmitter/receiver and the RIS, the RIS aperture, the radiation patterns of antennas, etc. The power scaling laws were studied in ref. [13] for asymptotically large RISs. The work in ref. [14] divided the RIS into multiple RIS tiles and derived the corresponding tile response functions with arbitrary transmission mode, and incident and reflection directions. Based on these, a physics-based end-to-end channel model was developed for the RIS-assisted wireless systems, taking into account the effect of transmission mode, incident angle, and reflection angle of all RIS tiles.

RIS channel estimation can be classified into three major categories, including model-based, data-driven, and the mixture of the former two. For the model-based RIS CE, various approaches are considered to estimate the individual channels, cascaded channel, or channel parameters. Under the assumption of passive RIS, the pilot signals received at the BS (over uplink transmissions) or MS (over downlink transmissions) include the information of channel coefficient matrices for both hops, i.e., BS-RIS and RIS-MS links, which in turn brings more difficulties on CSI acquisition. In the literature, such kind of RIS CE can be done by leveraging the bilinear generalized approximate message passing (BiG-AMP) for sparse matrix factorization and the Riemannian manifold gradient-based algorithm for matrix completion [15]. The framework of two-stage RIS-aided channel estimation (TRICE) was proposed in ref. [16] to estimate the cascaded channel matrix, followed by parallel factor (PARAFAC) tensor decomposition [17] to obtain the two individual channels. Compressive sensing (CS) techniques, e.g., orthogonal matching pursuit (OMP) and generalized approximate message passing (GAMP), were also applied in the RIS-assisted networks [18] to estimate the cascaded channel. Different from the previous works, the authors in ref. [19] focus on estimating the channel parameters in two stages, by adopting off-grid CS technique, i.e., atomic norm minimization (AMN). The availability of channel sparsity is essential in the aforementioned approaches. Extension from single-user scenario to multi-user scenario, the different properties of the channels, i.e., BS-RIS and RIS-MS channels, are leveraged in the design of the CE algorithm [20]. The properties include the changing rate and channel sparsity. The former determines the frequency of CE, and the latter determines the training overhead of channel estimation. Note that all the MSs share the same BS-RIS channel, which can be considered to reduce the training overhead. In terms of data-driven approaches, the authors in ref. [21] designed a twin convolutional neural network (CNN) architecture to estimate both direct and cascaded channels from received pilot signals. By leveraging the advantages of both model-based and data-driven approaches, deep unfolding-based RIS channel estimation exhibits excellent performance [22]. **Figure 3** depicts the RIS channel estimation results from different approaches, where three training overhead values, i.e., $K = 24,28,32$, are considered. The numbers of transmit antennas and RIS elements are 16 and 32, respectively, while the MS is assumed to have a single antenna. Deep unfolding outperforms the ANM thanks to its mixture nature of data-driven and model-based approaches [22].

The HMIMO channel modeling focuses on the radiative near-field propagation due to the electromagnetically large antenna arrays employed at the transmitter and/or the receiver. The model incorporating arbitrary scattering propagation conditions was

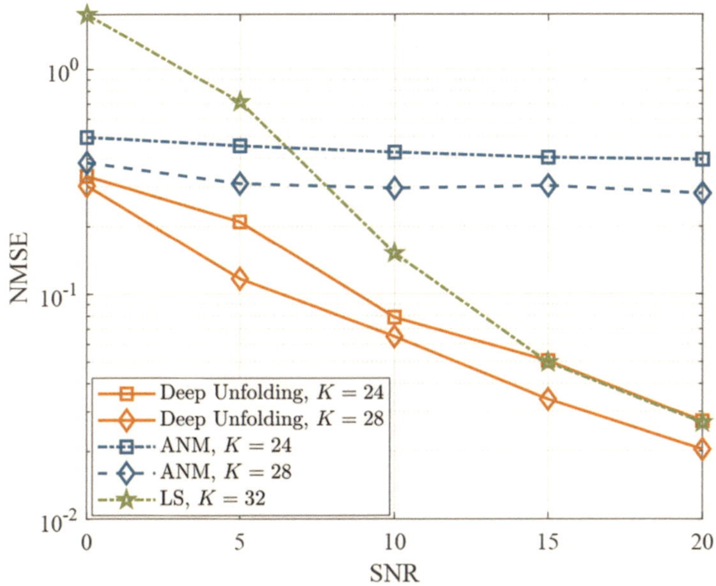


**Figure 3.**
*Comparison of RIS CE with different schemes.*

proposed by adopting the first principles of wave propagation and Fourier plane-wave series expansion of the channel response [23]. The equivalent wavenumber domain channel from the spatial domain was obtained via transformation. Meanwhile, the Fourier plane-wave series expansion of the channel response was studied in ref. [23]. The equivalent wavenumber domain channel from the spatial domain was obtained via transformation. Meanwhile, the physically-meaningful stochastic channel model of non-isotropic radio waves propagation was also obtained for the far-field case [24]. The mathematically tractable HMIMO channel model is essential for algorithm development, such as holographic beamforming, detailed in the next section.

The holographic MIMO channel estimation was investigated in ref. [25], where the authors proposed a subspace-based channel estimation approach for the far-field propagation condition. Such an approach only requires the information of the subspace of the spatial correlation matrix while attaining the performance of minimum mean square error (MMSE) estimator in the high SNR regime. It is well known that the MMSE estimator requires complicated matrix inverse operation and full knowledge of the spatial correlation matrix. An extension to near-field channel estimation can be found in ref. [26], where the polar domain sparsity other than angular domain sparsity along with off-grid CS technique, i.e., polar-domain simultaneous iterative gridless weighted (P-SIGW) scheme, were considered.

## 4. RIS beamforming vs. HMIMO beamforming

The RIS can perform beam focusing and offer beamforming gain in order to compensate for the severe path loss. For the SISO system, the optimal RIS beamforming vector can be found based on the CSI of the BS-RIS and RIS-MS channels. For more complicated scenarios, e.g., MIMO systems, finding the optimal RIS design is not straightforward because of the strict RIS hardware constraints. In the literature, researchers focused on the design of RIS beamforming and TX beamforming/precoding simultaneously, termed as joint active and passive beamforming. References [6, 27] are among the first ones dealing with such a challenging problem. In these works, perfect or imperfect CSI information was assumed. In other words, the joint design is considered after the CE phase, detailed in Section 3. Another alternative way to do this is to create a beam codebook and conduct beam scanning to find the optimal beam pair, one used at the BS and the other used at the RIS. There are several challenges raised in the codebook design. The large-sized RIS requires huge computation for the beam codebook design, aiming at the collection of beams having full coverage of the space. Second, the beam may have many sidelobes, resulting from the RIS hardware constraints. Third, the beam scanning process is supposed to be time inefficient when both nodes adopt large-sized codebooks.

Unlike RIS beamforming, HMIMO beamforming is much more tricky due to the closely packed patch antennas. Even the simple linear precoding schemes, such as zero-forcing (ZF), with continuous-aperture surfaces are impractical due to a large number of patch antennas, not to mention MMSE beamforming, which can be accounted for expensive matrix inversion operation. To solve this, ref. [28] leveraged a novel low-hardware complexity ZF precoding scheme that is based on a Neumann series (NS) expansion, which replaces the expensive matrix inversion operation while being similar in terms of achievable sum rate with conventional ZF, as shown in **Figure 4**. For the holographic beamforming, the authors in refs. [10, 29] also studied discrete amplitude-controlled holographic beamforming and analyzed the effect of radiation amplitude discretization on the sum rate for the downlink multi-user communication system. In order to realize holographic beamforming, the holographic interference principle that the holographic transceivers record the interference between the reference wave and arbitrary desired object waves, known as an interference pattern, is considered. By coupling the reference wave with the interference

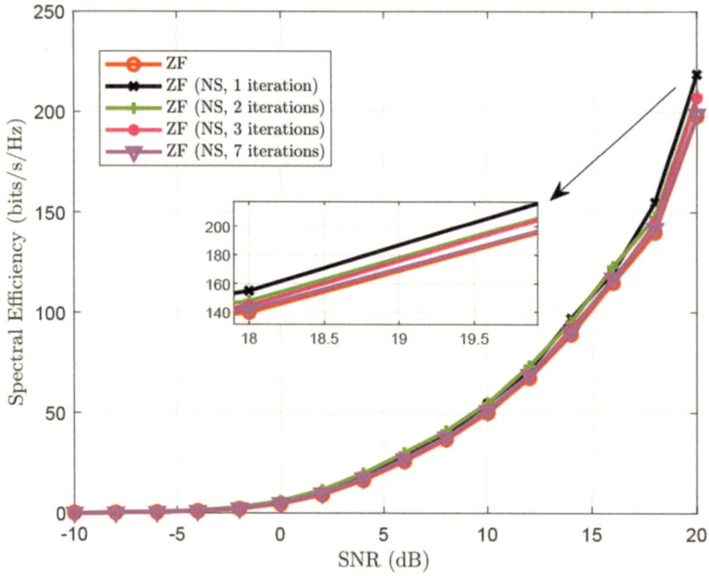


**Figure 4.**
*Comparison of ZF precoding schemes and NS-based ZF precoding for HMIMO systems with 729 transmit patch antennas and 144 receive patch antennas (spacing is $\lambda/3$) [28].*

pattern, the holographic transceivers are capable of performing beamforming toward the desired direction by radiation amplitude control of the reference wave propagating along the metasurface. Di [30] studied joint hybrid digital beamforming and holographic beamforming for wideband OFDM transmissions while compensating for the beam squint loss via linear additivity of holographic interference patterns.

The above work mainly dealt with beamforming in the spatial domain, however, the polarization domain should not be ignored in RIS and HMIMO systems as well. The adoption of the dual-polarization (DP) or tri-polarization (TP) feature is expected to further improve the performance without enlarging antenna array size, enabling multiple independent information to be sent in two or three polarization directions, thus offering polarization diversity in addition to spatial diversity to improve spectral efficiency. However, the cross-polarization in polarization systems also brings new interference and degrades system performance, thus, the beamforming in polarization domain is required. To exploit polarization diversity, a few recent works discussed the deployment of DP RIS systems [31–33]. The work in ref. [33] proposed a RIS-based wireless communication structure to control the reflected beam and polarization state to maximize the received signal power. de Sena et al. [31] also designed a transmission scheme in RIS-assisted systems. Although HMIMO can also integrate polarization techniques, there is still a little difference between RIS and HMIMO. Specifically, due to the large size of HMIMO and higher frequencies, the communication range shifts from the traditional far-field region to the near-field zone [13], and the achievable polarization diversity also increases from two to three, i.e., TP HMIMO is available. The difficulty of polarization interference increases in TP HMIMO since the number of cross-polarization components is one in DP RIS and two in TP HMIMO. To fully exploit polarization diversity and remove both spatial and polarization interference, a two-layer precoding design was investigated for multi-user TP HMIMO systems, which is compared with the user-cluster-based scheme, i.e., different users are assigned to different polarizations [34], as shown in **Figure 5**. A complete list of works on holographic beamforming can be found in ref. [9].

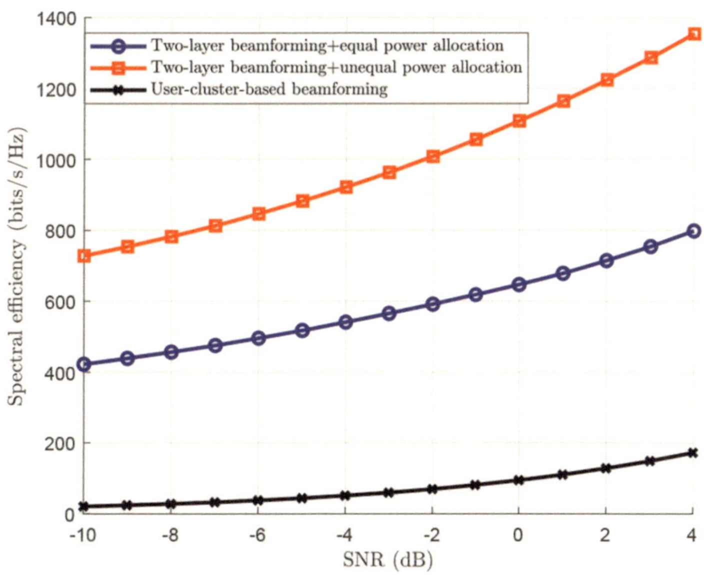


**Figure 5.**
*Spectral efficiency of the user-cluster-based and two-layer beamforming schemes.*

To be specific, the beamforming in the spatial domain mainly removes interference resulting from the mutual coupling or inter-users, while the beamforming in the polarization domain is designed to remove polarization interference caused by cross-polarization components.

## 5. Performance analyses

The phase shift and power allocation schemes are designed in RIS-assisted systems for higher energy efficiency and lower transmit power. Huang et al. [6] developed energy-efficient designs based on alternating maximization, gradient descent search, and sequential fractional programming methods, and these RIS-based resource allocation methods could provide up to 300% higher EE in comparison with the use of regular multi-antenna amplify-and-forward relaying, as shown in **Figure 6**. In addition, Yang et al. [35] adopted a dual method to solve the problem of resource allocation for multiuser communication networks with a RIS-assisted wireless transmitter. In this network, the sum transmit power of the network is minimized by controlling the phase beamforming of the RIS and transmit power of the base station, which could reduce up to 94% and 27% sum transmit power compared to the maximum ratio transmission (MRT) beamforming and ZF beamforming techniques, respectively.

Beamforming design is important in enlarging coverage, enhancing capacity, and removing inter-user interference. For example, Huang et al. [36] proposed a joint design of digital beamforming matrix at the BS and analog beamforming matrices at the RISs for the multi-hop RIS-assisted communication network to improve the coverage range at THz-band frequencies, leveraging deep reinforcement learning (DRL)

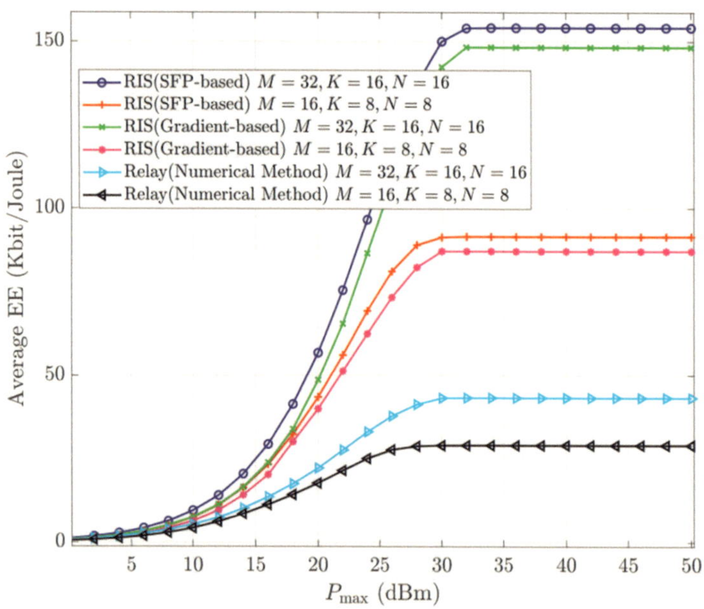


**Figure 6.**
*Average EE using either RIS or AF relay versus the maximum transmit power constraint $P_{max}$ a) $M = 32$ BS antennas, $K = 16$ users, $N = 16$ RIS elements; and b) $M = 16$ BS antennas, $K = 8$ users, $N = 8$ RIS elements [6].*

to combat the propagation loss. Simulation results showed that the two-hop scheme is able to improve 50% more coverage range of THz communications compared with the zero-forcing beamforming without RIS and 14% more transmission distances than that of the single-hop scheme. The authors in ref. [37] jointly optimized the active beamforming of the power station and the passive beamforming of the RIS in an iterative behavior using Lagrange dual theory to improve system EE, simulation results showed that a higher EE is achieved compared to the throughput-based maximization algorithm. The study in ref. [38] investigated the approximations of the average rate of user equipment and a RIS configuration algorithm to improve the average sum rate with low complexity in RIS-assisted multiple-input single-output (MISO) systems. The study in ref. [39] leveraged DRL to jointly design of transmit beamforming matrix at the base station and the phase shift matrix at the RIS, and the result showed its comparable sum-rate performance with the classic weighted minimum mean square error algorithm.

The above methods require the full knowledge of instantaneous CSI, which requires burdensome overhead. Therefore, some works designed beamforming schemes with imperfect CSI. For example, Gao et al. [40] studied the robust beamforming design for RIS-assisted communication systems from a multi-antenna access point to a single-antenna user under imperfect CSI. By decoupling the non-convex optimization problem into two subproblems, the transmit beamforming at the access point is optimized and discrete phase shifts of RIS is designed to minimize the transmission power of access point (AP), subject to a signal-to-noise ratio constraint at the user. Simulation results showed that the proposed scheme can approach the performance of the perfect CSI counterpart and substantially outperform traditional non-robust methods. Gan et al. [41] investigated the ergodic capacity using the alternating direction method of multipliers, fractional programming, and alternating optimization methods, in RIS-assisted multi-user MISO wireless systems, considering statistical CSI instead of instantaneous CSI. Simulation results showed that such statistical CSI design achieved decent performance compared with the instantaneous CSI-based design, especially in the low and moderate SNR regimes. Gan et al. [42] proposed a low-complexity algorithm via the two-timescale transmission protocol in cell-free systems through statistical CSI at RISs and instantaneous CSI at BSs, where the joint beamforming at BSs and RISs is facilitated via alternating optimization framework to maximize the average weighted sum-rate. A power gain on the order of $\mathcal{O}(M)$ is achieved without LoS components, with $M$ being the BS antenna's number.

The DoF is also an important performance indicator. In addition to the inherent DoF limit of the RIS, rotating the RIS rather than moving it over a wide area can also obtain a considerable improvement. For example, Cheng et al. [43] considered the extra DoF offered by the rotation of the RIS plane and investigated its potential in improving the performance of RIS-assisted wireless communication systems by considering the radiation pattern. The results showed that the maximum capacities are obtained by rotating RIS, as shown in **Figure 7**. Compared with RIS systems, the strong mutual coupling generated from the sub-wavelength spacing between adjacent antennas is inevitable in HMIMO communications, resulting in distorted radiation patterns and low radiation efficiencies. However, ignoring the mutual coupling would not seriously affect the DoF of the HMIMO, i.e., the DoF reaches its limit when the antenna number is larger than $2L_x/\lambda_0 + 1$, where $L_x$ is the array size, and $\lambda_0$ is the wavelength [44]. For an antenna number larger than $2L_x/\lambda_0 + 1$, the DoF ceases to increase while the radiation efficiencies keep decreasing, resulting in a reduced capacity.

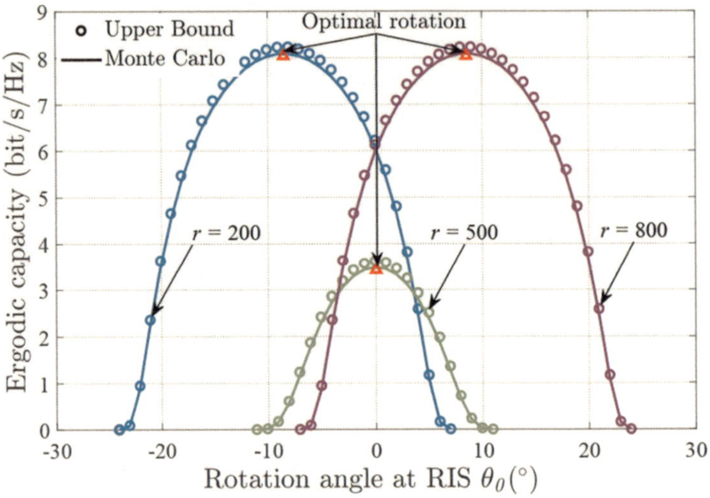


**Figure 7.**
*Impact of RIS rotation angle $\theta_0$ on the ergodic capacity [43].*

## 6. Challenges and outlook

In this chapter, we cover various aspects of RIS and HMIMO, such as channel modeling, estimation, beamforming control, etc. Both of the techniques share some common challenges before their implementations and applications to future wireless systems. For instance, the difficulty of channel estimation increases as the number of meta-atoms increases. It might be even more difficult to estimate RIS channels due to the inherent passive nature of the RIS. The HMIMO channel estimation requires a larger training overhead compared to the current massive MIMO CE due to a further increase in the number of elements. The fundamental study of the performance limits needs a full understanding of the electromagnetic (EM) theory and physics, which also applies to channel modeling.

### 6.1 CE for multi-hop RIS-empowered systems

RIS plays an important role in transmitting signals for unfavorable scenarios, especially when the direct links are blocked due to walls or obstacles. Most of the current works mainly focus on single-hop RIS-assisted systems, however, in practical scenarios where the receiver is quite far away from the transmitter, employing multi-hop RIS for signal relaying becomes imperative. In such a case, the desired signal will pass through more than one RIS, thus a high-order cascaded channel is generated. Unfortunately, the current channel acquisition methods, including model-free based schemes and model-based approaches, are only applicable to single-hop RIS-assisted wireless communications, and CE in multi-hop RIS is complex due to the involvement of higher-order channels. Plus the incapability of signal processing at the RIS part, the CE for multi-hop channels at the receiver/transmitter is much more challenging. Some existing works may enlighten the possible solutions to this challenge, for instance, the involved channels can be represented as variables in a factor graph, the relationship among these channels is denoted as factors, then the effective message-passing algorithms in multi-layers can be derived for the posterior probability of unknown channels. Nevertheless, the inherent ambiguities in such a factor graph should be carefully addressed.

In addition, most of the current works design beamforming for RIS-assisted wireless communications with perfect CSI, however, CE techniques may not perfectly estimate all involved channels, and imperfect CSI has negative impacts on beamforming design. Specifically, the estimation error in cascaded channels in RIS-aided communications may be larger due to error propagation. Consequently, taking estimation errors in beamforming design is necessary, i.e., a more robust beamforming scheme should be designed.

### 6.2 CE-implicit schemes for RIS-assisted communications

The training overhead in the CE process is normally large, therefore, designing RIS-assisted systems in the absence of explicit channel information could save temporal and spatial resources greatly. Fortunately, it is feasible to design such beamforming schemes without explicit CE for various RIS-empowered wireless communications. This approach optimizes system parameters without relying on the traditional explicit CE paradigm, saving training overhead and avoiding power allocation to the training part as well. For example, the explicit CE can be bypassed using machine learning methods to achieve a superior transmission rate or facilitate the phase matrix design using statistical parameters instead of instantaneous CSI.

### 6.3 Low-complexity beamforming for HMIMO systems

The large number of closely packed patch antennas increases the complexity of the beamforming design for HMIMO systems. For instance, the traditional ZF and MMSE beamforming schemes are impractical to be directly applied in hardware design. Therefore, low-complexity beamforming is imperative in practical applications. One beamforming approach is to replace matrix inversion with polynomial functions, as introduced in this chapter. However, such methods rely on the specific channel structure and may diverge under some parameter settings. Consequently, an effective and robust beamforming technique is expected for HMIMO systems, in order to eliminate both spatial and polarization interference or enhance signal strength in the desired direction.

### 6.4 Optimal design for HMIMO

Although the continuous HMIMO can achieve the spatially continuous aperture, it is infeasible to construct such a continuous structure in practical applications. Therefore, the discrete HMIMO that incorporates a large number of patch antennas is the most viable approach. Although increasing the number of patch antennas would bring performance benefits, this improvement reaches the plateau for the specific number of patch antennas, i.e., the optimal number of patch antennas, which can be accounted for by mutual coupling effects. For instance, the more patch antennas placed in a fixed area, the stronger mutual coupling generated, resulting deformed radiation pattern and reduced antenna efficiency. Consequently, the performance gain brought by the larger number of antennas ceases eventually. Based upon this observation, the optimal configuration to achieve the best performance of HMIMO systems is required to be investigated.

Albeit the aforementioned challenges for both RIS and HMIMO techniques, HMIMO other than RIS will be widely recognized as a beyond massive MIMO technique. We will witness the growing trend of paradigm shift from RIS to HMIMO in the near future.

## Abbreviations

| | |
|---|---|
| MIMO | multiple-input multiple-output |
| HMIMO | holographic MIMO |
| RIS | reconfigurable intelligent surface |
| mmWave | millimeter wave |
| THz | terahertz |
| CSI | channel state information |
| EE | energy efficiency |
| RF | radio frequency |
| SE | spectrum efficiency |
| AF | amplify-and-forward |
| DF | decode-and-forward |
| CF | compute-and-forward |
| BS | base station |
| MS | mobile station |
| LoS | line-of-sight |
| AoD | angle of departure |
| AoA | angle of arrival |
| ISAC | integrated sensing and communication |
| DoFs | degrees of freedom |
| OAM | orbital angular momentum |
| KPI | key performance indicator |
| PIN | positive intrinsic-negative |
| BiG-AMP | bilinear generalized approximate message passing |
| PARAFAC | parallel factor |
| OMP | orthogonal matching pursuit |
| GAMP | generalized approximate message passing |
| ANM | atomic norm minimization |
| MMSE | minimum mean square error |
| P-SIGW | polar-domain simultaneous iterative gridless weighted |
| CNN | convolutional neural network |
| CS | compressive sensing |
| DP | dual-polarization |
| TP | tri-polarization |
| DRL | deep reinforcement learning |
| MRT | maximum ratio transmission |
| AP | access point |
| MISO | multiple-input single-output |
| EM | electromagnetic |

## Author details

Jiguang He[1*], Chongwen Huang[2], Li Wei[3], Yuan Xu[2], Ahmed Al Hammadi[1] and Merouane Debbah[1]

1 Technology Innovation Institute, Masdar City, Abu Dhabi, United Arab Emirates

2 College of Information Science and Electronic Engineering, Zhejiang University, Hangzhou, China

3 Engineering Product Development (EPD) Pillar, Singapore University of Technology and Design, Singapore

*Address all correspondence to: jiguang.he@tii.ae

## References

[1] Rappaport TS et al. Millimeter wave Mobile communications for 5G cellular: It will work! IEEE Access. 2013;**1**: 335-349. DOI: 10.1109/ACCESS.2013.2260813

[2] Ayach OE, Rajagopal S, Abu-Surra S, Pi Z, Heath RW. Spatially sparse precoding in Millimeter wave MIMO systems. IEEE Transactions on Wireless Communications. 2014;**13**(3):1499-1513. DOI: 10.1109/TWC.2014.011714.130846

[3] Hans R, et al. A high-level comparison of recent technologies for massive MIMO architectures. arXiv preprint arXiv:2212.11842. 2022

[4] Wymeersch H et al. Integration of communication and sensing in 6G: A joint industrial and academic perspective. In: IEEE 32nd Annual International Symposium on Personal, Indoor and Mobile Radio Communications (PIMRC). Helsinki, Finland: IEEE; 2021. pp. 1-7. DOI: 10.1109/PIMRC50174.2021.9569364

[5] Pizzo A, Marzetta TL, Sanguinetti L. Degrees of freedom of holographic MIMO channels. In: IEEE 21st International Workshop on Signal Processing Advances in Wireless Communications (SPAWC). Atlanta, GA, USA: IEEE; 2020. pp. 1-5

[6] Huang C, Zappone A, Alexandropoulos GC, Debbah M, Yuen C. Reconfigurable intelligent surfaces for energy efficiency in wireless communication. IEEE Transactions on Wireless Communications. 2019;**18**(8): 4157-4170 Available from: https://ieeexplore.ieee.org/stamp/stamp.jsp?tp=&arnumber=8741198

[7] Rana B, Cho S-S, Hong I-P. Review paper on hardware of reconfigurable intelligent surfaces. IEEE Access. 2023; **11**:29614-29634 Available from: https://ieeexplore.ieee.org/abstract/document/10080950

[8] Huang C, Hu S, Alexandropoulos GC, Zappone A, Yuen C, Zhang R, et al. Holographic MIMO surfaces for 6G wireless networks: Opportunities, challenges, and trends. IEEE Wireless Communications. 2020;**27**(5):118-125 Available from: https://ieeexplore.ieee.org/stamp/stamp.jsp?tp=&arnumber=9136592

[9] Tierui G, Ioanna V, Ran J, Chongwen H, Alexandropoulos George C, Li W. Holographic MIMO communications: Theoretical foundations, enabling technologies, and future directions. 2022. Available from: https://arxiv.org/abs/2212.01257 [Accessed: 02 Dec 2022]

[10] Hu X, Deng R, Di B, Zhang H, Song L. Holographic beamforming for ultra massive MIMO with limited radiation amplitudes: How many quantized bits do we need? IEEE Communications Letters. 2022;**26**(6): 1403-1407. DOI: 10.1109/LCOMM.2022.3151801

[11] Deng R, Zhang Y, Zhang H, Di B, Zhang H, Song L. Reconfigurable holographic surface: A new paradigm to implement holographic radio. IEEE Vehicular Technology Magazine. 2023; **18**(1):20-28 Available from: https://ieeexplore.ieee.org/abstract/document/10018594

[12] Tang W et al. Wireless communications with reconfigurable intelligent surface: Path loss Modeling and experimental measurement. IEEE Transactions on Wireless Communications. 2021;**20**(1):421-439. DOI: 10.1109/TWC.2020.3024887

[13] Björnson E, Sanguinetti L. Power scaling Laws and Near-field Behaviors of massive MIMO and intelligent reflecting surfaces. IEEE Open Journal of the Communications Society. 2020;**1**: 1306-1324. DOI: 10.1109/ OJCOMS.2020.3020925

[14] Najafi M, Jamali V, Schober R, Poor HV. Physics-based Modeling and scalable optimization of large intelligent reflecting surfaces. IEEE Transactions on Communications. 2021;**69**(4):2673-2691. DOI: 10.1109/TCOMM.2020.3047098

[15] He Z, Yuan X. Cascaded Channel estimation for large intelligent Metasurface assisted massive MIMO. IEEE Wireless Communications Letters. 2020;**9**(2):210-214. DOI: 10.1109/ LWC.2019.2948632

[16] Ardah K, Gherekhloo S, de Almeida ALF, Haardt M. TRICE: A channel estimation framework for RIS-aided millimeter-wave MIMO systems. IEEE Signal Processing Letters. 2021;**28**: 513-517. DOI: 10.1109/LSP.2021.3059363

[17] de Araújo GT, de Almeida ALF. PARAFAC-Based Channel estimation for intelligent reflective surface assisted MIMO system. In: IEEE 11th Sensor Array and Multichannel Signal Processing Workshop (SAM). Hangzhou, China: IEEE; 2020. pp. 1-5. DOI: 10.1109/SAM48682.2020.9104260

[18] Wang P, Fang J, Duan H, Li H. Compressed Channel estimation for intelligent reflecting surface-assisted millimeter wave systems. IEEE Signal Processing Letters. 2020;**27**:905-909. DOI: 10.1109/LSP.2020.2998357

[19] He J, Wymeersch H, Juntti M. Channel estimation for RIS-aided mm wave MIMO systems via atomic norm minimization. IEEE Transactions on Wireless Communications. 2021;**20**(9): 5786-5797. DOI: 10.1109/TWC.2021. 3070064

[20] Hu C, Dai L, Han S, Wang X. Two-timescale channel estimation for reconfigurable intelligent surface aided wireless communications. IEEE Transactions on Communications. 2021; **69**(11):7736-7747. DOI: 10.1109/ TCOMM.2021.3072729

[21] Elbir AM, Papazafeiropoulos A, Kourtessis P, Chatzinotas S. Deep Channel learning for large intelligent surfaces aided mm-wave massive MIMO systems. IEEE Wireless Communications Letters. 2020;**9**(9):1447-1451. DOI: 10.1109/LWC.2020.2993699

[22] He J, Wymeersch H, Di Renzo M, Juntti M. Learning to estimate RIS-aided mm wave channels. IEEE Wireless Communications Letters. 2022;**11**(4): 841-845. DOI: 10.1109/ LWC.2022.3147250

[23] Pizzo A, Sanguinetti L, Marzetta TL. Fourier plane-wave series expansion for holographic MIMO communications. IEEE Transactions on Wireless Communications. 2022;**21**(9):6890-6905. DOI: 10.1109/TWC.2022.3152965

[24] Pizzo A, Marzetta TL, Sanguinetti L. Spatially-stationary model for holographic MIMO small-scale fading. IEEE Journal on Selected Areas in Communications. 2020;**38**(9):1964-1979. DOI: 10.1109/JSAC.2020.3000877

[25] Demir ÖT, Björnson E, Sanguinetti L. Channel modeling and channel estimation for holographic massive MIMO with planar arrays. IEEE Wireless Communications Letters. 2022; **11**(5):997-1001. DOI: 10.1109/ LWC.2022.3152600

[26] Cui M, Dai L. Channel estimation for extremely large-scale MIMO: Far-field or

near-field? IEEE Transactions on Communications. 2022;**70**(4):2663-2677. DOI: 10.1109/TCOMM.2022.3146400

[27] Wu Q, Zhang R. Intelligent reflecting surface enhanced wireless network via joint active and passive beamforming. IEEE Transactions on Wireless Communications. 2019;**18**(11): 5394-5409. DOI: 10.1109/TWC.2019.2936025

[28] Wei L, Huang C, Alexandropoulos GC, Sha WEI, Zhang Z, Debbah M, et al. Multi-user holographic MIMO surfaces: Channel Modeling and spectral efficiency analysis. IEEE Journal of Selected Topics in Signal Processing. 2022; **16**(5):1112-1124. Available from: https://ieeexplore.ieee.org/abstract/document/9779586

[29] Deng R, Di B, Zhang H, Tan Y, Song L. Reconfigurable holographic surface-enabled multi-user wireless communications: Amplitude-controlled holographic beamforming. IEEE Transactions on Wireless Communications. 2022;**21**(8): 6003-6017. DOI: 10.1109/TWC.2022.3144992

[30] Di B. Reconfigurable holographic metasurface aided wideband OFDM communications against beam squint. IEEE Transactions on Vehicular Technology. 2021;**70**(5):5099-5103. DOI: 10.1109/TVT.2021.3070361

[31] de Sena AS, Nardelli PHJ, da Costa DB, Lima FRM, Yang L, Popovski P, et al. IRS-assisted massive MIMO-NOMA networks: Exploiting wave polarization. IEEE Transactions on Wireless Communications. 2021;**20**(11):7166-7183 Available from: https://ieeexplore.ieee.org/abstract/document/9440813

[32] Han Y, Li X, Tang W, Jin S, Cheng Q, Cui TJ. Dual-polarized RIS-assisted mobile communications. IEEE Transactions on Wireless Communications. 2022;**21**(1):591-606 Available from: https://ieeexplore.ieee.org/document/9497725

[33] Sugiura S, Kawai Y, Matsui T, Lee T, Iizuka H. Joint beam and polarization forming of intelligent reflecting surfaces for wireless communications. IEEE Transactions on Vehicular Technology. 2021;**70**(2):1648-1657 Available from: https://ieeexplore.ieee.org/document/9339948

[34] Wei L et al. Tri-polarized holographic MIMO surfaces for near-field communications: Channel modeling and precoding design. IEEE Transactions on Wireless Communications. 2023. Available from: https://ieeexplore.ieee.org/abstract/document/10103817. DOI: 10.1109/TWC.2023.3266298

[35] Yang Z, Xu W, Huang C, Shi J, Shikh-Bahaei M. Beamforming Design for Multiuser Transmission through Reconfigurable Intelligent Surface. IEEE Transactions on Communications. 2021; **69**(1):589-601 Available from: https://ieeexplore.ieee.org/document/9211520

[36] Huang C, Yang Z, Alexandropoulos GC, Xiong K, Wei L, Yuen C, et al. Multi-hop RIS-empowered terahertz communications: A DRL-based hybrid beamforming design. IEEE Journal on Selected Areas in Communications. 2021;**39**(6):1663-1677 Available from: https://ieeexplore.ieee.org/abstract/document/9410457

[37] Xu Y, Gao Z, Wang Z, Huang C, Yang Z, Yuen C. RIS-enhanced WPCNs: Joint radio resource allocation and passive beamforming optimization. IEEE Transactions on Vehicular Technology. 2021;**70**(8):7980-7991 Available from: https://ieeexplore.ieee.org/abstract/document/9485102

[38] Xiao G, Yang T, Huang C, Wu X, Feng H, Hu B. Average rate approximation and maximization for RIS-assisted multi-user MISO system. IEEE Wireless Communications Letters. 2022;**11**(1):173-177 Available from: https://ieeexplore.ieee.org/document/9590494

[39] Huang C, Mo R, Yuen C. Reconfigurable intelligent surface assisted multiuser MISO systems exploiting deep reinforcement learning. IEEE Journal on Selected Areas in Communications. 2020;**38**(8):1839-1850 Available from: https://ieeexplore.ieee.org/document/9110869

[40] Gao H, Cui K, Huang C, Yuen C. Robust beamforming for RIS-assisted wireless communications with discrete phase shifts. IEEE Wireless Communications Letters. 2021;**10**(12): 2619-2623 Available from: https://ieeexplore.ieee.org/abstract/document/9521553

[41] Gan X, Zhong C, Huang C, Zhang Z. RIS-assisted multi-user MISO communications exploiting statistical CSI. IEEE Transactions on Communications. 2021;**69**(10): 6781-6792 Available from: https://ieeexplore.ieee.org/abstract/document/9500188

[42] Gan X, Zhong C, Huang C, Yang Z, Zhang Z. Multiple RISs assisted cell-free networks with two-timescale CSI: Performance analysis and system design. IEEE Transactions on Communications. 2022;**70**(11):7696-7710 Available from: https://ieeexplore.ieee.org/document/9899454

[43] Cheng Y, Peng W, Huang C, Alexandropoulos GC, Yuen C, Debbah M. RIS-aided wireless communications: Extra degrees of freedom via rotation and location optimization. IEEE Transactions on Wireless Communications. 2022;**21**(8): 6656-6671 Available from: https://ieeexplore.ieee.org/document/9722711

[44] Yuan SSA, Chen X, Huang C, Sha WEI. Effects of mutual coupling on degree of freedom and antenna efficiency in holographic MIMO communications. IEEE Open Journal of Antennas and Propagation. 2023;**4**: 237-244 Available from: https://ieeexplore.ieee.org/abstract/document/10045715

Chapter 3

# New Results on Single User Massive MIMO

*Kasturi Vasudevan, Surendra Kota, Lov Kumar and Himanshu Bhusan Mishra*


## Abstract

Achieving high bit rates is the main goal of wireless technologies like 5G and beyond. This translates to obtaining high spectral efficiencies using large number of antennas at the transmitter and receiver (single user massive multiple input multiple output or SU-MMIMO). It is possible to have a large number of antennas in the mobile handset at mm-wave frequencies in the range 30–300 GHz due to the small antenna size. In this work, we investigate the bit-error-rate (BER) performance of SU-MMIMO in two scenarios (a) using serially concatenated turbo code (SCTC) in uncorrelated channel and (b) parallel concatenated turbo code (PCTC) in correlated channel. Computer simulation results indicate that the BER is quite insensitive to re-transmissions and wide variations in the number of transmit and receive antennas. Moreover, we have obtained a BER of $10^{-5}$ at an average signal-to-interference plus noise ratio (SINR) per bit of just 1.25 dB with 512 transmit and receive antennas ($512 \times 512$ SU-MMIMO system) with a spectral efficiency of 256 bits/transmission or 256 bits/sec/Hz in an uncorrelated channel. Similar BER results have been obtained for SU-MMIMO using PCTC in correlated channel. A semi-analytic approach to estimating the BER of a turbo code has been derived.


**Keywords:** single user massive multiple input multiple output (SU-MMIMO), Rayleigh fading, serially concatenated turbo code (SCTC), parallel concatenated turbo code (PCTC), spectral efficiency (SE), signal-to-interference plus noise ratio (SINR) per bit, spatial multiplexing, bit-error-rate (BER)

## 1. Introduction

As wireless technologies evolve beyond 5G [1–3], there is a growing need to attain peak data rates of about gigabits per second per user, which is required for high definition video, remote surgery, autonomous vehicles, gaming and so on, while at the same time consuming minimum transmit power. This can only be achieved by using multiple antennas at the transmitter and receiver [4–8], small constellations like quadrature shift keying (QPSK) and powerful error correcting codes like turbo or low density parity check (LDPC) codes. Having a large number of antennas in the mobile handset is feasible in mm-wave frequencies [9–12] (30–300 GHz) due to the small antenna size. The main concern about mm wave communications has been its rather high attenuation in outdoor environments with rain and snow [13]. Therefore, at least in the initial stages, mm wave could be deployed indoors. The second issue relates to the poor penetration characteristics of mm wave through walls, doors, windows and other materials. This points towards to usage of mm wave [9] in a single room, say a big auditorium or underground parking and so on. Reconfigurable intelligent surface (RIS) [11–17] could be used to boost the propagation of mm waves, both indoors and outdoors. Most of the massive MIMO systems discussed in the literature are multi-user (MU) [18–26], that is, the base station has a large number of antennas and the mobile handset has only a single antenna ($N_t = 1$). A large number of users are served simultaneously by the base station. A comparison between MU-MMIMO and SU-MMIMO is given in **Table 1** [27, 28].

The base station in MU-MMIMO uses beamforming to improve the signal-to-noise ratio at the mobile handset. On the other hand, SU-MMIMO uses spatial multiplexing to improve the spectral efficiency in the downlink and uplink. The comparison between beamforming and spatial multiplexing is given in **Table 2** [27, 28]. The total transmit power of SU-MMIMO using uncoded QPSK versus MU-MMIMO using $M$-ary QAM is shown in **Table 3**. The minimum Euclidean distance between symbols of all constellations is taken to be 2. The peak-to-average power ratio (PAPR) for SU-MMIMO using QPSK is compared with MU-MMIMO using $M$-ary QAM in **Table 4** [27]. Of course in the case of frequency selective fading channels, OFDM needs to be used, which would result in PAPR greater than 0 dB even for QPSK signaling. It is clear from **Tables 1–4** that technologies that use SU-MMIMO have a lot to gain.

Moreover, since all transmit antennas use the same carrier frequency, there is no increase in bandwidth. SU-MMIMO with equal number of transmit and receive

| MU-MIMO | SU-MMIMO |
|---|---|
| Beamforming possible in downlink | Beamforming possible in uplink & downlink |
| Spatial multiplexing not possible | Spatial multiplexing possible in uplink & downlink |
| Low spectral efficiency per user | High spectral efficiency per user |
| High directivity in downlink in beamforming mode | High directivity in uplink & downlink in beamforming mode |

**Table 1.**
*Comparison of MU-MMIMO and SU-MMIMO.*

| Beamforming | Spatial multiplexing |
|---|---|
| High directivity | Little or no directivity |
| Line-of-sight communication required | Rich scattering channel required |
| Low spectral efficiency per user since the same signal is transmitted from each antenna element | High spectral efficiency per user since different signals are transmitted from each antenna element |
| Spectral efficiency can be improved by increasing the constellation size resulting in high PAPR | QPSK constellations with PAPR 0 dB can be used |
| Difficult to turbo/LDPC code large constellations | Easy to turbo/LDPC code QPSK |
| Large BER at average SINR per bit close to 0 dB | Small BER at average SINR per bit close to 0 dB |

**Table 2.**
*Comparison of beamforming and spatial multiplexing.*

| Spectral efficiency (bits/sec/Hz) | QPSK | | M-ary QAM | |
|---|---|---|---|---|
| | Transmit antennas $N_t$ | Total average transmit power | $M$-ary QAM | $N_t = 1$ average transmit power |
| 4 | 2 | 4 | 16-QAM | 10 |
| 6 | 3 | 6 | 64-QAM | 42 |
| 8 | 4 | 8 | 256-QAM | 170 |
| 10 | 5 | 10 | 1024-QAM | 682 |

**Table 3.**
*SU-MMIMO using QPSK vs. MU-MMIMO using* M-*ary.*

| Spectral efficiency (bits/sec/Hz) | QPSK | | M-ary $Nt = 1$ | |
|---|---|---|---|---|
| | Transmit antennas $Nt$ | PAPR (dB) | $M$ | PAPR (dB) |
| 4 | 2 | 0 | 16-QAM | 2.5 |
| 6 | 3 | 0 | 64-QAM | 3.7 |
| 8 | 4 | 0 | 256-QAM | 4.23 |
| 10 | 5 | 0 | 1024-QAM | 4.5 |

**Table 4.**
*PAPR of SU-MMIMO using QPSK vs. MU-MMIMO using* M-*ary.*

antennas is given in [29, 30]. The probability of erasure in MIMO-OFDM is presented in [31]. A practical SU-MMIMO receiver with estimated channel, carrier frequency offset and timing is described in [32, 33]. SU-MMIMO with unequal number of transmit and receive antennas and precoding is discussed in [34, 35] and the case without precoding in [36, 37]. All the earlier research on SU-MMIMO involved the use of a parallel concatenated turbo code (PCTC) and uncorrelated channel. In this work, we investigate the performance of SU-MMIMO using (a) serial concatenated turbo code (SCTC) in uncorrelated channel and (b) PCTC in correlated channel. Throughout this article we assume that the channel is known perfectly at the receiver. Perfect carrier and timing synchronization is also assumed.

This work is organized as follows. Section II discusses SU-MMIMO with SCTC in uncorrelated channel, the procedure for bit-error-rate (BER) estimation and computer simulation results. Section III deals with SU-MMIMO using PCTC in correlated channel with and without precoding along with computer simulation results. Section IV presents the conclusions and scope for future work.

## 2. SU-MMIMO with SCTC

### 2.1 System model

Consider the block diagram in **Figure 1** [36, 38]. The input bits $a_i$, $1 \leq i \leq L_{d1}$ is passed through an outer rate-1/2 recursive systematic convolutional (RSC) encoder to obtain the coded bit stream $b_i$, $1 \leq i \leq L_d$, where

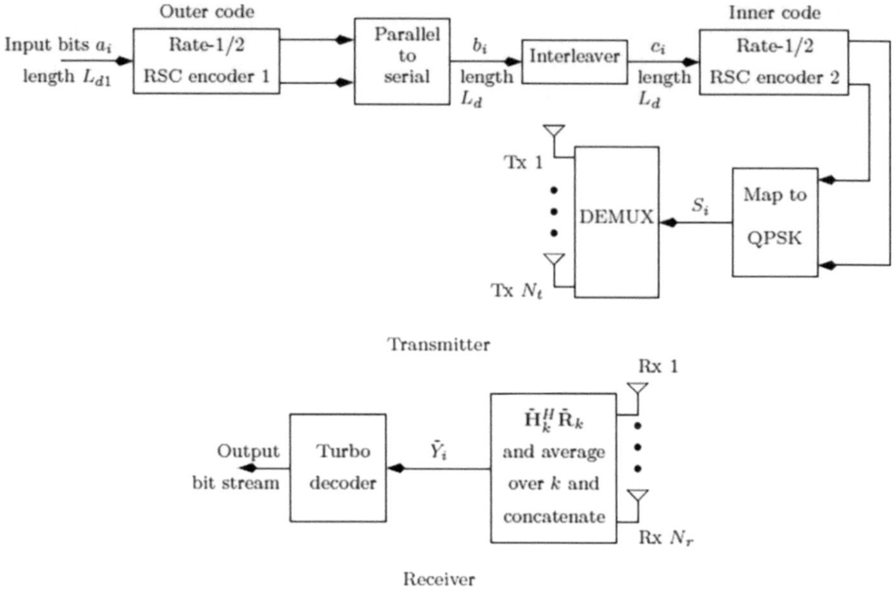


**Figure 1.**
*SU-MMIMO with serially concatenated turbo code.*

$$L_d = 2L_{d1}. \tag{1}$$

Now $b_i$ is input to an interleaver to generate $c_i$, $1 \le i \le L_d$. Next $c_i$ is passed through an inner rate-1/2 RSC encoder and mapped to symbols $S_i$, $1 \le i \le L_d$, in a quadrature phase shift keyed (QPSK) constellation having symbol coordinates $\pm 1 \pm \mathrm{j}$, where $\mathrm{j} = \sqrt{-1}$. Throughout this article we assume that bit "0" maps to $+1$ and bit "1" maps to $-1$. The set of $L_d$ QPSK symbols constitute a "frame" and are transmitted using $N_t$ antennas. We assume that

$$\frac{L_d}{N_t} = \text{an integer} \tag{2}$$

so that all symbols in the frame are transmitted using $N_t$ antennas. The set of QPSK symbols transmitted simultaneously using $N_t$ antennas constitute a "block". The generator matrix for both the inner and outer rate-1/2 RSC encoder is given by

$$\mathbf{G}(D) = \begin{bmatrix} 1 & \dfrac{1 + D^2}{1 + D + D^2} \end{bmatrix}. \tag{3}$$

Hence, both encoders have $S_E = 4$ states in the trellis. Assuming uncorrelated Rayleigh flat fading, the received signal for the $k^{th}$ re-transmission ($0 \le k \le N_{rt} - 1$, $k$ is an integer) is given by (2) of [36], which is repeated here for convenience

$$\tilde{\mathbf{R}}_k = \tilde{\mathbf{H}}_k \mathbf{S} + \tilde{\mathbf{W}}_k \tag{4}$$

where $\mathbf{S} \in \mathbb{C}^{N_t \times 1}$ whose elements are drawn from the QPSK constellation, $\tilde{\mathbf{H}}_k \in \mathbb{C}^{N_r \times N_t}$ whose elements are mutually independent and $\mathcal{CN}(0, 2\sigma_H^2)$ and

$\tilde{\mathbf{W}}_k \in \mathbb{C}^{N_r \times 1}$ is the additive white Gaussian noise (AWGN) vector whose elements are mutually independent and $\mathcal{CN}(0, 2\sigma_W^2)$. Note that $\sigma_H^2, \sigma_W^2$ denote the variance per dimension (real part or imaginary part) and $N_r$ is the number of receive antennas. We assume that $\tilde{\mathbf{H}}_k$ and $\tilde{\mathbf{W}}_k$ are independent across blocks and re-transmissions, hence (4) in [29] is valid with $N$ replaced by $N_t$. Recall that (see also (16) of [36])

$$N_{\text{tot}} = N_t + N_r. \tag{5}$$

Following the procedure given in Section 4 of [36] we get (see (36) of [36])

$$\tilde{Y}_i = F_i S_i + \tilde{U}_i \qquad \text{for } 1 \le i \le N_t. \tag{6}$$

After concatenation over blocks, $\tilde{Y}_i$ in (6) for $1 \le i \le L_d$ is sent to the turbo decoder (see also the sentence after (25) in [29]). For the sake of consistency with earlier work [38], we re-index $i$ as $0 \le i \le L_d - 1$ and use the same index $i$ for $a_i$, $b_i$, $c_i$ and $Y_i$ without any ambiguity. In the next subsection, we discuss the turbo decoding (BCJR) algorithm [39, 40] for the inner code.

### 2.2 BCJR for the inner code

Let $\mathcal{D}_n$ denote the set of states that diverge from state $n$ in the trellis [38, 40]. Similarly, let $\mathcal{C}_n$ denote the set of states that converge to state $n$. Let $\alpha_{i,n}$ denote the forward sum-of-products (SOP) at time $i$, $0 \le i \le L_d - 2$, at state $n$, $0 \le n \le S_E - 1$. Then the forward SOP can be recursively computed as follows (see also (30) of [38]):

$$\alpha'_{i+1,n} = \sum_{m \in \mathcal{C}_n} \alpha_{i,m} \gamma_{i,m,n} P(c_{i,m,n}); \qquad \alpha_{0,n} = 1; \qquad \alpha_{i+1,n} = \alpha'_{i+1,n} \Big/ \left( \sum_{n=0}^{S_E-1} \alpha'_{i+1,n} \right) \tag{7}$$

where $P(c_{i,m,n})$ denotes the *a priori* probability of the systematic bit corresponding to the transition from encoder state $m$ to $n$, at time $i$ (this is set to 0.5 at the beginning of the first iteration). The last equation in (7) is required to prevent numerical instabilities [40]. We have

$$\gamma_{i,m,n} = \exp\left( -\frac{\left(\tilde{Y}_i - S_{m,n}\right)^2}{2\sigma_U^2} \right) \tag{8}$$

where $\tilde{Y}_i$ is given by (6), $S_{m,n}$ is the QPSK symbol corresponding to the transition from encoder state $m$ to $n$ and $\sigma_U^2$ is given by (38) of [36] which is repeated here for convenience:

$$E\left[\left|\tilde{U}_i\right|^2\right] = \frac{8\sigma_H^4 N_r (N_t - 1) + 4\sigma_W^2 \sigma_H^2 N_r}{N_{rt}} \triangleq \sigma_U^2. \tag{9}$$

Robust turbo decoding (see Section 4.2 of [41]) can be employed to compute $\gamma_{i,m,n}$ in (8). Similarly, let $\beta_{i,m}$ denote the backward SOP at time $i$, $1 \le i \le L_d - 1$, at state $m$, $0 \le m \le S_E - 1$. Then the backward SOP can be recursively computed as (see also (33) of [38]):

$$\beta'_{i,m} = \sum_{n \in \mathcal{D}_m} \beta_{i+1,n}\gamma_{i,m,n}P(c_{i,m,n}); \beta_{L_{d,m}} = 1; \beta_{i,m} = \beta'_{i,m} \Big/ \left( \sum_{m=0}^{S_E-1} \beta'_{i,m} \right). \tag{10}$$

Let $\rho^+(n)$ denote the state that is reached from encoder state $n$ when the input symbol is +1. Similarly let $\rho^-(n)$ denote the state that can be reached from encoder state $n$ when the input symbol is −1. Then for $0 \le i \le L_d - 1$ we compute

$$C_{i+} = \sum_{n=0}^{S_E-1} \alpha_{i,n}\gamma_{i,n,\rho^+(n)}\beta_{i+1,\rho^+(n)}; \quad C_{i-} = \sum_{n=0}^{S_E-1} \alpha_{i,n}\gamma_{i,n,\rho^-(n)}\beta_{i+1,\rho^-(n)}. \tag{11}$$

Finally, the extrinsic information that is fed to the BCJR algorithm for the outer code is computed as, for $0 \le i \le L_d - 1$, (see (36) of [38]):

$$E(c_i = +1) = C_{i+} \big/ (C_{i+} + C_{i-}); E(c_i = -1) = C_{i-}/(C_{i+} + C_{i-}). \tag{12}$$

Next, we describe the BCJR for the outer code.

### 2.3 BCJR for the outer code

Let $\alpha_{i,n}$ denote the forward SOP at time $i$, $0 \le i \le L_{d1} - 2$, at state $n$, $0 \le n \le S_E - 1$. Then the forward SOP is recursively computed as follows:

$$\alpha'_{i+1,n} = \sum_{m \in \mathcal{C}_n} \alpha_{i,m}\gamma_{sys,i,m,n}\gamma_{par,i,m,n}P(a_{i,m,n}); \alpha_{0,n} = 1; \alpha_{i+1,n} = \alpha'_{i+1,n} \Big/ \left( \sum_{n=0}^{S_E-1} \alpha'_{i+1,n} \right) \tag{13}$$

where $P(a_{i,m,n})$ denotes the *a priori* probability of the systematic bit corresponding to the transition from state $m$ to state $n$, at time $i$. In the absence of any other information, we assume $(a_{i,m,n}) = 0.5$ [42]. We also have for $0 \le i \le L_{d1} - 1$ (similar to (38) of [38])

$$\gamma_{\text{sys},i,m,n} = \begin{cases} E(c_{\pi(2i)} = +1) & \text{if } \mathcal{H}_1 \\ E(c_{\pi(2i)} = -1) & \text{if } \mathcal{H}_2 \end{cases}; \quad \gamma_{\text{par},i,m,n} = \begin{cases} E(c_{\pi(2i+1)} = +1) & \text{if } \mathcal{H}_3 \\ E(c_{\pi(2i+1)} = -1) & \text{if } \mathcal{H}_4 \end{cases} \tag{14}$$

where $\pi(\cdot)$ denotes the interleaver map and
$\mathcal{H}_1$ : systematic bit from state $m$ to $n$ is + 1; $\mathcal{H}_2$ :
systematic bit from state $m$ to $n$ is − 1

$$\mathcal{H}_3 : \text{parity bit from state } m \text{ to } n \text{ is } + 1; \mathcal{H}_4 : \text{parity bit from state } m \text{ to } n \text{ is } - 1. \tag{15}$$

Observe that in (14) and (15) it is assumed that after the parallel-to-serial conversion in **Figure 1**, $b_{2i}$ corresponds to the systematic (data) bits and $b_{2i+1}$ corresponds to the parity bits for $0 \le i \le L_{d1} - 1$. Similarly, let $\beta_{i,m}$ denote the backward SOP at time $i$, $1 \le i \le L_{d1} - 1$, at state $m$, $0 \le m \le S_E - 1$. Then the backward SOP can be recursively computed as:

$$\beta'_{i,m} = \sum_{n \in \mathcal{D}_m} \beta_{i+1,n}\gamma_{sys,i,m,n}\gamma_{par,i,m,n}P(a_{i,m,n}); \beta_{L_{d1,m}} = 1; \beta_{i,m} = \beta'_{i,m} \Big/ \left( \sum_{m=0}^{S_E-1} \beta'_{i,m} \right). \tag{16}$$

Next, for $0 \le i \le L_{d1} - 1$ we compute

$$B_{2i+} = \sum_{n=0}^{S_E-1} \alpha_{i,n} \gamma_{\text{par},i,n,\rho^+(n)} \beta_{i+1,\rho^+(n)}; B_{2i-} = \sum_{n=0}^{S_E-1} \alpha_{i,n} \gamma_{\text{par},i,n,\rho^-(n)} \beta_{i+1,\rho^-(n)}. \tag{17}$$

Let $\mu^+(n)$, and $\mu^-(n)$ denote the states that are reached from state $n$ when the parity bit is $+1$ and $-1$respectively. Similarly for $0 \le i \le L_{d1} - 1$ compute

$$B_{2i+1+} = \sum_{n=0}^{S_E-1} \alpha_{i,n} \gamma_{\text{sys},i,n,\mu^+(n)} \beta_{i+1,\mu^+(n)}; B_{2i+1-} = \sum_{n=0}^{S_E-1} \alpha_{i,n} \gamma_{\text{sys},i,n,\mu^-(n)} \beta_{i+1,\mu^-(n)}. \tag{18}$$

The extrinsic information that is sent to the inner decoder for $0 \le i \le L_d - 1$ is computed as

$$E(b_i = +1) = {}^{B_{i+}}/_{(B_{i+} + B_{i-})}; E(b_i = -1) = B_{i-}/(B_{i+} + B_{i-}) \tag{19}$$

where $B_{i+}, B_{i-}$ are given by (17)or (18) depending on whether $i$ is even or odd respectively. Note that $P(c_{i,m,n})$ for $0 \le i \le L_d - 1$ in (7) and (10) is equal to

$$P(c_{i,m,n}) = \begin{cases} E\left(b_{\pi^{-1}(i)} = +1\right) & \text{if } \mathcal{H}_1 \\ E\left(b_{\pi^{-1}(i)} = -1\right) & \text{if } \mathcal{H}_2 \end{cases} \tag{20}$$

where $\pi^{-1}(\cdot)$ denotes the inverse interleaver map. Note that $c_{i,m,n}$ are the systematic (data) bits for the inner encoder.

After the convergence of the BCJR algorithm in the last iteration, the final *a posteriori* probabilities of $a_i$ for $0 \le i \le L_{d1} - 1$ is given by

$$P(a_i = +1) = E(b_{2i} = +1)E\left(c_{\pi(2i)} = +1\right); P(a_i = -1) = E(b_{2i} = -1)E\left(c_{\pi(2i)} = -1\right) \tag{21}$$

where $E(c_i = \pm 1)$ and $E(b_i = \pm 1)$ are given by (12) and (19) respectively. Finally note that for $0 \le i \le L_{d1} - 1$

$$a_i = b_{2i} = c_{\pi(2i)}. \tag{22}$$

In the next section we present the estimation of the bit-error-rate (BER) of the SCTC.

### 2.4 Estimation of BER

The estimation of BER of SCTC is based on the following propositions:

**Proposition1.** *The extrinsic information as computed in (12) and (19) lies in the range* $[0, 1]$ *(0 and 1 included). The extrinsic information in the range* $(0, 1)$*, 0 and 1 excluded, is Gaussian distributed* [43] *for each frame.*

This is illustrated in **Figure 2** for different values of the frame length $L_{d1}$, over many frames ($F$). We find that for large values of $L_{d1}$, the histogram better approximates the Gaussian characteristic. It may be noted that the extrinsic information at the output of one decoder is equal to the *a priori* probabilities for the other decoder.

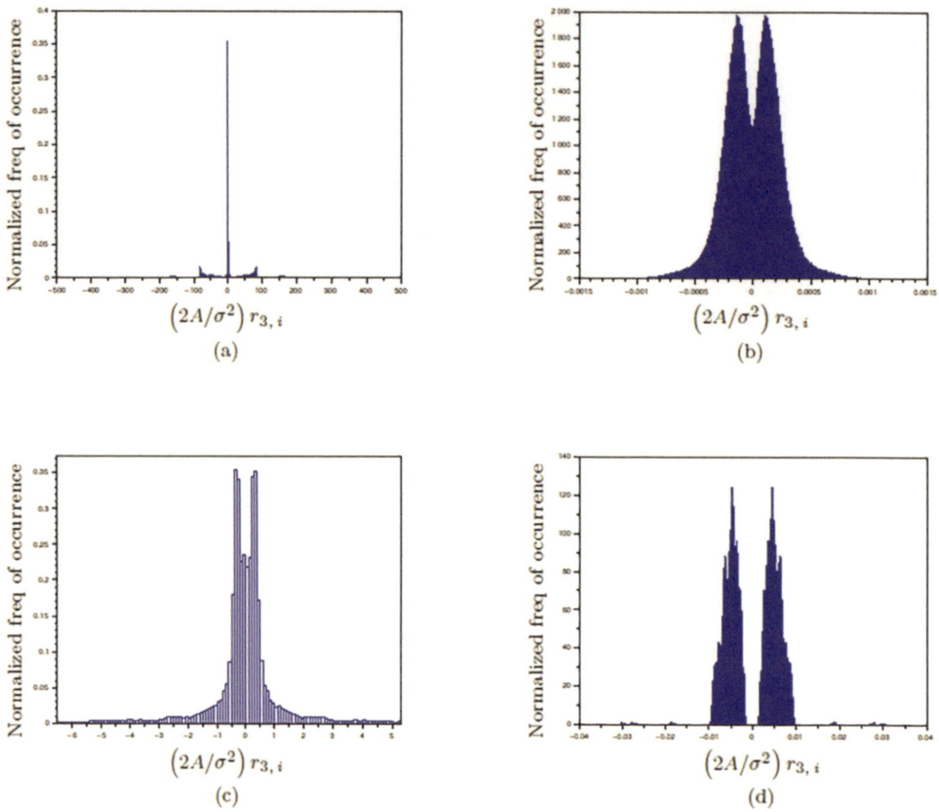


**Figure 2.**
*Normalized histogram for $N_{tot}$ = 1024, $N_t$ = 512, $N_{rt}$ = 2 (a) $L_{d1}$ = 1024, $SNR_{av,\,b}$ = 1.25 dB, F = $10^5$ frames (b) $L_{d1}$ = 50,176, $SNR_{av,\,b}$ = 0.3 dB, F = 2000 frames (c) expanded view of (around) $r_{3,i}$ = 0 and (d) $L_{d1}$ = 50,176, $SNR_{av,\,b}$ = 0.5 dB, F = 2000 frames.*

**Proposition 2.** *After convergence of the BCJR algorithm in the final iteration, the extrinsic information at a decoder output has the same mean and variance as that of the a priori probability at its input.*

**Proposition 3.** *The mean and variance of the Gaussian distribution may vary from frame to frame.*

This is illustrated in **Figure 3** over two frames, that is, $F = 2$.

$$P(e) = \frac{1}{2}\operatorname{erfc}\left(\sqrt{\frac{A^2}{\sigma^2}}\right). \tag{23}$$

Based on *Propositions 1 & 2* and (22), after convergence of the BCJR algorithm, we can write for $0 \le i \le L_{d1} - 1$

$$E(b_{2i} = +1) = \frac{1}{\sigma\sqrt{2\pi}} \mathrm{e}^{-(\mathrm{r}_{1,\mathrm{i}}-\mathrm{A})^2/(2\sigma^2)}; E\left(c_{\pi(2i)} = +1\right) = \frac{1}{\sigma\sqrt{2\pi}} \mathrm{e}^{-(\mathrm{r}_{2,\mathrm{i}}-\mathrm{A})^2/(2\sigma^2)} \tag{24}$$

where it is assumed that bit "0" maps to $A$ and bit "1" maps to $-A$ and

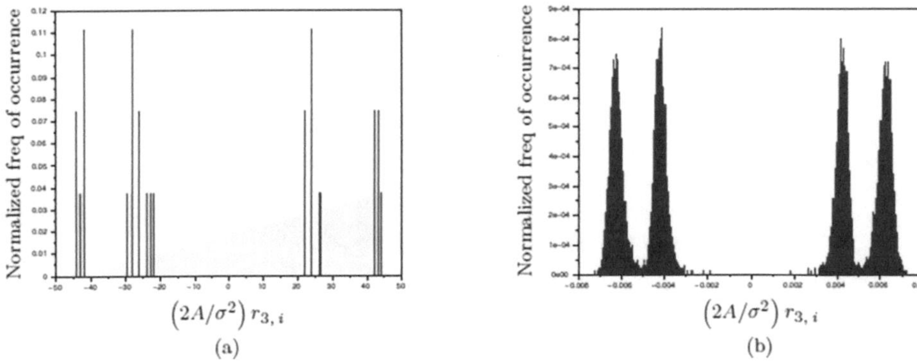


**Figure 3.**
*Normalized histogram over two frames (F = 2) for $N_{tot}$ = 1024, $N_t$ = 512, $N_{rt}$ = 2 (a) $L_{d1}$ = 1024, $SNR_{av,\,b}$ = 1.25 dB and (d) $L_{d1}$ = 50,176, $SNR_{av,\,b}$ = 0.5 dB.*

$$r_{1,i} = \pm A + w_{1,i}; r_{2,i} = \pm A + w_{2,i} \tag{25}$$

where $w_{1,i}, w_{2,i}$ denote real-valued samples of zero-mean additive white Gaussian noise (AWGN) with variance $\sigma^2$. Similarly we have

$$E(b_{2i} = -1) = \frac{1}{\sigma\sqrt{2\pi}} \mathrm{e}^{-(r_{1,i}+A)^2/(2\sigma^2)}; E\left(c_{\pi(2i)} = -1\right) = \frac{1}{\sigma\sqrt{2\pi}} \mathrm{e}^{-(r_{2,i}+A)^2/(2\sigma^2)}. \tag{26}$$

Clearly

$$\ln\left(\frac{E(b_{2i} = +1)}{E(b_{2i} = -1)}\right) = \frac{2A}{\sigma^2} r_{1,i}; \quad \ln\left(\frac{E\left(c_{\pi(2i)} = +1\right)}{E\left(c_{\pi(2i)} = -1\right)}\right) = \frac{2A}{\sigma^2} r_{2,i}. \tag{27}$$

From (21) and (26) we have for $0 \le i \le L_{d1} - 1$

$$\ln\left(\frac{P(a_i = +1)}{P(a_i = -1)}\right) = \frac{2A}{\sigma^2}(r_{1,i} + r_{2,i}) \triangleq \frac{2A}{\sigma^2} r_{3,i}. \tag{28}$$

Consider the average

$$\mathcal{Y} = \frac{2A}{\sigma^2 L_{d2}} \sum_{i=0}^{L_{d2}-1} a_i r_{3,i} = \frac{4A^2}{\sigma^2} + \mathcal{Z} \tag{29}$$

where

$$\mathcal{Z} = \frac{2A}{\sigma^2 L_{d2}} \sum_{i=0}^{L_{d2}-1} a_i (w_{1,i} + w_{2,i});. \quad L_{d2} < L_{d1} \tag{30}$$

Note that the average in (28) is done over less than $L_{d1}$ terms to avoid situations like

$$P(a_i = \pm 1) = 1 \text{ or } 0. \tag{31}$$

In fact, only those time instants $i$ have been considered in the summation of (28) for which

$$P(a_i = \pm 1) > \mathrm{e}^{-500}. \tag{32}$$

Now

$$E[\mathcal{Z}] = 0; E\left[\mathcal{Z}^2\right] = \frac{4A^2}{\sigma^4 L_{d2}^2} \sum_{i=0}^{L_{d2}-1} 2\sigma^2 = \frac{8A^2}{\sigma^2 L_{d2}} \tag{33}$$

where we have used the fact that $w_{1,i}, w_{2,i}$ are independent. Now, we know that the probability of error for the BPSK signal in (27), that is equal to [40].

$$r_{3,i} = r_{1,i} + r_{2,i} = \pm 2A + w_{1,i} + w_{2,i} \tag{34}$$

Therefore from (28), (32) and (34) we have

$$P_f(e) \approx \frac{1}{2} \operatorname{erfc}\left(\sqrt{\frac{|\mathcal{Y}|}{4}}\right) \tag{35}$$

where $P_f(e)$ denotes the probability of bit error for frame "$f$" and

$$E\left[\mathcal{Z}^2\right] \to 0 \qquad \text{for } L_{d2} \gg 1. \tag{36}$$

Observe that it is necessary to take the absolute value of $\mathcal{Y}$ in (35) since there is a possibility that it can be negative. The average probability of bit error over $F$ frames is given by

$$P(e) = \frac{1}{F} \sum_{f=0}^{F-1} P_f(e). \tag{37}$$

In the next section we present computer simulation results for SU-MMIMO using SCTC in uncorrelated channel.

### 2.5 Simulation results

The simulation parameters are given in **Table 5**. We can make the following observations from **Figures 4–6** [36]:

The theoretical prediction of BER closely matches with simulations.

- For $N_{\text{tot}}$ = 32, 1024, the BER is quite insensitive to wide variations in the total number of antennas $N_{\text{tot}}$, transmit antennas $N_t$ and retransmissions $N_{rt}$.
- For $N_{\text{tot}}$ = 2, the BER improves significantly with increasing retransmissions.

In **Figure 4(c)** we observe that there is more than 1 dB improvement in SINR compared to **Figures 4–6(a, b)**. However, large values of $L_{d1}$ may introduce more latency which is contrary to the requirements of 5G and beyond. In the next section we present SU-MMIMO using PCTC in correlated channel.

| Parameter | Value(s) | | | | | |
|---|---|---|---|---|---|---|
| Modulation | QPSK | | | | | |
| Total Antennas ($N_{tot} = N_t + N_r$) | 1024 | | 32 | | | 2 |
| Transmit antennas ($N_t$) | 400 | 512 | 7 | 12 | 16 | 1 |
| Frame length ($L_{d1}$) | 1200 50,400 | 1024 50,176 | 1001 | 1008 | 1024 | 1001 |
| Frames simulated ($F$) | $10^4$. $10^5$ for $L_{d1}$ range 1001 to 1200, 200, 2000 for $L_{d1}$ = 50,176, 50,400 | | | | | |
| Spectral eff. For $N_{rt}$ = 1 (bits/sec/Hz) | 200 | 256 | 3.5 | 6 | 8 | 0.5 |

**Table 5.**
*Simulation parameters for results in **Figures 4–6**.*

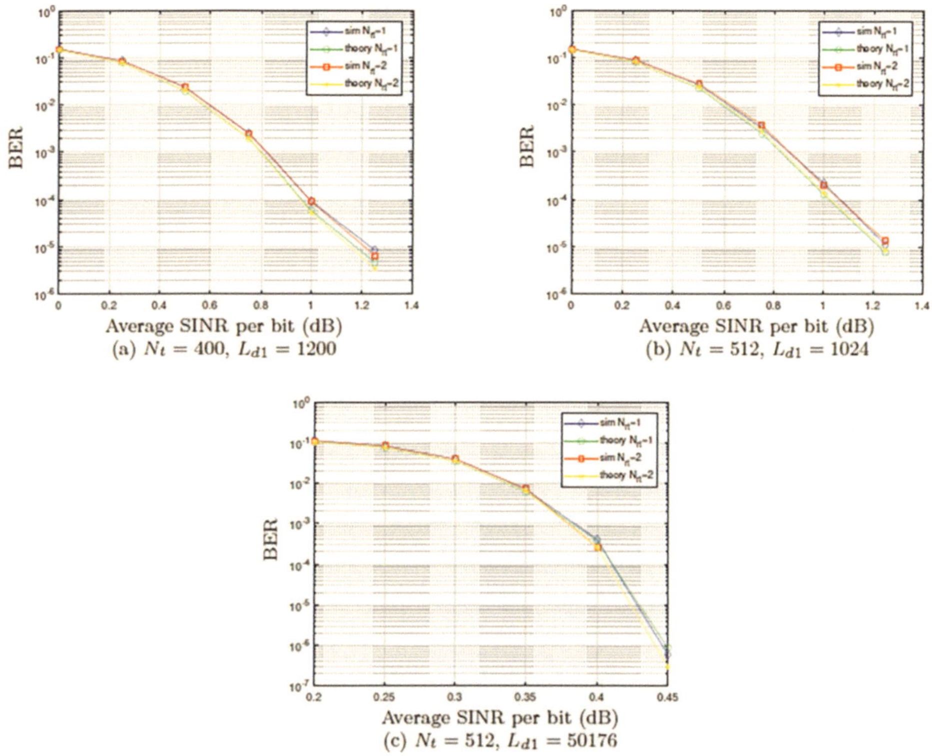


**Figure 4.**
*Simulation results for* $N_{tot}$ = *1024.*

## 3. SU-MMIMO using PCTC in correlated channel

### 3.1 System model

The block diagram of the system is identical to **Figure 2** in [36] and the received signal is given by (4). Note that in (4), the channel autocorrelation matrix is given by

$$\mathbf{R}_{\tilde{\mathbf{H}}\tilde{\mathbf{H}}} = \frac{1}{2}E\left[\tilde{\mathbf{H}}_k^H\tilde{\mathbf{H}}_k\right] = N_r\mathbf{I}_{N_t} \tag{38}$$

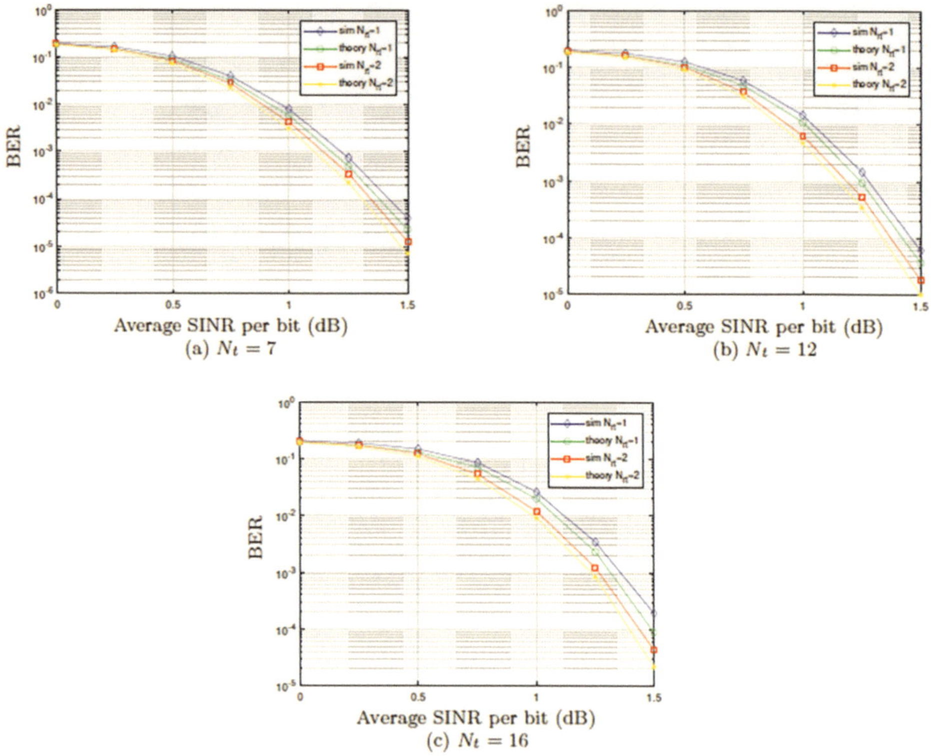


**Figure 5.**
*Simulation results for $N_{tot}$ = 32.*

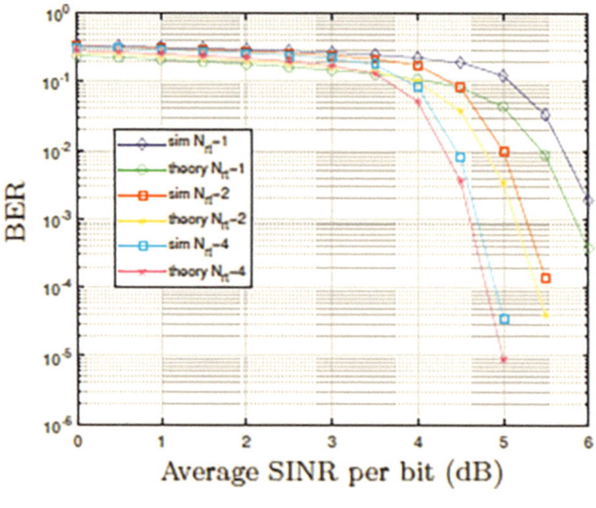


**Figure 6.**
*Simulation results for $N_{tot}$ = 2, $N_t$ = 1.*

where the superscript "$H$" denotes Hermitian and $\mathbf{I}_{N_t}$ denotes the $N_t \times N_t$ identity matrix. In this section, we investigate the situation where $\mathbf{R}_{\tilde{\mathbf{H}}\tilde{\mathbf{H}}}$ is not an identity matrix, but is a valid autocorrelation matrix [40]. As mentioned in [36], the elements of $\tilde{\mathbf{H}}_k$ – given by $\tilde{H}_{k,i,j}$ for the $k^{th}$ re-transmission, $i^{th}$ row, $j^{th}$ column of $\tilde{\mathbf{H}}_k$ – are zero-mean, complex Gaussian random variables with variance per dimension equal to $\sigma_H^2$. The in-phase and quadrature components of $\tilde{H}_{k,i,j}$ – denoted by $H_{k,i,j,I}$ and $H_{k,i,j,Q}$ respectively – are statistically independent. Moreover, we assume that the rows of $\tilde{\mathbf{H}}_k$ are statistically independent. Following the procedure in [36] for the case without precoding, we now find the expression for the average SINR per bit before and after averaging over re-transmissions ($k$). All symbols and notations have the usual meaning, as given in [36].

### 3.2 SINR analysis

The $i^{th}$ element of $\tilde{\mathbf{H}}_k^H \tilde{\mathbf{R}}_k$ is given by (25) of [36] which is repeated here for convenience

$$\tilde{Y}_{k,i} = \tilde{F}_{k,i,i} S_i + \tilde{I}_{k,i} + \tilde{V}_{k,i} \quad \text{for } 1 \le i \le N_t \tag{39}$$

where

$$\tilde{V}_{k,i} = \sum_{j=1}^{N_r} \tilde{H}_{k,j,i}^* \tilde{W}_{k,j}; \tilde{I}_{k,i} = \sum_{\substack{j=1 \\ j \neq i}}^{N_t} \tilde{F}_{k,i,j} S_j; \tilde{F}_{k,i,j} = \sum_{l=1}^{N_r} \tilde{H}_{k,l,i}^* \tilde{H}_{k,l,j}. \tag{40}$$

We have

$$\begin{aligned} E\left[\tilde{F}_{k,i,i}^2\right] &= E\left[\sum_{l=1}^{N_r} \left|\tilde{H}_{k,l,i}\right|^2 \sum_{m=1}^{N_r} \left|\tilde{H}_{k,m,i}\right|^2\right] \\ &= E\left[\sum_{l=1}^{N_r} \left(H_{k,l,i,I}^2 + H_{k,l,i,Q}^2\right) \sum_{m=1}^{N_r} \left(H_{k,m,i,I}^2 + H_{k,m,i,Q}^2\right)\right] = 4\sigma_H^4 N_r (N_r + 1) \end{aligned} \tag{41}$$

which is identical to (27) in [36] and we have used the following properties

1. The in-phase and quadrature components of $\tilde{H}_{k,i,j}$ are independent.
2. The rows of $\tilde{\mathbf{H}}_k$ are independent.
3. For zero-mean, real-valued Gaussian random variable $X$, . with variance equal to $\sigma_X^2$, $E[X^4] = 3\sigma_X^4$.

The interference power is

$$E\left[\left|\tilde{I}_{k,i}\right|^2\right] = E\left[\sum_{\substack{j=1 \\ j \neq i}}^{N_t} \tilde{F}_{k,i,j} S_j \sum_{\substack{l=1 \\ l \neq i}}^{N_t} \tilde{F}_{k,i,l}^* S_l^*\right] = \sum_{\substack{j=1 \\ j \neq i}}^{N_t} \sum_{\substack{l=1 \\ l \neq i}}^{N_t} E\left[\tilde{F}_{k,i,j} \tilde{F}_{k,i,l}^*\right] E\left[S_j S_l^*\right] = P_{\text{av}} \sum_{\substack{j=1 \\ j \neq i}}^{N_t} E\left[\left|\tilde{F}_{k,i,j}\right|^2\right]. \tag{42}$$

where we have used (9) in. Similarly the noise power is

$$
\begin{aligned}
E\left[\left|\tilde{V}_{k,i}\right|^2\right] &= E\left[\sum_{j=1}^{N_r} \tilde{H}_{k,j,i}^{*}\tilde{W}_{k,j}\sum_{m=1}^{N_r}\tilde{H}_{k,m,i}\tilde{W}_{k,m}^{*}\right] \\
&= \sum_{j=1}^{N_r}\sum_{m=1}^{N_r} E\left[\tilde{H}_{k,j,i}^{*}\tilde{H}_{k,m,i}\right]E\left[\tilde{W}_{k,m}^{*}\tilde{W}_{k,j}\right] \\
&= \sum_{j=1}^{N_r}\sum_{m=1}^{N_r} 2\sigma_H^2\delta_K(j-m)2\sigma_W^2(j-m) = 4N_r\sigma_H^2\sigma_W^2
\end{aligned} \tag{43}
$$

which is identical to (29) in [36] and we have used the following properties:

1. Rows of $\tilde{\mathbf{H}}_k$ are independent.

2. Sifting property of the Kronecker delta function.

3. Noise and channel coefficients are independent.

Now in (42)

$$
\begin{aligned}
E\left[\left|\tilde{F}_{k,i,j}\right|^2\right] &= E\left[\sum_{l=1}^{N_r}\tilde{H}_{k,l,i}^{*}\tilde{H}_{k,l,j}\sum_{m=1}^{N_r}\tilde{H}_{k,m,i}\tilde{H}_{k,m,j}^{*}\right] \\
&= \sum_{l=1}^{N_r} E\left[\tilde{H}_{k,l,i}^{*}\tilde{H}_{k,l,j}\left(\tilde{H}_{k,l,i}\tilde{H}_{k,l,j}^{*} + \sum_{\substack{m=1\\ m\neq l}}^{N_r}\tilde{H}_{k,m,i}\tilde{H}_{k,m,j}^{*}\right)\right] \\
&= \sum_{l=1}^{N_r} E\left[\left|\tilde{H}_{k,l,i}\right|^2\left|\tilde{H}_{k,l,j}\right|^2 + \left(\sum_{\substack{m=1\\ m\neq l}}^{N_r}\tilde{H}_{k,l,i}^{*}\tilde{H}_{k,l,j}\tilde{H}_{k,m,i}\tilde{H}_{k,m,j}^{*}\right)\right].
\end{aligned} \tag{44}
$$

Now the first summation in (44) is equal to

$$
E_1 = E\left[\left|\tilde{H}_{k,l,i}\right|^2\left|\tilde{H}_{k,l,j}\right|^2\right] = E\left[\left(H_{k,l,i,I}^2 + H_{k,l,i,Q}^2\right)\left(H_{k,l,j,I}^2 + H_{k,l,j,Q}^2\right)\right] = 4\sigma_H^4 + 4R_{\tilde{H}\tilde{H},j-i}^2 \tag{45}
$$

where we have used the property that for real-valued, zero-mean Gaussian random variables $X_i$, $1 \leq i \leq 4$ [44, 45]

$$
E[X_1X_2X_3X_4] = C_{12}C_{34} + C_{13}C_{24} + C_{14}C_{23} \tag{46}
$$

where

$$C_{ij} = E\left[X_i X_j\right] \qquad \text{for } 1 \le i, j \le 4 \tag{47}$$

and

$$R_{\tilde{H}\tilde{H},j-i} = E\left[H_{k,l,i,I} H_{k,l,j,I}\right] = E\left[H_{k,l,i,Q} H_{k,l,j,Q}\right] = \frac{1}{2} E\left[\tilde{H}^*_{k,l,i} \tilde{H}_{k,l,j}\right] = R_{\tilde{H}\tilde{H},i-j} \tag{48}$$

is the real-valued autocorrelation of $\tilde{H}_{k,m,n}$ and we have made the assumption that the in-phase and quadrature components of $\tilde{H}_{k,m,n}$ are independent. The second summation in (44) can be written as

$$\begin{aligned} E_2 &= \sum_{\substack{m=1 \\ m \neq l}}^{N_r} E\left[\tilde{H}^*_{k,l,i} \tilde{H}_{k,l,j} \tilde{H}_{k,m,i} \tilde{H}^*_{k,m,j}\right] \\ &= \sum_{\substack{m=1 \\ m \neq l}}^{N_r} E\left[\tilde{H}^*_{k,l,i} \tilde{H}_{k,l,j}\right] E\left[\tilde{H}_{k,m,i} \tilde{H}^*_{k,m,j}\right] = \sum_{\substack{m=1 \\ m \neq l}}^{N_r} 4R^2_{\tilde{H}\tilde{H},j-i} = 4(N_r - 1) R^2_{\tilde{H}\tilde{H},j-i} \end{aligned} \tag{49}$$

where we have used the property that the rows of $\tilde{\mathbf{H}}_k$ are independent. Therefore (44) becomes

$$E\left[\left|\tilde{F}_{k,i,j}\right|^2\right] = N_r(E_1 + E_2) = 4N_r\left[\sigma_H^4 + R^2_{\tilde{H}\tilde{H},j-i} + (N_r - 1)R^2_{\tilde{H}\tilde{H},j-i}\right] = 4N_r\left[\sigma_H^4 + N_r R^2_{\tilde{H}\tilde{H},j-i}\right]. \tag{50}$$

The total power of interference plus noise is

$$E\left[\left|\tilde{I}_{k,i} + \tilde{V}_{k,i}\right|^2\right] = E\left[\left|\tilde{I}_{k,i}\right|^2\right] + E\left[\left|\tilde{V}_{k,i}\right|^2\right] = 4P_{\text{av}} N_r \sum_{\substack{j=1 \\ j \neq i}}^{N_t} \left[\sigma_H^4 + N_r R^2_{\tilde{H}\tilde{H},j-i}\right] + 4N_r \sigma_H^2 \sigma_W^2 \tag{51}$$

where we have made the assumption that noise and symbols are independent. The average SINR per bit for the $i^{th}$ transmit antenna is similar to (31) of [36] which is repeated here for convenience

$$\text{SINR}_{\text{av},b,i} = \frac{E\left[\left|\tilde{F}_{k,i,i} S_i\right|^2\right] \times 2N_{rt}}{E\left[\left|\tilde{I}_{k,i} + \tilde{V}_{k,i}\right|^2\right]} \qquad \text{for } 1 \le i \le N_t \tag{52}$$

into which (41) and (51) have to be substituted. The upper bound on the average SINR per bit for the $i^{th}$ transmit antenna is obtained by setting $\sigma_W^2 = 0$ in (51), (52) and is given by, for $1 \le i \le N_t$

$$\text{SINR}_{\text{av},b,\text{UB},i} = \frac{\sigma_H^4 (1 + N_r) \times 2N_{rt}}{\sum_{\substack{j=1 \\ j \neq i}}^{N_t} \left[\sigma_H^4 + N_r R^2_{\tilde{H}\tilde{H},j-i}\right]}. \tag{53}$$

Observe that in contrast to (31) and (32) in [36], the average SINR per bit and its upper bound depend on the transmit antenna. Let us now compute the average SINR per bit after averaging over retransmissions. The received signal after averaging over retransmissions is given by (6) with (see also (20) of [36])

$$F_i = \frac{1}{N_{rt}} \sum_{k=0}^{N_{rt}-1} \tilde{F}_{k,i,i}$$

$$\tilde{U} = \frac{1}{N_{rt}} \sum_{k=0}^{N_{rt}-1} \left(\tilde{I}_{k,i} + \tilde{V}_{k,i}\right) = \frac{1}{N_{rt}} \sum_{k=0}^{N_{rt}-1} \tilde{U}_{k,i}' \qquad \text{(say)} \tag{54}$$

where $\tilde{F}_{k,i,i}$, $\tilde{I}_{k,i}$ and $\tilde{V}_{k,i}$ are given in (39). The power of the signal component of (6) is

$$\begin{aligned} E\left[|S_i|^2 F_i^2\right] &= P_{\text{av}} E\left[F_i^2\right] = \frac{P_{\text{av}}}{N_{rt}^2} E\left[\sum_{k=0}^{N_{rt}-1} \tilde{F}_{k,i,i} \sum_{l=0}^{N_{rt}-1} \tilde{F}_{l,i,i}\right] \\ &= \frac{P_{\text{av}}}{N_{rt}^2} \sum_{k=0}^{N_{rt}-1} \left[\sum_{\substack{l=0 \\ l\neq k}}^{N_{rt}-1} E\left[\tilde{F}_{k,i,i}\right] E\left[\tilde{F}_{l,i,i}\right] + E\left[\left|\tilde{F}_{k,i,i}\right|^2\right]\right] \end{aligned} \tag{55}$$

where we have used the fact that the channel is independent across retransmissions, therefore

$$E\left[\tilde{F}_{k,i,i}\tilde{F}_{l,i,i}\right] = E\left[\tilde{F}_{k,i,i}\right] E\left[\tilde{F}_{l,i,i}\right] \qquad \text{for } k \neq l. \tag{56}$$

Now

$$E\left[\tilde{F}_{k,i,i}\right] = E\left[\sum_{l=1}^{N_r} \left|\tilde{H}_{k,l,i}\right|^2\right] = 2N_r\sigma_H^2. \tag{57}$$

Substituting (41) and (57) in (55) we get

$$E\left[|S_i|^2 F_i^2\right] = \frac{4N_r P_{\text{av}} \sigma_H^4}{N_{rt}} \left(1 + N_r N_{rt}\right). \tag{58}$$

The power of the interference component in (6) and (54) is

$$E\left[\left|\tilde{U}_i\right|^2\right] = \frac{1}{N_{rt}^2} E\left[\sum_{k=0}^{N_{rt}-1} \left(\tilde{I}_{k,i} + \tilde{V}_{k,i}\right) \sum_{l=0}^{N_{rt}-1} \left(\tilde{I}_{l,i}^* + \tilde{V}_{l,i}^*\right)\right] = \frac{1}{N_{rt}^2} \sum_{k=0}^{N_{rt}-1} \sum_{l=0}^{N_{rt}-1} E\left[\tilde{I}_{k,i}\tilde{I}_{l,i}^*\right] + E\left[\tilde{V}_{k,i}\tilde{V}_{l,i}^*\right] \tag{59}$$

where we have used the following properties from (40)

$$E\left[\tilde{I}_{k,i}\right] = E\left[\tilde{V}_{k,i}\right] = 0; E\left[\tilde{I}_{k,i}\tilde{V}_{l,i}^*\right] = E\left[\tilde{V}_{k,i}\tilde{I}_{l,i}^*\right] = 0 \qquad \text{for all } k, l \tag{60}$$

since $S_j$ and $\tilde{W}_{k,j}$ are mutually independent with zero-mean. Now

$$
\begin{aligned}
E\left[\tilde{I}_{k,i}\tilde{I}_{l,i}^{*}\right] &= E\left[\sum_{\substack{j=1\\ j\neq i}}^{N_t}\tilde{F}_{k,i,j}S_j\sum_{\substack{n=1\\ n\neq i}}^{N_t}\tilde{F}_{l,i,n}^{*}S_n^{*}\right] = \sum_{\substack{j=1\\ j\neq i}}^{N_t}\sum_{\substack{n=1\\ n\neq i}}^{N_t}E\left[\tilde{F}_{k,i,j}\tilde{F}_{l,i,n}^{*}\right]E\left[S_jS_n^{*}\right]\\
&= \sum_{\substack{j=1\\ j\neq i}}^{N_t}\sum_{\substack{n=1\\ n\neq i}}^{N_t}E\left[\tilde{F}_{k,i,j}\tilde{F}_{l,i,n}^{*}\right]P_{\mathrm{av}}\delta_K(j-n)\\
&= P_{\mathrm{av}}\sum_{\substack{j=1\\ j\neq i}}^{N_t}E\left[\tilde{F}_{k,i,j}\tilde{F}_{l,i,j}^{*}\right]
\end{aligned}
\tag{61}
$$

where we have used the property that the symbols are uncorrelated and $\delta_K(\cdot)$ is the Kronecker delta function [40]. When $k = l$, (61) is given by (42) and (50). When $k \neq l$, (61) is given by

$$
E\left[\tilde{I}_{k,i}\tilde{I}_{l,i}^{*}\right] = P_{\mathrm{av}}\sum_{\substack{j=1\\ j\neq i}}^{N_t}E\left[\tilde{F}_{k,i,j}\right]E\left[\tilde{F}_{l,i,j}^{*}\right] = P_{\mathrm{av}}\sum_{\substack{j=1\\ j\neq i}}^{N_t}4N_r^2R_{\tilde{H}\tilde{H},j-i}^2
\tag{62}
$$

where we have used (40) and (48). Similarly, we have

$$
E\left[\tilde{V}_{k,i}\tilde{V}_{l,i}^{*}\right] = 4N_r\sigma_H^2\sigma_W^2\delta_K(k-l)
\tag{63}
$$

where we have used (43). Substituting (42), (50), (62) and (63) in (59) we get

$$
\begin{aligned}
E\left[\left|\tilde{U}_i\right|^2\right] &= \frac{1}{N_{rt}^2}\left[4P_{\mathrm{av}}N_rN_{rt}\sum_{\substack{j=1\\ j\neq i}}^{N_t}\left(\sigma_H^4+N_rR_{\tilde{H}\tilde{H},j-i}^2\right)+4P_{\mathrm{av}}N_r^2N_{rt}(N_{rt}-1)\sum_{\substack{j=1\\ j\neq i}}^{N_t}R_{\tilde{H}\tilde{H},j-i}^2\right]\\
&\quad+\frac{4N_r}{N_{rt}}\sigma_H^2\sigma_W^2\\
&= \frac{1}{N_{rt}}\left[4P_{\mathrm{av}}N_r\sum_{\substack{j=1\\ j\neq i}}^{N_t}\left(\sigma_H^4+N_rR_{\tilde{H}\tilde{H},j-i}^2\right)+4P_{\mathrm{av}}N_r^2(N_{rt}-1)\sum_{\substack{j=1\\ j\neq i}}^{N_t}R_{\tilde{H}\tilde{H},j-i}^2\right]\\
&\quad+\frac{4N_r}{N_{rt}}\sigma_H^2\sigma_W^2.
\end{aligned}
\tag{64}
$$

The average SINR per bit for the $i^{th}$ transmit antenna, after averaging over retransmissions (also referred to as "combining" [36]) is given by

$$\mathrm{SINR}_{\mathrm{av},b,\mathrm{C},i} = \frac{2P_{\mathrm{av}}E\left[F_i^2\right]}{E\left[\left|\tilde{U}_i\right|^2\right]} \tag{65}$$

into which (58) and (64) have to be substituted. The upper bound on the average SINR per bit after "combining" for the $i^{th}$ transmit antenna is given by

$$\mathrm{SINR}_{\mathrm{av},b,\mathrm{C},\mathrm{UB},i} = \mathrm{SINR}_{\mathrm{av},b,\mathrm{C},i}\big|_{\sigma_W^2=0}. \tag{66}$$

The plots of the average SINR per bit for the $i^{th}$ transmit antenna before and after "combining" are shown in **Figures** 7 and **8** respectively for $N_{\mathrm{tot}} = 1024$ and $N_{rt} = 2$. The channel correlation is given by

$$R_{\tilde{H}\tilde{H},j-i} = 0.9^{|j-i|}\sigma_H^2 \tag{67}$$

in (48), which is obtained by passing samples of white Gaussian noise through a unit-energy, first-order infinite impulse response (IIR) lowpass filter with $a = -0.9$ (see (30) of [46]).

We observe in **Figures** 7 and **8** that

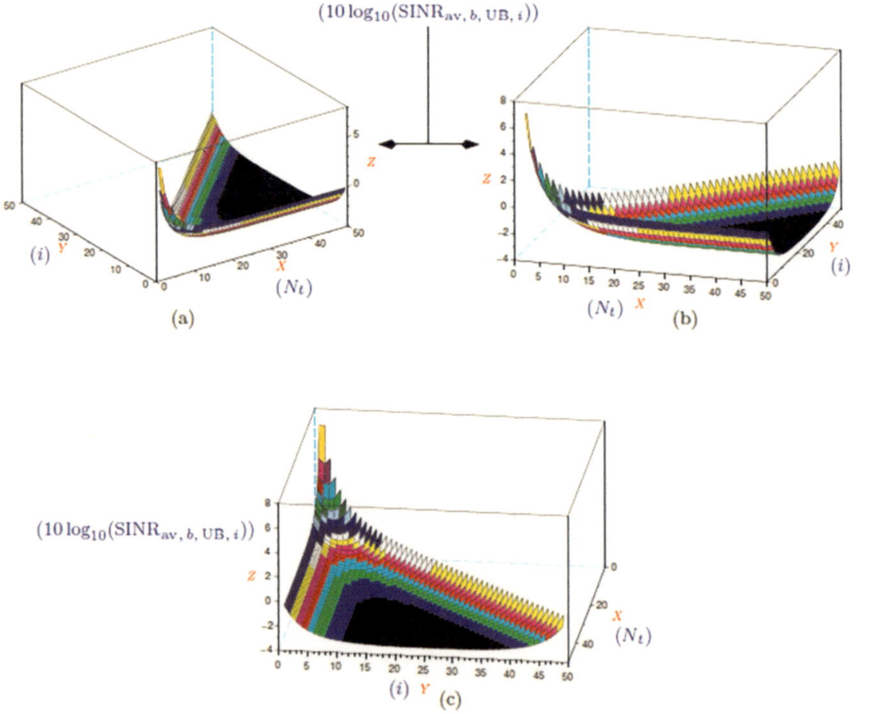


**Figure 7.**
*Plot of $SINR_{av,b,UB,i}$ for $N_{tot}$ = 1024, $N_{rt}$ = 2. (a) Back view. (b) Sideview. (c) Front view.*

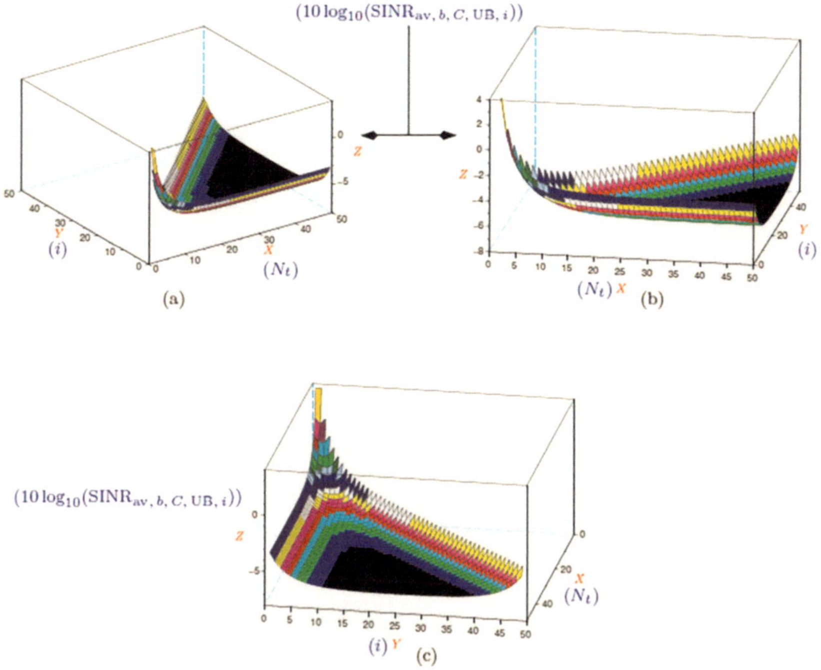


**Figure 8.**
*Plot of* $SINR_{av,b,C,UB,i}$ *for* $N_{tot} = 1024$, $N_{rt} = 2$. *(a) Back view. (b) Side view. (c) Front view.*

The upper bound on the average SINR per bit decreases rapidly with increasing transmit antennas $N_t$ and falls below 0 dB for $N_t > 5$ (see **Figures 7(b)** and **8(b)**). Since the spectral efficiency of the system is $N_t/(2N_{rt})$ bits/sec/Hz (see (33) of [36]), the system would be of no practical use, since the BER would be close to 0.5 for $N_t > 5$.

The upper bound on the average SINR per bit after "combining" is *less* than that before "combining". Therefore retransmissions are ineffective.

In view of the above observation, it becomes necessary to design a better receiver using precoding. This is presented in the next section.

### 3.3 Precoding

Similar to (4) consider the modified received signal given by

$$\tilde{\mathbf{R}}_k = \tilde{\mathbf{H}}_k \tilde{\mathbf{B}} \mathbf{S} + \tilde{\mathbf{W}}_k \tag{68}$$

where

$$\tilde{\mathbf{B}} = \begin{bmatrix} 1 & 0 & \cdots & 0 \\ \tilde{a}_{1,1} & 1 & \cdots & 0 \\ \vdots & \cdots & \cdots & \vdots \\ \tilde{a}_{N_t-1,N_t-1} & \cdots & \tilde{a}_{N_t-1,1} & 1 \end{bmatrix}^T \triangleq \tilde{\mathbf{A}}^T \tag{69}$$

where $(\cdot)^T$ denotes transpose. In (69), $\tilde{\mathbf{A}}$ is an $N_t \times N_t$ lower triangular matrix with diagonal elements equal to unity and $\tilde{a}_{i,j}$ denotes the $j^{th}$ coefficient of the optimum $i^{th}$-order forward prediction filter [40] and $\tilde{\mathbf{B}}$ is the precoding matrix. Let

$$\tilde{\mathbf{Y}}_k = \tilde{\mathbf{B}}^H \tilde{\mathbf{H}}_k^H \tilde{\mathbf{R}}_k = \tilde{\mathbf{B}}^H \tilde{\mathbf{H}}_k^H \tilde{\mathbf{H}}_k \tilde{\mathbf{B}} \mathbf{S} + \tilde{\mathbf{B}}^H \tilde{\mathbf{H}}_k^H \tilde{\mathbf{W}}_k. \tag{70}$$

Define

$$\tilde{\mathbf{Z}}_k = \tilde{\mathbf{H}}_k \tilde{\mathbf{B}} = \begin{bmatrix} \tilde{Z}_{k,1,1} & \cdots & \tilde{Z}_{k,1,N_t} \\ \vdots & \cdots & \vdots \\ \tilde{Z}_{k,N_r,1} & \cdots & \tilde{Z}_{k,N_r,N_t} \end{bmatrix}. \tag{71}$$

Now [40]

$$\frac{1}{2} E\left[\tilde{\mathbf{Z}}_k^H \tilde{\mathbf{Z}}_k\right] = N_r \begin{bmatrix} \sigma_{Z,1}^2 & 0 & \cdots & 0 \\ 0 & \sigma_{Z,2}^2 & \cdots & 0 \\ \vdots & \cdots & \cdots & \vdots \\ 0 & \cdots & 0 & \sigma_{Z,N_t}^2 \end{bmatrix} \stackrel{\Delta}{=} \tilde{\mathbf{R}}_{\tilde{Z}\tilde{Z}} \tag{72}$$

is an $N_t \times N_t$ diagonal matrix and $\sigma_{Z,i}^2$ denotes the variance per dimension of the optimum $(i-1)^{th}$-order forward prediction filter. Note that [40]

$$\sigma_{Z,1}^2 = \sigma_H^2; \sigma_{Z,i}^2 \geq \sigma_{Z,j}^2 \qquad \text{for } i < j. \tag{73}$$

Let

$$\tilde{\mathbf{V}}_k = \tilde{\mathbf{Z}}_k^H \tilde{\mathbf{W}}_k = \begin{bmatrix} \tilde{V}_{k,1} & \cdots & \tilde{V}_{k,N_t} \end{bmatrix}^T \tag{74}$$

which is an $N_t \times 1$ vector. Now

$$E\left[\tilde{V}_{k,i}\tilde{V}_{k,m}^*\right] = E\left[\sum_{j=1}^{N_r} \tilde{Z}_{k,j,i}^* \tilde{W}_{k,j} \sum_{l=1}^{N_r} \tilde{Z}_{k,l,m} \tilde{W}_{k,l}^*\right] = \sum_{j=1}^{N_r}\sum_{l=1}^{N_r} E\left[\tilde{Z}_{k,l,m}\tilde{Z}_{k,j,i}^*\right] E\left[\tilde{W}_{k,j}\tilde{W}_{k,l}^*\right]$$

$$= \sum_{j=1}^{N_r}\sum_{l=1}^{N_r} 2\sigma_{Z,i}^2 \delta_K(i-m)\delta_K(j-l) \times 2\sigma_W^2 \delta_K(j-l) = 4N_r \sigma_{Z,i}^2 \sigma_W^2 \delta_K(i-m) \tag{75}$$

where we have used (72). Let

$$\tilde{\mathbf{F}}_k = \tilde{\mathbf{Z}}_k^H \tilde{\mathbf{Z}}_k \tag{76}$$

which is an $N_t \times N_t$ matrix. Substituting (76) in (70) we get

$$\tilde{\mathbf{Y}}_k = \tilde{\mathbf{F}}_k \mathbf{S} + \tilde{\mathbf{V}}_k. \tag{77}$$

Similar to (39), the $i^{th}$ element of $\tilde{\mathbf{Y}}_k$ in (77) is given by

$$\tilde{y}_{k,i} = \tilde{F}_{k,i,i} S_i + \tilde{I}_{k,i} + \tilde{V}_{k,i} \text{for } 1 \leq i \leq N_t \tag{78}$$

where

$$\tilde{V}_{k,i} = \sum_{j=1}^{N_r} \tilde{Z}^*_{k,j,i} \tilde{W}_{k,j}; \tilde{I}_{k,i} = \sum_{\substack{j=1 \\ j \neq i}}^{N_t} \tilde{F}_{k,i,j} S_j; \tilde{F}_{k,i,j} \sum_{l=1}^{N_r} \tilde{Z}^*_{k,l,i} \tilde{Z}_{k,l,j}. \tag{79}$$

Note that from (72) and (76) we have

$$E\left[\tilde{F}_{k,i,i}\right] = 2N_r \sigma^2_{Z,i} \tag{80}$$

Now

$$\begin{aligned} E\left[\tilde{F}^2_{k,i,i}\right] &= E\left[\sum_{l=1}^{N_r} \left|\tilde{Z}_{k,l,i}\right|^2 \sum_{m=1}^{N_r} \left|\tilde{Z}_{k,m,i}\right|^2\right] = \sum_{l=1}^{N_r} \left|\tilde{Z}_{k,l,i}\right|^4 + \\ &+ \sum_{\substack{m=1 \\ m \neq l}}^{N_r} E\left[\left|\tilde{Z}_{k,l,i}\right|^2\right] E\left[\left|\tilde{Z}_{k,m,i}\right|^2\right] = 4N_r(N_r+1)\sigma^2_{Z,i}. \end{aligned} \tag{81}$$

Similarly

$$E\left[\left|\tilde{I}_{k,i}\right|^2\right] = E\left[\sum_{\substack{j=1 \\ j \neq i}}^{N_t} \tilde{F}_{k,i,j} S_j \sum_{\substack{l=1 \\ l \neq i}}^{N_t} \tilde{F}^*_{k,i,l} S^*_l\right] = P_{\text{av}} \sum_{\substack{j=1 \\ j \neq i}}^{N_t} E\left[\left|\tilde{F}_{k,i,j}\right|^2\right]. \tag{82}$$

Now

$$E\left[\left|\tilde{F}_{k,i,j}\right|^2\right] = E\left[\sum_{l=1}^{N_r} \tilde{Z}^*_{k,l,i} \tilde{Z}_{k,l,j} \sum_{m=1}^{N_r} \tilde{Z}_{k,m,i} \tilde{Z}^*_{k,m,j}\right] = \sum_{l=1}^{N_r} \sum_{m=1}^{N_r} 4\sigma^2_{Z,i}\sigma^2_{Z,j}\delta_K(l-m) = 4N_r\sigma^2_{Z,i}\sigma^2_{Z,j} \tag{83}$$

where we have used (72). Substituting (83) in (82) we get

$$E\left[\left|\tilde{I}_{k,i}\right|^2\right] = 4P_{\text{av}} N_r \sigma^2_{Z,i} \sum_{\substack{j=1 \\ j \neq i}}^{N_t} \sigma^2_{Z,j}. \tag{84}$$

Note that

$$E\left[\left|\tilde{I}_{k,i} + \tilde{V}_{k,i}\right|^2\right] = E\left[\left|\tilde{I}_{k,i}\right|^2\right] + E\left[\left|\tilde{V}_{k,i}\right|^2\right]. \tag{85}$$

The average SINR per bit for the $i^{th}$ transmit antenna is given by (52) and is equal to

$$\text{SINR}_{\text{av},b,i} = \frac{E\left[\left|\tilde{F}_{k,i,i} S_i\right|^2 \times 2N_{rt}\right]}{E\left[\left|\tilde{I}_{k,i} + \tilde{V}_{k,i}\right|^2\right]} = \frac{P_{\text{av}}(N_r+1)\,\sigma^2_{Z,i} \times 2N_{rt}}{P_{\text{av}} \sum_{\substack{j=1 \\ j \neq i}}^{N_t} \sigma^2_{Z,j} + \sigma^2_W} \tag{86}$$

where we have used (75), (81) and (84). The upper bound on the average SINR per bit for the $i^{th}$ transmit antenna is obtained by setting $\sigma_W^2 = 0$ in (86) and is equal to

$$\mathrm{SINR}_{\mathrm{av},b,\mathrm{UB},i} = \frac{(N_r + 1)\sigma_{Z,i}^2 \times 2N_{rt}}{\sum_{\substack{j=1 \\ j \neq i}}^{N_t} \sigma_{Z,j}^2} \tag{87}$$

which is illustrated in **Figure 9** for $N_{\mathrm{tot}} = 1024$ and $N_{rt} = 2$. The value of the upper bound on the average SINR per bit for $N_t = i = 50$ is 18.6 dB. The channel correlation is given by (67). Note that a first-order prediction filter completely decorrelates the channel with [40]

$$\tilde{a}_{i,1} = -0.9 \text{ for } 1 \le i \le N_t - 1; \tilde{a}_{i,j} = 0 \text{ for } 2 \le i \le N_t - 1, 2 \le j \le i. \tag{88}$$

We also have [40]

$$\sigma_{Z,i}^2 = \sigma_{Z,2}^2 = \left(1 - |-0.9|^2\right)\sigma_{Z,1}^2 = 0.19\sigma_{Z,1}^2 \text{for } i > 2. \tag{89}$$

Therefore we see in **Figure 9** that the first transmit antenna $i = 1$ has a high $\mathrm{SINR}_{\mathrm{av},b,\mathrm{UB},i}$ due to low interference power from remaining transmit antennas,

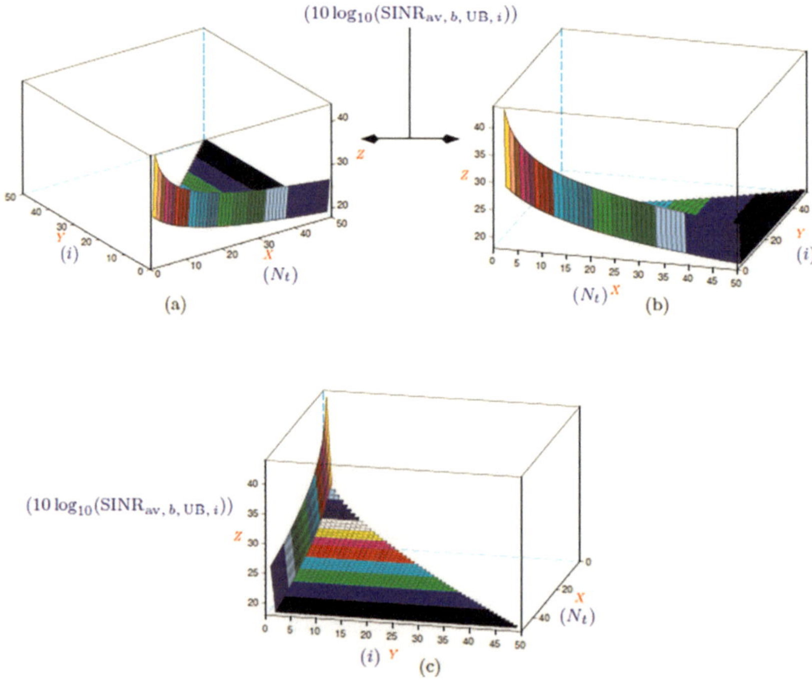


**Figure 9.**
*Plot of $\mathrm{SINR}_{av,b,UB,i}$ for $N_{tot} = 1024$, $N_{rt} = 2$ after precoding. (a) Back view. (b) Sideview. (c) Front view.*

whereas for $i \neq 1$ the $\text{SINR}_{\text{av},b,\text{UB},i}$ is low due to high interference power from the first transmit antenna ($i = 1$). The received signal after "combining" is given by (6) and (54). Note that from (54) and (79)

$$E\left[\tilde{F}_i^2\right] = \frac{1}{N_{rt}^2} E\left[\sum_{k=0}^{N_{rt}-1} \tilde{F}_{k,i,i} \sum_{l=0}^{N_{rt}-1} \tilde{F}_{l,i,i}\right] = \frac{1}{N_{rt}^2} \sum_{k=0}^{N_{rt}-1} E\left[\left|\tilde{F}_{k,i,i}\right|^2\right] + \\ + \sum_{\substack{l=0 \\ l \neq k}}^{N_{rt}-1} E\left[\tilde{F}_{k,i,i}\tilde{F}_{l,i,i}\right] = \frac{4N_r\sigma_{Z,i}^2}{N_{rt}^2} \sum_{k=0}^{N_{rt}-1} (N_r + 1) + (N_{rt} - 1)N_r \\ = \frac{4N_r\sigma_{Z,i}^2}{N_{rt}^2}(1 + N_r N_{rt}) \tag{90}$$

where we have used (56), (80) and (81). Similarly from (54), (75), (84) and (85) we have

$$E\left[\tilde{U}_i^2\right] = \frac{1}{N_{rt}^2} E\left[\sum_{k=0}^{N_{rt}-1} \tilde{U}'_{k,i} \sum_{l=0}^{N_{rt}-1} \left(\tilde{U}'_{l,i}\right)^*\right] = \frac{1}{N_{rt}^2} \sum_{k=0}^{N_{rt}-1} \sum_{l=0}^{N_{rt}-1} E\left[\tilde{U}'_{k,i}\left(\tilde{U}'_{l,i}\right)^*\right] \\ = \frac{1}{N_{rt}^2} \sum_{k=0}^{N_{rt}-1} \sum_{l=0}^{N_{rt}-1} E\left[\left|\tilde{U}'_{k,i}\right|^2\right] \delta_K(k-l) \\ = \frac{1}{N_{rt}} E\left[\left|\tilde{U}'_{k,i}\right|^2\right] = \frac{1}{N_{rt}}\left[E\left[\left|\tilde{I}_{k,i}\right|^2\right] + E\left[\left|\tilde{V}_{k,i}\right|^2\right]\right] = \frac{4N_r\sigma_{Z,i}^2}{N_{rt}}\left[P_{\text{av}} \sum_{\substack{j=1 \\ j \neq i}}^{N_t} \sigma_{Z,j}^2 + \sigma_W^2\right]. \tag{91}$$

Substituting (90) and (91) in (65) we have, after simplification, for $1 \leq i \leq N_t$

$$\text{SINR}_{\text{av},b,C,i} = \frac{2P_{\text{av}}E\left[F_i^2\right]}{E\left[\left|\tilde{U}_i\right|^2\right]} = \frac{(N_r N_{rt} + 1)\sigma_{Z,i}^2 \times 2P_{\text{av}}}{P_{\text{av}} \sum_{\substack{j=1 \\ j \neq i}}^{N_t} \sigma_{Z,j}^2 + \sigma_W^2}. \tag{92}$$

The upper bound on the average SINR per bit for the $i^{th}$ transmit antenna is obtained by substituting (92) in (66) and is equal to

$$\text{SINR}_{\text{av},b,C,\text{UB},i} = \frac{(N_r N_{rt} + 1)\sigma_{Z,i}^2 \times 2}{\sum_{\substack{j=1 \\ j \neq i}}^{N_t} \sigma_{Z,j}^2} \approx \text{SINR}_{\text{av},b,\text{UB},i} \tag{93}$$

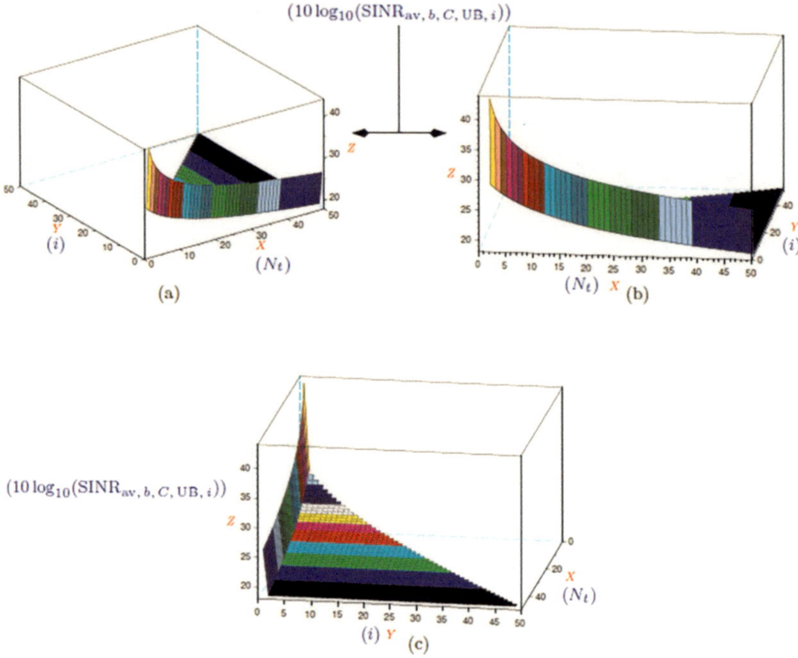


**Figure 10.**
*Plot of $SINR_{av,b,C,UB,i}$ for $N_{tot}$ = 1024, $N_{rt}$ = 2 after precoding. (a) Back view. (b) Side view. (c) Front view.*

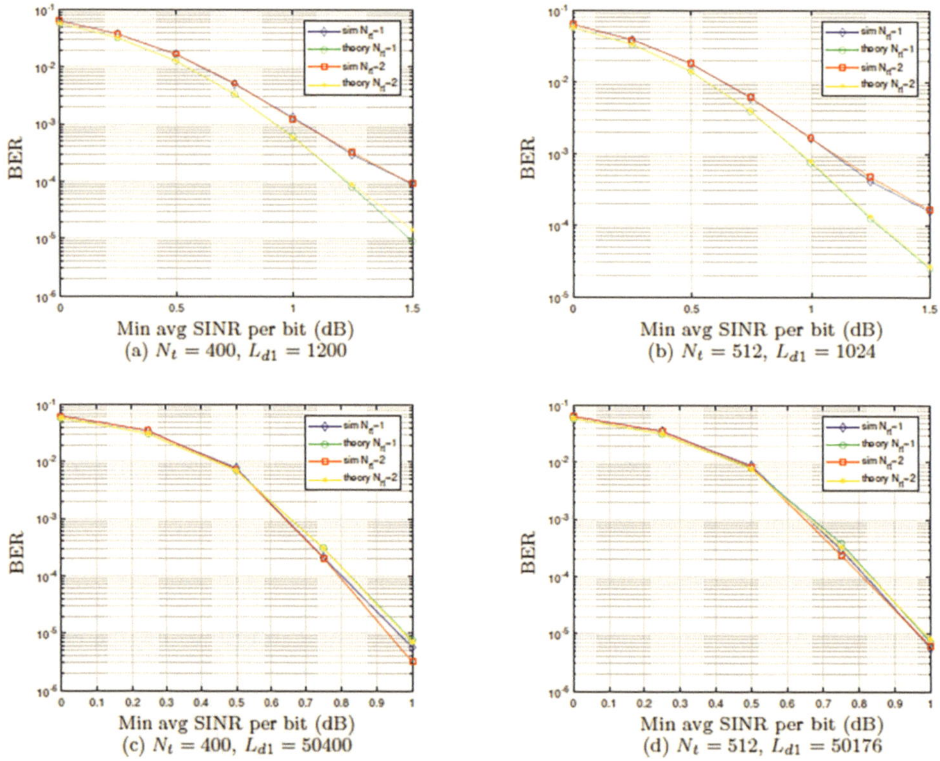


**Figure 11.**
*Simulation results with precoding for $N_{tot}$ = 1024.*

for $1 \le i \le N_t$, $N_r \gg 1$. This is illustrated in **Figure 10** for $N_{\text{tot}} = 1024$ and $N_{rt} = 2$. We again observe that the first transmit antenna ($i = 1$) has a high upper bound on the average SINR per bit, after "combining", compared to the remaining transmit antennas. The value of the upper bound on the average SINR per bit after "combining" for $N_t = i = 50$, $N_{\text{tot}} = 1024$ is 18.6 dB. After concatenation, $\tilde{Y}_i$ for $0 \le i \le L_d - 1$, in (6) and (54) is given to the turbo decoder [29, 40]. Let (see (26) of [29]):

$$\gamma_{1,i,m,n} = \exp\left[-\frac{\left|\tilde{Y}_i - F_i S_{m,n}\right|^2}{2\sigma_{U,i}^2}\right]; \gamma_{2,i,m,n} = \exp\left[-\frac{\left|\tilde{Y}_{i1} - F_{i1} S_{m,n}\right|^2}{2\sigma_{U,i}^2}\right] \tag{94}$$

$$\tilde{\mathbf{Y}}_1 = \left[\tilde{Y}_1 \cdots\cdots \tilde{Y}_{L_{d1}-1}\right]; \tilde{\mathbf{Y}}_2 = \left[\tilde{Y}_{L_{d1}} \cdots\cdots \tilde{Y}_{L_d-1}\right]. \tag{95}$$

Thenwhere

$$i1 = i + L_{d1} \qquad \text{for } 0 \le i \le L_{d1} - 1. \tag{96}$$

The rest of the turbo decoding algorithm is similar to that discussed in [29, 40] will not be repeated here. In the next subsection we present the computer simulation results for correlated channel with precoding and PCTC.

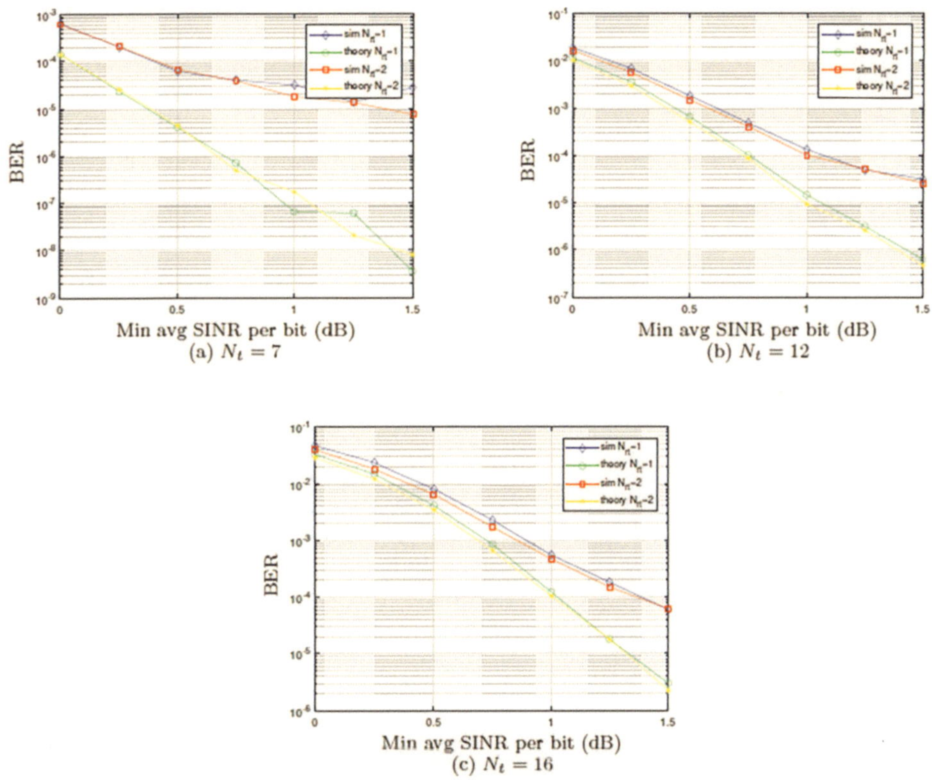


**Figure 12.**
*Simulation results with precoding for $N_{tot}$ = 32.*

### 3.4 Simulation results

The channel correlation is given by (67). The BER results for $N_{\text{tot}} = 1024$ with precoding are depicted in **Figure 11**. Incidentally, the value of the upper bound on the average SINR per bit before and after "combining" for $N_t = i = 512$, $N_{\text{tot}} = 1024$ is 6 dB. The BER results for $N_{\text{tot}} = 32$ with precoding are depicted in **Figure 12**. Note that since the average SINR per bit depends on the transmit antenna, the *minimum* average SINR per bit is indicated along the $x$-axis of **Figures 11** and **12**. We also observe from **Figures 11(a,b)** and **12** that there is a large difference between theory and simulations. This is probably because, the average SINR per bit is not identical for all transmit antennas. In particular, we observe from **Figures 9** and **10** that the first transmit antenna has a large average SINR per bit compared to the remaining antennas. However, in **Figure 11(c,d)** there is a close match between theory and simulations. This could be attributed to having a large number of blocks in a frame, as given by (2), resulting in better statistical properties. Even though the number of blocks is large in 12, the number of transmit antennas is small, resulting in inferior statistical properties. In order to improve the accuracy of the BER estimate for $N_{\text{tot}} = 32$, we propose to transmit "dummy data" from the first transmit antenna and "actual data" from the remaining antennas. The BER results shown in **Figure 13** indicates a good match between theory and practice. However, comparison of **Figures 11** and **14** demonstrates that "dummy data" is ineffective for large number of transmit antennas.

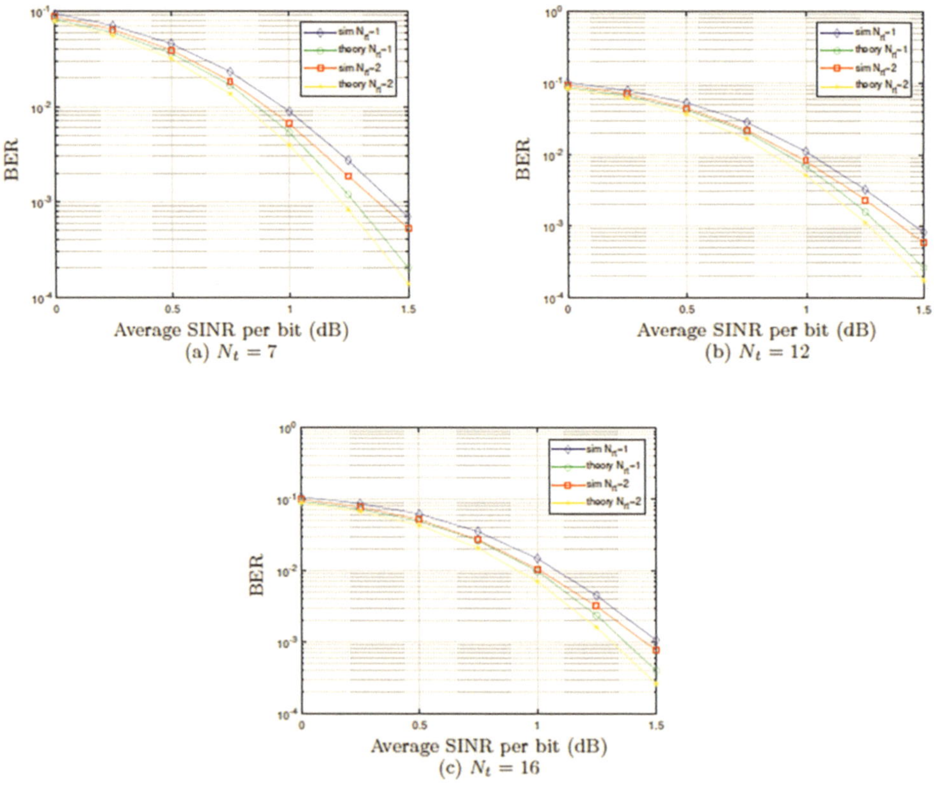


**Figure 13.**
*Simulation results with precoding and dummy data for $N_{tot}$ = 32.*

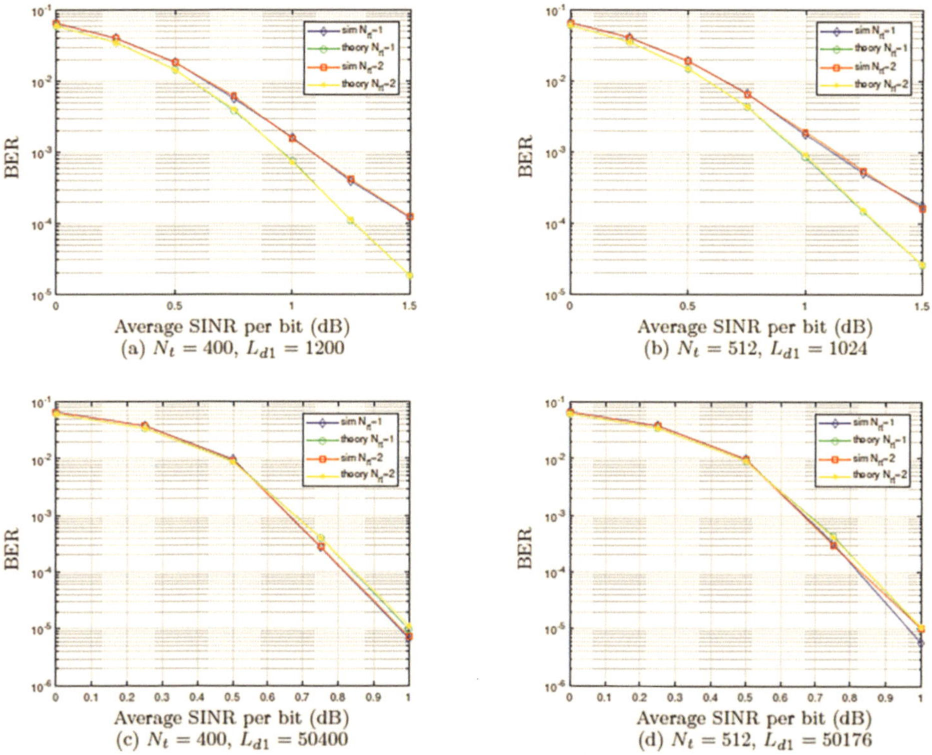


**Figure 14.**
*Simulation results with precoding and dummy data for $N_{tot}$ = 1024.*

## 4. Conclusions and future work

This article presents the advantages of single-user massive multiple input multiple output (SU-MMIMO) over multi-user (MU) MMIMO systems. The bit-error-rate (BER) performance of SU-MMIMO using serially concatenated turbo codes (SCTC) over uncorrelated channel is presented. A semi-analytic approach to estimating the BER of a turbo code is derived. A detailed signal-to-interference-plus-noise ratio analysis for SU-MMIMO over correlated channel is presented. The BER performance of SU-MMIMO with parallel concatenated turbo code (PCTC) over correlated channel is studied. Future work could involve estimating the MMIMO channel, since the present work assumes perfect knowledge of the channel.

## Author details

Kasturi Vasudevan[1*], Surendra Kota[1], Lov Kumar[1] and Himanshu Bhusan Mishra[2]

1 Department of Electrical Engineering, Indian Institute of Technology, Kanpur, India

2 Department of Electronics Engineering, Indian Institute of Technology (Indian School of Mines), Dhanbad, India

*Address all correspondence to: vasu@iitk.ac.in

## References

[1] Chowdhury MZ, Shahjalal M, Ahmed S, Jang YM. 6G wireless communication systems: Applications, requirements, technologies, challenges, and research directions. IEEE Open Journal of the Communications Society. 2020;**1**:957-975

[2] Pereira de Figueiredo FA. An overview of massive MIMO for 5G and 6G. IEEE Latin America Transactions. 2022;**20**(6):931-940

[3] Masaracchia A, Sharma V, Canberk B, Dobre OA, Duong TQ. Digital twin for 6G: Taxonomy, research challenges, and the road ahead. IEEE Open Journal of the Communications Society. 2022;**3**: 2137-2150

[4] Kshetrimayum RS, Mishra M, Aïssa S, Koul SK, Sharawi MS. Diversity order and measure of MIMO antennas in single-user, multiuser, and massive MIMO wireless communications. IEEE Antennas and Wireless Propagation Letters. 2023;**22**(1):19-23

[5] Gkonis PK. A survey on machine learning techniques for massive MIMO configurations: Application areas, performance limitations and future challenges. IEEE Access. 2023;**11**:67-88

[6] Zakavi MJ, Nezamalhosseini SA, Chen LR. Multiuser massive MIMO-OFDM for visible light communication systems. IEEE Access. 2023;**11**:2259-2273

[7] Ning B, Tian Z, Mei W, Chen Z, Han C, Li S, et al. Beamforming technologies for ultra-massive MIMO in terahertz communications. IEEE Open Journal of the Communications Society. 2023;**4**:614-658

[8] Molodtsov V, Bychkov R, Osinsky A, Yarotsky D, Ivanov A. Beamspace selection in multi-user massive MIMO. IEEE Access. 2023;**11**:18761-18771

[9] Elayan H, Amin O, Shihada B, Shubair RM, Alouini MS. Terahertz band: The last piece of RF spectrum puzzle for communication systems. IEEE Open Journal of the Communications Society. 2020;**1**:1-32

[10] Izadinasab K, Shaban AW, Damen O. Detection for hybrid beamforming millimeter wave massive MIMO systems. IEEE Communications Letters. 2021;**25**(4):1168-1172

[11] Kebede T, Wondie Y, Steinbrunn J, Kassa HB, Kornegay KT. Precoding and beamforming techniques in mm wave massive MIMO: Performance assessment. IEEE Access. 2022;**10**: 16365-16387

[12] Wei C, Yang Z, Dang J, Li P, Wang H, Yu X. Accurate wideband channel estimation for Thz massive MIMO systems. IEEE Communications Letters. 2023;**27**(1): 293-297

[13] Zhang P, Li J, Wang H, You X. Millimeter-Wave Space-Time Propagation Characteristics in Urban Macrocell Scenarios. In: ICC 2019 - 2019 IEEE International Conference on Communications (ICC), Shanghai, China. 2019. pp. 1-6. DOI: 10.1109/ICC.2019.8761087

[14] Nadeem Q-U-A, Alwazani H, Kammoun A, Chaaban A, Debbah M, Alouini M-S. Intelligent reflecting surface assisted multi-user MISO communication: Channel estimation and beamforming design. IEEE Open Journal of the Communications Society. 2020;**1**: 661-680

[15] Liu Y, Liu X, Mu X, Hou T, Xu J, Di Renzo M, et al. Reconfigurable intelligent surfaces: Principles and opportunities. IEEE Communications Surveys & Tutorials. 2021;**23**(3): 1546-1577

[16] Björnson E, Wymeersch H, Matthiesen B, Popovski P, Sanguinetti L, de Carvalho E. Reconfigurable intelligent surfaces: A signal processing perspective with wireless applications. IEEE Signal Processing Magazine. 2022;**39**(2):135-158

[17] Zhang S, Zhang L, Lu Y, Ding T. Research on the propagation characteristics of millimeter wave signals in complicated enclosed spaces. In: Proceedings of the 5th China Aeronautical Science and Technology Conference. Singapore: Springer Singapore; 2022. pp. 1015-1021

[18] Lu L, Li GY, Swindlehurst AL, Ashikhmin A, Zhang R. An overview of massive MIMO: Benefits and challenges. IEEE Journal of Selected Topics in Signal Processing. 2014;**8**(5):742-758

[19] Wu M, Yin B, Wang G, Dick C, Cavallaro JR, Studer C. Large-scale MIMO detection for 3GPP LTE: Algorithms and FPGA implementations. IEEE Journal of Selected Topics in Signal Processing. 2014;**8**(5):916-929

[20] Ma J, Ping L. Data-aided channel estimation in large antenna systems. IEEE Transactions on Signal Processing. 2014;**62**(12):3111-3124

[21] Ciuonzo D, Rossi PS, Dey S. Massive MIMO channel aware decision fusion. IEEE Transactions on Signal Processing. 2015;**63**(3):604-619

[22] Wang S, Li Y, Wang J. Multiuser detection in massive spatial modulation MIMO with low-resolution ADCs. IEEE Transactions on Wireless Communications. 2015;**14**(4):2156-2168

[23] Peng Y, Li Y, Wang P. An enhanced channel estimation method for millimeter wave systems with massive antenna arrays. IEEE Communications Letters. 2015;**19**(9):1592-1595

[24] Qin X, Yan Z, He G. A near-optimal detection scheme based on joint steepest descent and Jacobi method for uplink massive MIMO systems. IEEE Communications Letters. 2016;**20**(2): 276-279

[25] Choi J, Mo J, Heath RW. Near maximum-likelihood detector and channel estimator for uplink multiuser massive MIMO systems with one-bit ADCs. IEEE Transactions on Communications. 2016;**64**(5):2005-2018

[26] Khwandah SA, Cosmas JP, Lazaridis PI, Zaharis ZD, Chochliouros IP. Massive MIMO systems for 5G communications. Wireless Personal Communications. 2021;**120**(3): 2101-2115

[27] Vasudevan K, Kota S, Pathak GK, Reddy APK, Kumar L. Turbo coded single user massive MIMO. In: 1st Massive MIMO Workshop. IEEE Future Networks; 2021

[28] Sun H, Ng C, Huo Y, Hu RQ, Wang N, Chen C-M, et al. Massive MIMO. In: International Network Generations Roadmap 2022 Edition. IEEE Future Networks; 2022

[29] Vasudevan K, Madhu K, Singh S. Data detection in single user massive MIMO using Re-transmissions. The Open Signal Processing Journal. 2019;**6**: 15-26

[30] Vasudevan K, Madhu K, and Singh S. Scilab Code for Data Detection

in Single User Massive MIMO using Retransmissions. 2019. Available from: https://www.codeocean.com/

[31] Vasudevan K, Reddy APK, Pathak GK, Singh S. On the probability of erasure for MIMO OFDM. Semiconductor Science and Information Devices. 2020;**2**(1):1-5

[32] Vasudevan K, Singh S, Reddy APK. Coherent receiver for turbo coded single-user massive MIMO-OFDM with retransmissions. In: Mohammady S, editor. Multiplexing. London: IntechOpen; 2019. pp. 1-21

[33] Vasudevan K, Singh S, and Reddy APK. Scilab Code for Coherent Receiver for Turbo Coded Single-User Massive MIMO-OFDM with Retransmissions. 2019. Available from: https://www.codeocean.com/

[34] Vasudevan K, Pathak GK, Reddy APK. Turbo coded single user massive MIMO with precoding. In: Proc. of the 1st IFSA Winter Conference on Automation, Robotics & Communications for Industry 4.0 (ARCI' 2021). Chamonix-Mont-Blanc: IFSA; 2021. pp. 6-11

[35] Vasudevan K, Pathak GK, and Reddy APK. Scilab Code for Turbo Coded Single User Massive MIMO with Precoding. 2021. Available from: https://www.codeocean.com/

[36] Vasudevan K, Reddy APK, Pathak GK, Albreem M. Turbo coded single user massive MIMO. Sensors & Transducers Journal. 2021;**252**(5):65-75

[37] Vasudevan K, Reddy APK, Pathak GK, and Albreem MAM. Scilab code for turbo coded single user massive MIMO. 2021. Available from: https://www.codeocean.com/

[38] Vasudevan K. Turbo equalization of serially concatenated turbo codes using a predictive DFE-based receiver. Signal, Image and Video Processing. 2007;**1**(3): 239-252

[39] Bahl L, Cocke J, Jelinek F, Raviv J. Optimal decoding of linear codes for minimizing symbol error rate. IEEE Transactions on Information Theory. 1974;**20**(2):284-287

[40] Vasudevan K. Digital Communications and Signal Processing, Second Edition (CDROM Included). Hyderabad: Universities Press (India); 2010. Available from: https://www.universitiespress.com

[41] Vasudevan K. Coherent detection of turbo-coded OFDM signals transmitted through frequency selective Rayleigh fading channels with receiver diversity and increased throughput. Wireless Personal Communications. 2015;**82**(3): 1623-1642. DOI: 10.1007/s11277-015-2303-8

[42] Tüchler M, Koetter R, Singer AC. Turbo equalization: Principles and new results. IEEE Transactions on Communications. 2002;**50**(5):754-767

[43] ten Brink S. Convergence behaviour of iteratively decoded parallel concatenated codes. IEEE Transactions on Communications. 2001;**49**(10): 1727-1737

[44] Papoulis A. Probability, Random Variables and Stochastic Processes. 3rd ed. New York: McGraw-Hill; 1991

[45] Vasudevan K. Analog Communications: Problems & Solutions. Springer: Ane Books; 2018

[46] Vasudevan K. Detection of signals in correlated interference using a predictive VA. Signal Processing Journal, Elsevier Science. 2004;**84**(12):2271-2286

Chapter 4

# RS-Based MIMO-NOMA Systems in Multicast Framework

*Sareh Majidi Ivari, Mohammad Reza Soleymani and Yousef R. Shayan*

## Abstract

This chapter presents a novel scheme that integrates the rate-splitting (RS) technique in Multiple Input Multiple Output (MIMO) systems with non-orthogonal multiple access (NOMA) to improve performance and capacity in wireless communication systems under imperfect channel state information at the transmitter (CSIT) and in overloaded regimes. The proposed approach addresses a general and realistic scenario, incorporating both unicast and multicast users, aiming to increase system throughput through the optimization of precoding vectors and power allocation. A generic power allocation optimization technique is introduced, which can be employed for maximizing both the minimum-rate and sum-rate, focusing on the rate of the weakest user within each group per cluster. To tackle the non-convex nature of the problems, the proposed technique leverages the WMMSE-rate relationship and an alternating optimization (AO) algorithm, transforming the problem into a convex one. The chapter provides a comprehensive analysis of the proposed scheme, offering a tutorial background and presenting novel insights for an enhanced understanding.

**Keywords:** MIMO, RS, NOMA, fairness, sum-rate

## 1. Introduction

In recent years, the demand for high-speed wireless communication has grown significantly, driven by the widespread use of smartphones, tablets, and other wireless devices [1]. Users now expect constant internet connectivity and access to high-quality voice and video services, putting tremendous pressure on wireless communication networks to keep up with the increasing demand.

One major challenge faced by wireless communication systems is the limited availability of radio spectrum. As more devices and users come online, the demand for radio spectrum increases. To address this challenge, new technologies have been developed to utilize the available spectrum more efficiently, such as the MIMO-NOMA scheme [2].

MIMO-NOMA is a promising technology that integrates multiple antenna technology (MIMO) with NOMA to enhance the efficiency and capacity of wireless communication systems [2]. MIMO techniques leverage the spatial dimension by transmitting multiple data streams simultaneously over the same frequency-time resource. In the NOMA scheme, the transmitter sends a superposition of messages for multiple users, and users apply successive interference cancelation (SIC) to remove messages intended for weaker users and decode their own message [3]. By combining MIMO and NOMA, the MIMO-NOMA scheme brings together the advantages of both technologies, allowing multiple users to transmit and receive data concurrently using multiple antennas, non-orthogonal power allocation, and advanced signal processing techniques [4].

The integration of MIMO and NOMA is particularly useful in scenarios where multiple users are located in the same spatial direction but at distinct propagation distances, such as urban areas, stadiums, or office buildings [5]. In such scenarios, traditional wireless communication techniques like Orthogonal Multiple Access (OMA) may not provide sufficient capacity to serve all users [6]. In contrast, MIMO-NOMA can serve multiple users using the same resources, thereby improving the overall system capacity [2]. For instance, in a crowded stadium, many users may want to use their mobile devices to access the internet or stream videos simultaneously. MIMO-NOMA can serve these users using the same frequency band and time slot, whereas traditional techniques would require each user to be served in a separate time slot or frequency band.

In MIMO-NOMA systems, the transmitter employs interference cancelation techniques to form spatially orthogonal beams, with each beam carrying information for multiple users or groups of users [7]. The conventional linear precoding techniques such as Zero-forcing Beamforming (ZFBF) and Minimum Mean Square Error (MMSE) are commonly used to achieve spatial orthogonality [8, 9]. In the conventional linear precoding, the interference is canceled at the transmitter, and the receiver treats it as background noise [10]. These techniques play a crucial role in enhancing capacity, improving spectral efficiency, and enhancing overall system performance in wireless communication systems. In this chapter, the MIMO-NOMA scheme based on these conventional linear precoding is denoted as Conv-based MIMO-NOMA.

However, the implementation of MIMO-NOMA faces challenges, especially in the presence of imperfect CSIT and overloaded regime [10]. Imperfect CSIT can arise due to various factors, including channel estimation errors, quantization, and feedback delays [11, 12]. The accuracy of the channel information plays a vital role in interference cancelation techniques, and imperfect CSIT can degrade the performance of linear precoding methods.

Furthermore, wireless networks often comprise a combination of unicast and multicast users, which poses additional challenges for interference cancelation methods [13]. Accommodating both unicast and multicast users requires efficient resource allocation and power control strategies to optimize system performance and ensure fairness among users. Additionally, the performance of linear precoding techniques can deteriorate in overloaded regimes, where the number of users or groups of users exceeds the available resources. This further emphasizes the need for advanced techniques that can overcome the limitations of traditional linear precoding methods and improve system performance in realistic scenarios.

To address these challenges, the rate-splitting (RS) technique has emerged as a generic and powerful solution for interference cancelation in MIMO-NOMA systems [13]. RS decomposes the transmitted signal into two parts: a common part decoded by all users and a private part intended for the specific user. This enables the base station to exploit multiuser interference and achieve higher spectral efficiency [14].

RS has demonstrated significant potential for improving the sum-rate in multiuser MIMO systems under perfect CSIT conditions [15, 16]. It allows the base station to exploit multiuser interference and achieve higher spectral efficiency by decomposing the signal into common and private parts. Several studies have shown that RS can increase system capacity, reduce interference, and improve MMF rate performance in scenarios with perfect CSIT [17, 18]. However, the assumption of perfect CSIT may not hold in real-world scenarios.

Researchers have examined the effects of imperfect CSIT on the sum-rate and MMF rate performance of RS-based MIMO systems and proposed robust RS algorithms to counteract the impacts of imperfect CSIT [16, 19]. However, most of these studies focus on RS in the unicast framework and do not consider realistic scenarios with both unicast and multicast users.

In addition to unicast transmission, RS has been studied in the context of multicast transmission in MIMO systems [20]. Multicast transmission presents unique challenges, as it involves simultaneously transmitting the same information to multiple users. Therefore, it has received more attention. The performance of RS has been investigated in terms of MMF rate in [20]. RS has also been explored in multibeam multicast satellite communication systems in terms of MMF rate [21, 22].

RS-based MIMO-NOMA in the uplink has been investigated in [23]. The MMF rate is optimized for the proposed system in the unicast framework. However, none of the aforementioned works consider RS in MIMO-NOMA in the downlink with realistic scenarios, considering both unicast and multicast users under imperfect CSIT.

This chapter investigates the use of RS in MIMO-NOMA in downlink under imperfect CSI and both unicast and multicast users. The objective is to investigate the potential of RS-based MIMO-NOMA to improve system throughput and user fairness in realistic scenarios with imperfect CSIT. The chapter provides a comprehensive guide for researchers, engineers, and students interested in understanding the principles and applications of RS-based MIMO-NOMA for future wireless communication systems.

### 1.1 Contributions and organization of the chapter

This chapter explores the use of RS and NOMA in multiuser MIMO systems in downlink and presents a comprehensive analysis of the proposed scheme while investigating the challenges and trade-offs involved in its implementation. The main contributions of this chapter include the first application of RS in multiuser MIMO-NOMA systems under imperfect CSIT assumption, where RS is used to cancel interference and combat the effects of imperfect CSIT.

The chapter also covers the derivation of achievable data rates for both the common and private parts of user groups in the proposed RS-based MIMO-NOMA system. Precoding vectors are designed for both the common and private parts to enhance performance. The common part's precoding vector is optimized to maximize the rate of the common message, while the private part's precoding vectors are designed to cancel interference in both unicast and multicast frameworks. A low-complexity technique for designing the private precoding vectors is proposed in multicast transmission to address the lack of spatial degrees of freedom. The proposed technique builds upon unicast linear precoding methods and employs a Singular Value Decomposition (SVD) mapper.

Furthermore, the chapter formulates the max-min fairness MMF rate and sum-rate optimization problems for the RS-based MIMO-NOMA system under imperfect CSIT using the Average Rate (AR) framework. The weighted minimum mean square error (WMMSE) approach is employed to transform the formulated MMF and sum rate problems into convex problems. First, the chapter derives a rate-WMMSE relationship, and then, using this relationship and a low-complexity solution based on alternating optimization (AO), the problems are transformed into equivalent convex problems.

Overall, this chapter provides a comprehensive analysis of the RS-based MIMO-NOMA system under imperfect CSIT, and the proposed solutions and derivations of achievable data rates and optimization problems offer valuable insights into the design of future MIMO-NOMA systems.

The chapter is organized as follows. It begins with an introduction to the system model, including the signal and CSIT models. The design of precoding vectors for both common and private parts is discussed in Section 3. Section 4 focuses on power allocation optimization problems to maximize the minimum rate and sum-rate. The performance of the proposed technique is evaluated through simulations in Section 5. Finally, the chapter concludes with a summary of the findings and potential future research directions in Section 6.

Notations: Throughout this chapter, the following notations are used. Boldface capital letters, boldface lowercase letters, and ordinary letters represent matrices, column vectors, and scalars, respectively. The real component of a complex number $x$ is denoted by $\mathfrak{R}(x)$. The operators $()^T$ and $()^H$ represent transposition and Hermitian transpose, respectively. $||$ and $||$ are abbreviations for absolute value and Euclidean norm, respectively. $\mathbb{E}(.)$ represents the expected value of a random variable.

## 2. System model

This chapter presents a comprehensive study of RS-based MIMO-NOMA for a realistic wireless communication framework that includes both multicast and unicast users under imperfect CSIT assumption. The system consists of a single base station with $N_t$ antennas serving $I$ single-antenna users, where $N_t \leq I$. The base station forms $K$ clusters and generates one beam per cluster. The users that are in the same spatial direction but with distinct propagation distances are grouped into a cluster. This helps enhance the channel gain and combat inter-cluster interference. The distinctive propagation distances also facilitate SIC at mobile users. The $k$-th cluster contains $G_k$ groups which has $M_{g_k}$ users, where $M_{g_k} \geq 1$. If $M_{g_k} = 1$, group $g_k$ contains only one unicast user, and if $M_{g_k} > 1$, it contains multicast users.

The system model notation is defined, where $\mathcal{I}$ represents the set of indices of all users, $\mathcal{K}$ represents the set of clusters, $\mathcal{I}_k$ represents the set of users in the $k$ th cluster, and $\mathcal{G}_k$ represents the set of groups of users in the $k$ th cluster. The proposed system model of the RS-based MIMO-NOMA is depicted in **Figure 1**. In this figure, the parameters are as follows, $K = 4, I_1 = 4, I_2 = 1, I_3 = 3, I_4 = 2$. In cluster 1 and cluster 3 there are multicast users, $G_1 = 2$ and $M_{1_1} = 2, M_{2_1} = 2$ $G_2 = 1, M_{1_2} = 1,$ $G_3 = 3, M_{1_3} = 1, M_{2_3} = 1, M_{3_3} = 1,$ $G_4 = 2, M_{1_4} = 1, M_{2_4} = 1$.

The base station uses MIMO-NOMA to transmit multiple data streams simultaneously to the $I$ users by encoding $I$ messages into $K$ streams from a single data source. To enhance the system capacity and user fairness, the base station employs RS to divide the data of each user into two parts: a common part and a private part. The common stream is decoded by all users, while the private part is intended only for the specific user. The private part of the $k$-th message is further split into $G_k$ sub-streams, and each sub-stream is assigned to a group of users in the $k$-th cluster using the principles of NOMA. This helps to cancel inter-cluster interference and enhance the spectral efficiency. The combination of RS and NOMA provides flexibility in the

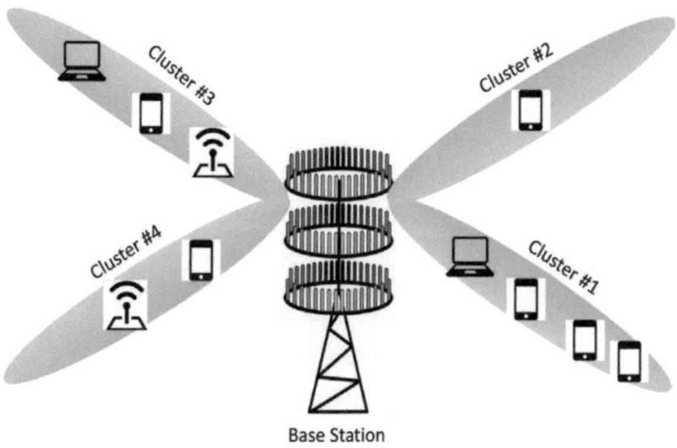


**Figure 1.**
*System model of the proposed RS-based MIMO-NOMA with multicast and unicast users.*

allocation of transmission power among users and the trade-off between system throughput and user fairness.

## 2.1 Signal model

In this section, we examine the transmitted and received signals to derive the signal-to-interference plus noise (SINR) and the achievable data rate. First, we need to discuss two main techniques of the proposed RS-based MIMO-NOMA: Rate Splitting and NOMA.

### *2.1.1 Rate-splitting approach*

Generally in the $L$-layer RS, the transmitter splits the message of each cluster- $k$ into $L$-sub-messages, $W_k^1, W_k^2, \ldots, W_k^L, \forall k \in \mathcal{K}$. Among the $L$ messages, one message is shared by all users, which is called the common part. In this chapter, we consider 1-layer RS, in which a message is split into two parts: a common and a private messages. The common part of all messages $W_1^c, W_2^c, \ldots, W_K^c$ is packed together and encoded into a common stream $s_c$ which is shared by all users. In the other hand, the private message of each message is encoded independently into private streams, $W_k^p \rightarrow s_k$.

As a result, the transmitted signal in time unit is $\mathbf{x}(t)$, where the time units are omitted for simplicity of expression. Therefore, the transmitted signal is

$$\mathbf{x} = \sqrt{p_c}\mathbf{w}_c s_c + \sum_{k=1}^{K} \mathbf{w}_k \sqrt{p_k} s_k \tag{1}$$

where $\mathbf{w}_c$ is the unit-norm precoding vector of the common message and $\mathbf{w}_k$ precodes the $k$-th message. $p_c$ is the power allocated to the common part. $p_k$ is the power allocated to the $k$-th cluster. The transmitted signal is constrained to

$$p_c + \sum_{k=1}^{K} p_k \|\mathbf{w}_k\|^2 \leq P_{\mathrm{T}} \tag{2}$$

where $P_{\mathrm{T}}$ is the maximum available power at the transmitter.

### 2.1.2 NOMA approach

Following the NOMA scheme in the power domain, different groups of users in a cluster are allocated different power levels according to their channel conditions to obtain the maximum gain in system performance. The transmitter sends all users information by sending the superposition of messages. Such power allocation is also beneficial to separate different groups of users. Therefore, users can apply SIC to cancel interference from the weaker groups of users in a cluster. However, the weak users perform single user detection (SUD) with considering the interference from the stronger users as the background noise. According to the NOMA scheme, the private stream can be contained information for more than one group of users. It means that the private stream $s_k$ consists of

$$s_k = \sum_{g_k=1}^{G_k} \sqrt{\alpha_{g_k}} s_{k_g}, \tag{3}$$

where $\alpha_{g_k}$ $(\sum_{g_k=1}^{G_k} \alpha_{g_k} = 1)$ denote fraction of the power allocated to $g$-th group in cluster $k$.

The received signal at user-$i$ is $y_i = \mathbf{h}_i \mathbf{x} + n_i, \forall i \in \mathcal{I}$. In terms of notation, $\mathbf{h}_i \in \mathbb{C}^{1\times N_t}$ is the channel vector between the transmitter and $i$-th user. This chapetr defines $\mu(i)$ as mapping a user index to its corresponding cluster and group indices, $\mu : i \rightarrow (k, g_k)$. Therefore, the received signal by $i$-th user which belongs to $k$-th cluster and $g_k$-th group is expressed as

$$y_i = \sqrt{p_c}\mathbf{h}_i \mathbf{w}_c s_c + \sum_{k=1}^{K} \sqrt{p_k}\mathbf{h}_i \mathbf{w}_k s_k + n_i, \tag{4}$$

where $n_i \sim \mathcal{CN}(0, \sigma_i^2)$ is the additive noise terms that contaminate the reception of $i$-th user. By substituting the Eq. (3) into the Eq. (4), the received signal is

$$y_i = \sqrt{p_c}\mathbf{h}_i \mathbf{w}_c s_c + \sqrt{\alpha_{g_k} p_k}\mathbf{h}_i \mathbf{w}_k s_{g_k} + \sum_{h_k > g_k}^{G_k} \sqrt{\alpha_{h_k} p_k}\mathbf{h}_i \mathbf{w}_k s_{h_k} + \sum_{j=1, j\neq k}^{K} \sqrt{p_j}\mathbf{h}_i \mathbf{w}_j s_j + n_i. \tag{5}$$

According to the RS technique, each user firstly decodes the common stream $s_c$ and treats the private streams as noise. Therefore, the SINR of the common part of user-$i$ is:

$$\gamma_{c,i} = \frac{p_c |\mathbf{h}_i \mathbf{w}_c|^2}{\sum_{j=1}^{K} p_j |\mathbf{h}_i \mathbf{w}_j|^2 + \sigma_i^2}, \tag{6}$$

and its corresponding rate is $R_{c,i} = \log_2(1 + \gamma_{c,i})$. In the RS scheme, the common message, $s_c$, is shared among all beams and groups, and each user should be able to decode $s_c$. Therefore, the common rate is defined as

$$R_c = \min_{i\in\mathcal{I}} R_{c,i} \triangleq \sum_{k=1}^{K} \sum_{g_k=1}^{G_k} C_{g_k}, \tag{7}$$

where $C_{g_k}$ denotes the portion of common rate of group $g_k$ in the $k$-th cluster.

After users decode and remove the common signal, $s_c$ through SIC, then each users decodes its private message. According to the NOMA scheme, users in group $g_k$ in cluster $k$, $\forall k \in \mathcal{K}, \forall g_k \in \mathcal{G}_k$, perform SIC to decode $s_{h_k}, \forall h_k < g_k$ and remove it from the received signal. Finally, users apply SUD to decode $s_{g_k}$ by considering all the other interference streams as noise. Therefore, the SINR of $i$-th user is determined by

$$\gamma_i = \frac{\alpha_{g_k} p_k |\mathbf{h}_i \mathbf{w}_k|^2}{1 + \sum_{h_k > g_k}^{G_k} \alpha_{h_k} p_k |\mathbf{h}_i \mathbf{w}_k|^2 + \sum_{j=1, j \neq k}^{K} p_j |\mathbf{h}_i \mathbf{w}_j|^2}. \tag{8}$$

In the multicast transmission, to guarantee all users can decode their messages, the user with the lowest SINR within a group dictates the rate of the corresponding group. Therefore, the achievable rate of group $g_k$ in cluster $k$, $r_{g_k}$, is defined by

$$r_{g_k} \triangleq \min_{i \in \mathcal{I}_{g_k}} R_i, \tag{9}$$

Therefore, the rate of users in group $g_k$ are composed of $C_{g_k}$ and $r_{g_k}$ and written as

$$R_{g_k} = C_{g_k} + r_{g_k}, \tag{10}$$

and the sum-rate is $R_{\text{sum-rate}} = R_c + \sum_{k=1}^{K} \sum_{g_k=1}^{G_k} r_{g_k}$.

### 2.2 CSIT uncertainty model

In this study, we assume that the receiver has perfect channel state information (CSI), while the transmitter has imperfect CSI due to limited feedback, such as quantized feedback with a fixed number of bits. The imperfect CSI of user $i$ is modeled as

$$\mathbf{h}_i = \hat{\mathbf{h}}_i + \tilde{\mathbf{h}}_i \tag{11}$$

where $\hat{\mathbf{h}}_i$ and $\tilde{\mathbf{h}}_i$ denote the estimated channel state and the corresponding channel estimation error at the transmitter, respectively. The uncertainty in CSIT (i.e., the channel estimation error) can be characterized by a conditional density $f\left(\mathbf{h}|\hat{\mathbf{h}}\right)$ that is known at the transmitter.

Consider $i$-th user and we define

$$\begin{aligned} \Upsilon_i &= \mathbb{E}|\mathbf{h}_i|^2, \\ \hat{\Upsilon}_i &= \mathbb{E}\left|\hat{\mathbf{h}}_i\right|^2, \\ \tilde{\Upsilon}_i &= \mathbb{E}\left|\tilde{\mathbf{h}}_i\right|^2. \end{aligned} \tag{12}$$

According to the orthogonal principles, $\hat{\mathbf{h}}_i$ and $\tilde{\mathbf{h}}_i$ are uncorrelated, and $\tilde{\mathbf{h}}_i$ has a zero mean. Therefore,

$$\Upsilon_i = \hat{\Upsilon}_i + \tilde{\Upsilon}_i. \tag{13}$$

We can consider

$$\tilde{\Upsilon}_i = \sigma_{e,i}^2 \Upsilon_i \tag{14}$$

where $\sigma_{e,i}^2 \in [0, 1]$ is the normalized CSIT error variance [16, 24]. Therefore, we have

$$\hat{\Upsilon}_i = \left(1 - \sigma_{e,i}^2\right) \Upsilon_i \tag{15}$$

A value of $\sigma_{e,i}^2 = 1$ corresponds to no instantaneous CSIT, while a value of $\sigma_{e,i}^2 = 0$ represents perfect instantaneous CSIT. For simplicity, we assume that all users have identical normalized CSIT error variances, that is, $\sigma_{e,i}^2 = \sigma_e^2, \forall i \in \mathcal{I}$.

The CSIT error variance scales with the signal-to-noise ratio (SNR) as $\sigma_e^2 = P_T^{-\eta}$, where $\eta \in [0, \infty)$ is the CSIT quality parameter. $\eta$ can be interpreted as a relation to the number of feedback bits, where $\eta = 0$ corresponds to a fixed number of feedback bits for all SNRs, and $\eta = \infty$ corresponds to an infinite number of feedback bits. The CSIT quality parameter is truncated such that $\beta \in [0, 1]$. In this context, $\eta = 1$ corresponds to perfect CSIT in the multiplexing gain sense [16, 24].

## 3. Precoder design

The proposed RS-based MIMO-NOMA system requires careful design of linear precoding vectors and power allocation to optimize performance and capacity. Firstly, in this section, we investigate the design of the linear precoding vectors for the private and common parts, denoted as $\mathbf{w}_k$ and $\mathbf{w}_c$, respectively. The linear precoding vectors $\mathbf{w}_k$ and $\mathbf{w}_c$ should be designed in a way that mitigates inter cluster interference and maximizes the achievable rate of the common message. In the following subsections, we investigate the design of the linear precoding vectors for the private and common parts.

### 3.1 Linear precoding vector of the private part, $\mathbf{w}_k$

Designing the linear precoding vector for the private part in the unicast framework under perfect CSIT is a relatively straightforward process. The optimal structure of $w_k$ is a generalization of regularized zero-forcing (RZF) precoding. However, in the presence of imperfect CSIT, the optimal precoders for private messages are still unknown and must be optimized numerically, as shown in [25]. This optimization process becomes particularly complex in large-scale systems. Despite this, RZF based on the channel estimates $\hat{\mathbf{H}}$ can be a suitable strategy for precoders of private messages, based on the findings of [26].

In the context of multicast transmission, designing the linear precoding vectors $\mathbf{w}_k$ is particularly challenging due to the matrix characterization of each cluster rather than a vector. To address this challenge, we propose a novel approach based on singular value decomposition (SVD) mapping. Specifically, we use the SVD mapping to transform the multicast transmission scenario into a set of parallel unicast channels, where the optimization problem is simplified. We then use the RZF technique to design the linear precoding vectors for the private messages in the unicast channels. This approach provides a low-complexity and efficient solution for designing the precoding vectors in the presence of imperfect CSIT.

The precoding vector using the ZBF is obtained as:

$$\mathbf{W}_{\mathbf{RZF}} = \frac{1}{\sqrt{\gamma_{\mathbf{RZF}}}} \left( \left( \hat{\mathbf{G}}^H \hat{\mathbf{G}} + \frac{K}{P_{\mathrm{T}}} \mathbf{I}K \right)^{-1} \hat{\mathbf{G}}^H \right), \tag{16}$$

where $\hat{\mathbf{G}}$ is the estimated composite channel matrix, and $\mathbf{I}_K$ is the $K$-dimensional identity matrix. To ensure that the power constraints are satisfied, the precoding matrix should be normalized by the factor $\gamma_{\mathbf{RZF}}$, which is defined as:

$$\gamma_{\mathbf{RZF}} = \max_k \left( \mathrm{diag}\left( \mathbf{W}_{\mathbf{RZF}} (\mathbf{W}_{\mathbf{RZF}})^H \right) \right). \tag{17}$$

Here, $\mathrm{diag}(\mathbf{A})$ denotes the diagonal elements of a matrix $\mathbf{A}$, and $(\mathbf{A})^H$ represents the conjugate transpose of $\mathbf{A}$.

The estimated composite channel matrix $\hat{\mathbf{G}} = [\hat{\mathbf{g}}_1, \hat{\mathbf{g}}_2, \dots, \hat{\mathbf{g}}_K]$ is obtained using the SVD mapping per beam [27]. In the SVD mapper, the estimated channel matrix of users in cluster $k$, denoted by $\hat{\mathbf{C}}_k = \left[ \hat{\mathbf{h}}_{\mathcal{I}_k(1)}^H, \dots, \hat{\mathbf{h}}_{\mathcal{I}_k(2M)}^H \right]$, is first subjected to SVD as follows:

$$\hat{\mathbf{C}}_k^H \hat{\mathbf{C}}_k = \mathbf{U}_k \Sigma_k \mathbf{V}_k^H, \tag{18}$$

where $\Sigma_k$ is the diagonal matrix of singular valusers, and $\mathbf{U}_k$ $(\mathbf{V}_k)$ gathers the left-singular vectors (right-singular vectors) [27]. Then, the right or left singular vector corresponding to the highest singular value is selected, which constructs the $\hat{\mathbf{g}}_k$ vector. The SVD mapper improves the energy spread over the users and the robustness to the CSIT uncertainty.

### 3.2 Linear precoding vector of the common part, $\mathbf{w}_c$

The precoding vector of the common message, $\mathbf{w}_c$, is designed to maximize the achievable rate of the common message. Therefore, the optimization problem is defined as

$$\overline{\mathcal{D}}_1 : \max_{\mathbf{w}_c \in \mathcal{N}} \min_{i \in \mathcal{I}} \frac{p_c |\mathbf{h}_i \mathbf{w}_c|^2}{\sum_{j=1}^K p_j |\mathbf{h}_i \mathbf{w}_j|^2 + \sigma_i^2} \tag{19}$$

$$\text{s.t.} \quad \|\mathbf{w}_c\|^2 = 1 \tag{20}$$

Since there is no interference in receiving the common message, the precoding vector for the common part can be designed as a linear combination of the channel vectors of all users, for a realization of $n \in \mathcal{N}$, the precoder of the common message is designed as

$$\mathbf{w}_c = \sum_{i \in \mathcal{I}} a_i \hat{\mathbf{h}}_i^H. \tag{21}$$

where $a_i$ is the weight for the channel vector of user $i$, and $\hat{\mathbf{h}}_i^H$ is the conjugate transpose of the normalized channel vector of user $i$. By assuming $\sigma_{e,i}^2 = \sigma_e^2, \|\mathbf{h}_i\|^2 = 1, \|\mathbf{h}_i\hat{\mathbf{h}}_j^H\|^2 = (1-\sigma_e^2)\varepsilon^2, \forall i \in \mathcal{I}, j \neq i$, and substituting (21) into (19) and (20), the problem $\overline{\mathcal{D}}_1$ is equivalently transformed to $\overline{\mathcal{D}}_2$

$$\overline{\mathcal{D}}_2 : \max_{a_i} \min_{i \in \mathcal{I}} \pi_i (1-\sigma_e^2) a_i^2 + \pi_i (1-\sigma_e^2) \varepsilon^2 \sum_{n=1, n \neq i}^{I} a_n^2 \tag{22}$$

$$\text{s.t.} \quad \sum_{i \in \mathcal{I}} a_i^2 = \frac{1}{N_t} \tag{23}$$

The goal is to find the optimal weights $a_i$ that maximize the minimum SINR of all users, subject to the constraint that the sum of the squared weights is equal to $1/(N_t)$. The optimal solution of problem $\overline{\mathcal{D}}_2$ is obtained when all terms are equal [28], that is, $\pi_i a_i^2 + \pi_i \varepsilon^2 \sum_{n=1, n \neq i}^{I} a_n^2 = \pi_j a_j^2 + \pi_j \varepsilon^2 \sum_{n=1, n \neq j}^{I} a_n^2, \forall i \neq j$. Therefore, the optimal precoding vector is achieved when all users experience the same common part SINR (6). In this chapter for simplicity and in order to obtain a more insightful and tractable asymptotic performance, we consider that $\pi_i = \pi_j, \forall i \neq j$ and $\varepsilon$ is very small, then the optimal $a_i$ is equal to $a_i^* = 1/\sqrt{N_t I}$, where $I$ is the total number of users.

## 4. Power allocation optimization

In this section, we examine the optimal power allocation for maximizing the MMF rate and sum-rate in the proposed RS-based MIMO-NOMA system under imperfect CSIT. To formulate the optimization problem under imperfect CSIT, we adopt a Stochastic Average Rate (AR) framework. Stochastic ARs are short-term metrics that represent the expected performance across the CSIT error distribution for a specific channel state estimate.

To define the AR framework, we first introduce three matrices: $\mathbf{H}$, $\hat{\mathbf{H}}$, and $\tilde{\mathbf{H}}$ which comprise the users' channel coefficients, users' channel coefficient estimations, and estimation errors. Given that the channel coefficients of users are independent and identically distributed (i.i.d.) and a sample index set $\mathcal{N} = \{1, 2, \dots, N\}$, we construct a realization sample containing $N$ i.i.d. realizations drawn from a conditional distribution $f(\mathbf{H}|\hat{\mathbf{H}})$. The realization sample can be expressed as:

$$\mathbb{H}^N \triangleq \left\{ \mathbf{H}^{(n)} = \hat{\mathbf{H}} + \tilde{\mathbf{H}}^{(n)} | \hat{\mathbf{H}}, n \in \mathcal{N} \right\}. \tag{24}$$

The realizations are accessible at the transmitter and can be utilized to approximate the ARs experienced by each user using Sample Average Functions (SAFs). As the number of samples (N) approaches infinity, $N \to \infty$, according to the strong law of large numbers, the ARs for user-$i$ are as follows:

$$\overline{R}_{c,i} = \lim_{N \to \infty} \overline{R}_{c,i}^{(N)} = \lim_{N \to \infty} \frac{1}{N} \sum_{n=1}^{N} R_{c,i}\left(\mathbf{H}^{(n)}\right), \tag{25}$$

$$\overline{R}_i = \lim_{N\to\infty} \overline{R}_i^{(N)} = \lim_{N\to\infty} \frac{1}{N}\sum_{n=1}^{N} R_i\left(\mathbf{H}^{(n)}\right) \tag{26}$$

where $R_{c,i}\left(\mathbf{H}^{(n)}\right)$ and $R_i\left(\mathbf{H}^{(n)}\right)$ are the achievable rates for $i$-th user based on the $n$-th realization in the sample set $\mathbb{H}^{(N)}$. The AR framework is then used to formulate the optimization problems for power allocation to maximize the MMF rate and sum-rate.

### 4.1 Problem statement

We define the optimization problems in this section. The AR framework is used to formulate the MMF and sum-rate optimization problems under imperfect CSIT.

#### *4.1.1 Max-min fairness analysis*

The MMF optimization problem using the AR framework can be formulated as

$$\overline{\mathcal{P}}_1 : \underset{\mathbf{p},\,\alpha,\,\overline{\mathbf{c}}}{\text{argmax}} \quad \min_{k\in\mathcal{K}} \min_{g_k\in\mathcal{G}_k} \left\{ \overline{C}_{g_k} + \min_{i\in\mathcal{I}_{g_k}} \overline{R}_i^{(N)} \right\} \tag{27}$$

s.t.

$$\overline{R}_{c,i}^{(N)} \geq \sum_{k=1}^{K}\sum_{g_k=1}^{G_k} \overline{C}_{g_k}, \forall i\in\mathcal{I} \tag{28}$$

$$\overline{C}_{g_k} \geq 0, \forall g_k \in \mathcal{G}_k, \forall k\in\mathcal{K} \tag{29}$$

$$\alpha_{g_k} \in [0,1], \sum_{g_k=1}^{G_k} \alpha_{g_k} = 1, \forall k\in K \tag{30}$$

$$p_c + \sum_{k=1}^{K} p_k \|\mathbf{w}_k\|^2 \leq P_{\mathrm{T}}, \forall k\in\mathcal{K} \tag{31}$$

here $\overline{\mathbf{c}} = \left[\overline{C}_{1,1}, \ldots, \overline{C}_{1,G}, \ldots, \overline{C}_{K,1}, \ldots, \overline{C}_{g_k}\right]$ is the vector of Average common-rate portions, and $\mathbf{p} = \{p_c, p_1, p_2, \ldots, p_K\}$, $\alpha = \{\alpha_1, \alpha_2, \ldots, \alpha_K\}$ are the vectors of powers and fraction of powers. The constraint (28) guarantees $s_c$ to be decoded by each user since the definition of the Average common rate is $\overline{R}_c = \sum_{k=1}^{K}\sum_{g_k=1}^{G_k}\overline{C}_{g_k} = \min_{i\in\mathcal{I}} \overline{R}_{c,i}$. Constraint (29) implies that each portion of the Average common rate is non-negative. Constraints (30) and (31) are the power constraint. By solving Problem $\overline{\mathcal{P}}_1$, variables ($\overline{\mathbf{c}}$, $\mathbf{p}$, $\alpha$) are jointly optimized. Note that by fixing $p_c = 0$ and $\overline{\mathbf{c}} = 0$, the RS scheme turns into Conv-based MIMO-NOMA.

#### *4.1.2 Sum-rate analysis*

The sum-rate optimization is another problem which is addressed in this chapter. The sum-rate maximization under imperfect CSIT is also formulated using the AR framework as

$$\overline{\mathcal{S}}_1 : \underset{\overline{R}_c,\mathbf{p},\alpha}{\text{argmax}} \quad \overline{R}_c + \sum_{k=1}^{K}\sum_{g_k=1}^{G_k} \min_{i\in I_{g_k}} \overline{R}_i^{(N)} \tag{32}$$

s.t.

$$\overline{R}_{c,i}^{[N]} \geq \overline{R}_c, \forall i \in \mathcal{I} \tag{33}$$

$$(24d), (24e) \tag{34}$$

where $\overline{R}_c$ is an auxiliary variable. The constraint (33) guarantees that all users can decode $s_c$.

Problems $\overline{\mathcal{P}}_1$ and $\overline{\mathcal{S}}_1$ are non-convex problems that are very challenging to solve because they contain superimposed rate expressions.

The weighted mean squared error (WMMSE) approach is a powerful technique for solving non-convex optimization problems with superimposed rate expressions. The idea behind this approach is to replace the original rate expressions with a set of WMMSE expressions that are easier to handle mathematically. Using the WMMSE expressions, the original problems $\overline{\mathcal{P}}_1$ and $\overline{\mathcal{S}}_1$ can be reformulated as block-wise convex optimization problems, which can be solved iteratively using interior-point methods. Specifically, the reformulated problems involve optimizing the WMMSE variables and the power allocation coefficients, subject to some convex constraints. The optimization procedure involves iteratively updating the WMMSE variables and the power allocation coefficients until convergence is achieved.

## 4.2 Rate-WMMSE relationship

To define the achievable data rate with set of WMMSE expression, first we establish the Rate-WMMSE relationship. The mean square errors (MSEs) of the estimate $\hat{s}_{c,i}$ of the common signal $s_c$ for user $i$ is given by:

$$\varepsilon_{c,i} = \mathbb{E}\left\{|\hat{s}_{c,i} - s_{c,i}|^2\right\} = \mathbb{E}\left\{\left|\hat{s}_{c,i} - q_{c,i}y_i\right|^2\right\}, \tag{35}$$

where $q_{c,i}$ is a scalar equalizer. Since the transmitter sends the superposition of $s_c$ and $s_{g_k}$, $\forall k \in \mathcal{K}, g_k \in \mathcal{G}_k$, user $i$ first decodes and removes $s_c$ from the received signal using SIC. Next, user $i$ which belongs to cluster $k$ and group $g_k$ decodes and removes the signals intended for the weaker groups in cluster $k$, $h_k < g_k$, through the SIC. Therefore, the MSE of the estimate $\hat{s}_{g_k}$ of the private signal $s_{g_k}$ for user $i$ in group $g_k$ of cluster $k$ is given by:

$$\varepsilon_i = \mathbb{E}\left\{|\hat{s}_i - s_i|^2\right\} = \mathbb{E}\left\{\left|\hat{s}_i - q_i\left(y_i - \sqrt{(1-t)P}\mathbf{h}_i\mathbf{w}_c s_c - \sum_{h_k=1}^{h_k<g_k}\sqrt{\alpha_{h_k}p_k}\mathbf{h}_i\mathbf{w}_k s_{h_k}\right)\right|^2\right\} \tag{36}$$

Here, $\mathbb{E}[\cdot]$ denotes the expectation operator, and $|\cdot|^2$ denotes the squared magnitude. With substituting the Eq. (5) into the Eq. (35) and (36), the MSEs of the common and private parts can be rewritten as

$$\varepsilon_{c,i} = \left|q_{c,i}\right|^2 T_{c,i} + 1 - 2\Re\left\{\sqrt{p_c} q_{c,i} \mathbf{h}_i \mathbf{w}_c\right\} \tag{37}$$

$$\varepsilon_i = \left|q_i\right|^2 T_i + 1 - 2\Re\left\{\sqrt{\alpha_{g_k} p_k} q_i \mathbf{h}_i \mathbf{w}_k\right\} \tag{38}$$

where

$$T_{c,i} = p_c|\mathbf{h}_i\mathbf{w}_c|^2 + \sum_{k=1}^{K} p_k|\mathbf{h}_i\mathbf{w}_k|^2 + \sigma_i^2, \tag{39}$$

$$T_i = p_k\alpha_{g_k}|\mathbf{h}_i\mathbf{w}_k|^2 + p_k\sum_{h_k>g_k}^{G_k}\alpha_{h_k}|\mathbf{h}_i\mathbf{w}_k|^2 + \sum_{j=1,j\neq k}^{K} p_j\left|\mathbf{h}_i\mathbf{w}_j\right|^2 + \sigma_i^2 \tag{40}$$

Moreover, we define the interference as

$$I_{c,i} = T_{c,i} - p_c|\mathbf{h}_i\mathbf{w}_c|^2, \tag{41}$$

$$I_i = T_i - p_k\alpha_{g_k}|\mathbf{h}_i\mathbf{w}_k|^2. \tag{42}$$

The optimum equalizers achieve by minimizing the MSEs over equalizers,

$$\frac{\partial\varepsilon_{c,i}}{q_{c,i}} = 0 \to q_{c,i}^{\mathrm{MMSE}} = \sqrt{p_c}\mathbf{h}_i\mathbf{w}_c T_i^{-1} \tag{43}$$

$$\frac{\partial\varepsilon_i}{q_i} = 0 \to q_i^{\mathrm{MMSE}} = \sqrt{p_k\alpha_{g_k}}\mathbf{h}_i\mathbf{w}_k T_i^{-1} \tag{44}$$

The minimum MSEs (MMSEs) with optimum equalizers are

$$\varepsilon_{c,i}^{\mathrm{MMSE}} = \min_{q_{c,i}} \varepsilon_{c,i} = T_{c,i}^{-1} I_{c,i}, \tag{45}$$

$$\varepsilon_i^{\mathrm{MMSE}} = \min_{q_i} \varepsilon_i = T_i^{-1} I_i. \tag{46}$$

Apparently, the SINRs can be expressed in the form of MMSEs, i.e., $\gamma = \left(1/\varepsilon^{\mathrm{MMSE}}\right) - 1$. Consequently, the corresponding rates are written as $R = -\log_2\left(\varepsilon^{\mathrm{MMSE}}\right)$.

Next, we define the augmented weighted MSEs (WMSEs) for the common and private parts. The term "augmented WMSE" is employed because it incorporates additional information or constraints into the standard WMSE, aiming to better capture the characteristics of the system under consideration, such as fairness or rate requirements, and facilitate the optimization process. This augmentation is particularly relevant in wireless communication system optimization problems, especially when dealing with RS or non-orthogonal multiple access techniques, to achieve more accurate and reliable results. The weighted WMSEs are given by:

$$\xi_{c,i} = u_{c,i}\varepsilon_{c,i} - \log_2(u_{c,i}), \xi_i = u_i\varepsilon_i - \log_2(u_i), \tag{47}$$

where $u_{c,i}, u_i > 0$ are weights associated with MSEs. In the following, we consider $\xi$ s as WMSEs and, for simplicity, drop the "augmented". After defining the augmented

WMSEs, they are minimized with respect to both equalizers and weights, yielding the following conditions:

$$\frac{\partial \xi_{c,i}\left(q_{c,i}^{\mathrm{MMSE}}\right)}{\partial q_{c,i}, u_{c,i}} = 0, \tag{48}$$

$$\frac{\partial \xi_{i}\left(q_{i}^{\mathrm{MMSE}}\right)}{\partial q_{i}, u_{i}} = 0. \tag{49}$$

Then the optimal equalizers are substituted into the WMSEs, and we obtain

$$\xi_{c,i}\left(q_{c,i}^{\mathrm{MMSE}}\right) = \min_{q_{c,i}} \xi_{c,i} = u_{c,i}\varepsilon_{c,i}^{\mathrm{MMSE}} - \log_2(u_{c,i}) \tag{50}$$

$$\xi_{i}\left(q_{i}^{\mathrm{MMSE}}\right) = \min_{q_{i}} \xi_{i} = u_{i}\varepsilon_{i}^{\mathrm{MMSE}} - \log_2(u_{i}) \tag{51}$$

As a result, the optimum weights can be determined as:

$$u_{c,i} = \left(\varepsilon_{c,i}^{\mathrm{MMSE}}\right)^{-1}, \tag{52}$$

$$u_{i} = \left(\varepsilon_{i}^{\mathrm{MMSE}}\right)^{-1}. \tag{53}$$

We substitute (52) and (53) into (50), (51), leading to the Rate-WMMSE relationship

$$\xi_{c,i}^{\mathrm{MMSE}} = \min_{q_{c,i}, u_{c,i}} \xi_{c,i} = 1 + \log_2 \varepsilon_{c,i}^{\mathrm{MMSE}} = 1 - R_{c,i} \tag{54}$$

$$\xi_{i}^{\mathrm{MMSE}} = \min_{q_{i}, u_{i}} \xi_{i} = 1 + \log_2 \varepsilon_{i}^{\mathrm{MMSE}} = 1 - R_{i}. \tag{55}$$

With considering imperfect CSIT, a Stochastic Average Rate-WMMSE relationship is developed, and the average WMMSEs are given by:

$$\bar{\xi}_{c,i}^{\mathrm{MMSE}(N)} = \frac{1}{N}\lim_{N\to\infty}\sum_{n=1}^{N} \xi_{c,i}^{\mathrm{MMSE}(n)} = 1 - \overline{R}_{c,i}^{(N)}, \tag{56}$$

$$\bar{\xi}_{i}^{\mathrm{MMSE}(N)} = \frac{1}{N}\lim_{N\to\infty}\sum_{n=1}^{N} \xi_{i}^{\mathrm{MMSE}(n)} = 1 - \overline{R}_{i}^{(N)} \tag{57}$$

where $\xi_{c,i}^{\mathrm{MMSE}(n)}$ and $\xi_{i}^{\mathrm{MMSE}(n)}$ are associated with the $n$-th realization in $\mathbb{H}^{(N)}$. The sets of optimum MMSE equalizers associated with (56) and (57) are defined as

$$\mathbf{g}_{c,i}^{\mathrm{MMSE}} = \left\{q_{c,i}^{\mathrm{MMSE}(n)} | n \in \mathcal{N}\right\}, \tag{58}$$

$$\mathbf{g}_{i}^{\mathrm{MMSE}} = \left\{q_{i}^{\mathrm{MMSE}(n)} | n \in \mathcal{N}\right\}. \tag{59}$$

Moreover, the sets of optimum weights are

$$\mathbf{u}_{c,i}^{\mathrm{MMSE}} = \left\{u_{c,i}^{\mathrm{MMSE}(n)} | n \in \mathcal{N}\right\}, \tag{60}$$

$$\mathbf{u}_i^{\text{MMSE}} = \left\{u_i^{\text{MMSE}(n)} | n \in \mathcal{N}\right\}. \tag{61}$$

Therefore, in each realization in $\mathbb{H}^{(N)}$, the optimum equalizer and weights are calculated. The composite set of optimum equalizer and weights are defined as

$$\mathbf{G}^{\text{MMSE}} = \left\{\mathbf{g}_{c,i}^{\text{MMSE}}, \mathbf{g}_i^{\text{MMSE}} | i \in \mathcal{I}, \right\} \tag{62}$$

$$\mathbf{U}^{\text{MMSE}} = \left\{\mathbf{u}_{c,i}^{\text{MMSE}}, \mathbf{u}_i^{\text{MMSE}} | i \in \mathcal{I}\right\} \tag{63}$$

Using the Rate-WMMSE relationship, the optimization problems are rewritten using the WMMSE variables in the following section.

### 4.3 WMMSE reformulation

In this section, we reformulate the optimization problems using the WMMSE expressions.

#### *4.3.1 Max-min fairness analysis*

Using the Rate-WMMSE relationship, and auxiliary variables, $\overline{z}$, $\mathbf{G}$, $\mathbf{U}$, $\overline{r}_g = \left\{\overline{r}_{1_g}, \dots, \overline{r}_{g_k}\right\}$, the problem $\overline{\mathcal{P}}_1$ can be transferred into an equivalent WMMSE problem, $\overline{\mathcal{P}}_2$:

$$\overline{\mathcal{P}}_2 : \underset{\mathbf{p}, \alpha, \overline{\mathbf{c}}, \overline{z}, \overline{r}_g}{\text{argmax}} \quad \overline{z} \tag{64}$$

s.t.

$$\overline{C}_{g_k} + \overline{r}_{g_k} \geq \overline{z}, \forall k \in \mathcal{K}, \forall g_k = \{1, \dots, G_k\} \tag{65}$$

$$1 - \overline{\xi}_i^{(N)} \geq \overline{r}_{g_k}, \forall i \in \mathcal{I}_{g_k}, \forall k \in \mathcal{K}, \forall g_k = \{1, \dots, G_k\} \tag{66}$$

$$1 - \overline{\xi}_{c,i}^{(N)} \geq \sum_{k=1}^{K} \sum_{g_k=1}^{G_k} \overline{C}_{g_k}, \forall i \in \mathcal{I} \tag{67}$$

$$(24d), (24e) \tag{68}$$

where $\overline{\xi}_{c,i}$ and $\overline{\xi}_i$ are given in (47). It is worth to mention if $\left(\mathbf{p}^*, \alpha^*, \overline{\mathbf{c}}^*, \overline{z}^*, \mathbf{G}^*, \overline{r}_g^*, \mathbf{U}^*\right)$ satisfies the KKT optimality conditions of $\overline{\mathcal{P}}_2$, $(\mathbf{p}^*, \alpha^*, \overline{\mathbf{c}}^*)$ will satisfy the KKT optimality conditions of $\overline{\mathcal{P}}_1$.

#### *4.3.2 Sum-rate analysis*

Motivated by the Rate-WMMSE relationships given in (56), (57), and the auxiliary variable, $\left(\overline{\xi}_c, \mathbf{U}, \mathbf{G}\right)$, the problem $\overline{\mathcal{S}}_1$ is equivalently transferred into the problem $\overline{\mathcal{S}}_2$. The problem is reformulated as

$$\overline{\mathcal{S}}_2 : \underset{\overline{\xi}_c, \mathbf{p}, \alpha_k}{\text{argmin}} \quad \overline{\xi}_c + \sum_{k=1}^{K} \sum_{g_k=1}^{G_k} \max_{i \in \mathcal{I}_{g_k}} \overline{\xi}_i^{(N)} \tag{69}$$

s.t.

$$\overline{\xi}_{c,i}^{[N]} \leq \overline{\xi}_c, \forall i \in \mathcal{I} \tag{70}$$

$$(24d), (24e) \tag{71}$$

where $\overline{\xi}_c$ refers to the common AWMSE. Noted problem $\overline{\mathcal{S}}_2$ and problem $\overline{\mathcal{S}}_1$ are equivalence. It means that for any point $\left(\mathbf{p}^*, \alpha^*, \overline{\xi}_c^*, \mathbf{G}^*, \mathbf{U}^*\right)$ satisfying the KKT optimality conditions of problem $\overline{\mathcal{S}}_2$, $(\mathbf{p}^*, \alpha^*)$ satisfies the KKT optimality conditions of problem $\overline{\mathcal{S}}_1$.

The problems $\overline{\mathcal{P}}_2$ and $\overline{\mathcal{S}}_2$ remain non-convex. However, they become convex when two out of the three variables, namely equalizer, weight, and power, are fixed. Taking into account this block-wise convexity property, we propose an Alternating Optimization algorithm to address the problems $\overline{\mathcal{P}}_2$ and $\overline{\mathcal{S}}_2$.

### 4.4 Alternating optimization algorithm

The problems $\overline{\mathcal{P}}_2$ and $\overline{\mathcal{S}}_2$ remain non-convex for the entire set of optimization variables, which include $\alpha$, $p$, $\overline{\mathbf{c}}$, $\mathbf{U}$, and $\mathbf{G}$. However, they exhibit block-wise convexity, which can be leveraged to propose an alternating optimization algorithm. Each iteration of the algorithm consists of two steps: (1) updating $\mathbf{U}$ and $\mathbf{G}$ based on the value of $\mathbf{p}$ and $\alpha$ from the previous iteration, and (2) updating $\mathbf{p}$, $\alpha$, and $\overline{\mathbf{c}}$ using $\mathbf{U}$ and $\mathbf{G}$ obtained in step 1. We now provide a detailed explanation of these two steps.

#### *4.4.1 Step 1: Updating* G, U

In $l$-th iteration, all the equalizers and weights are updated according to the $\mathbf{p}, \alpha$ form the previous round, $l-1$, $\mathbf{G}\left(p^{[l-1]}, \alpha^{[l-1]}\right)$, $\mathbf{U}\left(p^{[l-1]}, \alpha^{[l-1]}\right)$. The corresponding SAFs $\overline{u}_{c,i}, \overline{u}_i, \overline{g}_{c,i}, \overline{g}_i$ are calculated by taking average over $N$ realization. To facilitate the next step, we introduce a set of variables are

$$t_{c,i} = u_{c,i}^{(n)} \left| q_{c,i}^{(n)} \right|^2, \quad t_i = u_i^{(n)} \left| q_i^{(n)} \right|^2, \tag{72}$$

$$\Psi_{c,i}^{(n)} = t_{c,i} \mathbf{h}_i^{(n)H} \mathbf{h}_i^{(n)}, \quad \Psi_i^{(n)} = t_i \mathbf{h}_i^{(n)H} \mathbf{h}_i^{(n)}, \tag{73}$$

$$f_{c,i}^{(n)} = u_{c,i}^{(n)} q_{c,i}^{(n)} \mathbf{h}_i^{(n)} \mathbf{w}_c^{(n)}, \quad f_i^{(n)} = u_i^{(n)} q_i^{(n)} \mathbf{h}_i^{(n)} \mathbf{w}_k^{(n)} \tag{74}$$

$$v_{c,i}^{(n)} = \log_2(u_{c,i}), \quad v_i^{(n)} = \log_2(u_i) \tag{75}$$

and the corresponding SAFs are calculated in the same way,

$$t_{c,i}^{(N)}, \Psi_{c,i}^{(N)}, f_{c,i}^{(N)}, v_{c,i}^{(N)}, t_i, \Psi_i^{(N)}, f_i^{(N)}, v_i^{(N)} \tag{76}$$

#### *4.4.2 Step 2: Updating* $\mathbf{p}, \alpha$

In the $l$-th iteration up to this step, we fix $\mathbf{G}$, $\mathbf{U}$, and the other introduced variables, which are obtained using the updated valusers of $(\mathbf{U}, \mathbf{G})$. With these updated variables, in this step, the problems $\overline{\mathcal{P}}_2$ and $\overline{\mathcal{S}}_2$ transform into problems $\overline{\mathcal{P}}_3^{[l]}$ and $\overline{\mathcal{S}}_3^{[l]}$, which are convex problems. These problems can be solved using interior-point methods, allowing for the optimization of $\mathbf{p}$, $\alpha_{\mathbf{k}}$, and the other auxiliary variables.

$$\overline{\mathcal{P}}_3^{[l]} : \underset{\mathbf{p}, \alpha, \overline{\mathbf{c}}, \overline{z}, \overline{r}_g,}{\operatorname{argmax}} \ \overline{z} \tag{77}$$

s.t.

$$\overline{C}_{k_g} + \overline{r}_{k_g} \geq \overline{z}, \quad \forall k \in \mathcal{K}, \forall g_k = \{1, \dots, G_k\} \tag{78}$$

$$\begin{aligned} 1 - \overline{r}_{g_k} \geq & \sum_{j=1, j\neq k}^{K} p_j \overline{\mathbf{w}}_j^H \overline{\Psi}_i^{(N)} \overline{\mathbf{w}}_j^{(N)} - 2\mathcal{R}\left\{\sqrt{\alpha_{g_k} p_k} \overline{f}_i^{(N)}\right\} + \overline{t}_i^{(N)} + \overline{u}_i^{(N)} - \overline{v}_i^{(N)} \\ & + \sum_{h \geq g} p_k \alpha_{h_k} \overline{\mathbf{w}}_k^H \overline{\Psi}_i^{(N)} \overline{\mathbf{w}}_k^{(N)}, \quad \forall i \in \mathcal{I}_{g_k}, \forall k \in \mathcal{K}, \forall g_k \in \mathcal{G}_k \end{aligned} \tag{79}$$

$$\begin{aligned} 1 - \sum_{k=1}^{K} \sum_{g_k=1}^{G_k} \overline{C}_{g_k} \geq & p_c \overline{\mathbf{w}}_c^H \overline{\Psi}_{c,i}^{(N)} \overline{\mathbf{w}}_c + \sum_{k=1}^{K} p_k \overline{\mathbf{w}}_k^H \overline{\Psi}_{c,i}^{(N)} \overline{\mathbf{w}}_k + \overline{t}_{c,i}^{(N)} - 2\mathcal{R}\left\{\sqrt{p_c} \overline{f}_{c,i}^{(N)}\right\} \\ & + \overline{u}_{c,i}^{(N)} - \overline{v}_{c,i}^{(N)}, \quad \forall i \in \mathcal{I} \end{aligned} \tag{80}$$

$$(24d), (24e) \tag{81}$$

and

$$\overline{\mathcal{S}}_3^{[l]} : \underset{\overline{\xi}_c, \mathbf{p}, \alpha_{\mathbf{k}}}{\operatorname{argmin}} \quad \overline{\xi}_c + \sum_{k=1}^{K} \sum_{g=1}^{G} \left\{ \max_{i \in \mathcal{I}_{g_k}} \overline{\xi}_i \right\} \tag{82}$$

s.t.

$$p_c \overline{\mathbf{w}}_c^H \overline{\Psi}_{c,i}^{(N)} \overline{\mathbf{w}}_c + \sum_{k=1}^{K} p_k \overline{\mathbf{w}}_k^H \overline{\Psi}_{c,i}^{(N)} \overline{\mathbf{w}}_k + \overline{t}_{c,i}^{(N)} - 2\mathcal{R}\left\{\sqrt{p_c} \overline{f}_{c,i}^{(N)}\right\} + \overline{u}_{c,i}^{(N)} - \overline{v}_{c,i}^{(N)} \leq \overline{\xi}_c, \forall i \in \mathcal{I} \tag{83}$$

$$(24d), (24e) \tag{84}$$

where $\overline{\xi}_i$ is

$$\begin{aligned} \overline{\xi}_i = & \sum_{j=1, j\neq k}^{K} p_j \overline{\mathbf{w}}_j^H \overline{\Psi}_i^{(N)} \overline{\mathbf{w}}_j^{(N)} + \overline{t}_i^{(N)} - 2\mathcal{R}\left\{\sqrt{\alpha_{g_k} p_k} \overline{f}_i^{(N)}\right\} + p_k \sum_{h_k \geq g_k} \alpha_{h_k} \overline{\mathbf{w}}_k^H \overline{\Psi}_i^{(N)} \overline{\mathbf{w}}_k^{(N)} \\ & + \overline{u}_i^{(N)} - \overline{v}_i^{(N)}, \forall i \in \mathcal{I}_{g_k}, \forall k \in \mathcal{K}, \forall g_k \in \mathcal{G}_k \end{aligned}$$

As the iteration procedure continusers, the objective function in $\mathcal{P}_3$ or $\mathcal{S}_3$ grows until convergence. The proposed alternating optimization approach alternately optimizes the variables of the corresponding WMMSE problem $\overline{\mathcal{P}}_3$ and $\overline{\mathcal{S}}_3$. The proposed algorithm is guaranteed to converge as the objective function is bounded above for the specified power limitations.

## 5. Illustrative results and discussions

In this section, we evaluate the performance of the proposed RS-based MIMO-NOMA scheme through numerical simulations and validate the effectiveness of the power allocation algorithm. Specifically, we investigate the achievable MMF rate and sum-rate in different scenarios by varying the SNR, the number of users per group per cluster ($M$) and the degree of CSIT uncertainty, $\eta$. We compare the performance of the proposed RS-based MIMO-NOMA with conventional MIMO-NOMA.

In the Conv-based MIMO-NOMA system, instead of the RS, the conventional linear precoding such as RZF is applied to cancel interbeam interference between clusters of users, and NOMA is applied to provide service for more than one group of users.

### 5.1 Simulation setup

To carry out our analysis, we consider a single-cell cellular network with a radius of 500 m, where the base station is located at the center. It is equipped with an array of $N_t = 64$ antennas and forms $K = 12$ clusters. Each cluster has two groups of users, $G_k = 2$, $k \in \mathcal{K}$, and all groups have the same cardinality, $M$. Users are randomly and uniformly distributed throughout the cell, excluding an inner circle of radius 50 meters.

The large-scale fading coefficient for user $i$ is expressed as $\beta_i = \frac{\overline{d}}{x_i^{\nu}}$, where $x_i$ indicates the distance between the $i$-th user and the base station. Here, the constant $\overline{d} = 10^{-5}$ serves the role of regulating the channel attenuation at a distance of 50 m, and $\nu$ symbolizes the path loss exponent, which is assumed to be 3.76 for this study. The large-scale fading ($\beta_i$) in this context follows a log-normal distribution with a standard deviation of 8 dB.

Furthermore, in this chapter, we consider the noise variance to be set at 1. As a result, the SNR is defined by the peak power, denoted as $p_{max}$.

### 5.2 MMF rate analysis results

In this section, we aim to compare the performance of our proposed RS-based MIMO-NOMA scheme with that of the Conv-based MIMO-NOMA technique, specifically focusing on the MMF rate. The primary objective of this comparison is to maximize the minimum achievable rate by optimizing power allocation. We evaluate the MMF rate performance under varying SNR conditions, while simultaneously adjusting the number of users per group and the degrees of CSIT uncertainty.

**Figure 2** presents a comparison of the MMF rate for the proposed RS-based MIMO-NOMA and conventional MIMO-NOMA systems as a function of SNR. **Figure 2a** compares the MMF rate for different numbers of users per group, considering cases with two and three users per group. The results reveal that the gain of the RS-based MIMO-NOMA over the conventional MIMO-NOMA systems expands as the number of users per group increases. This gain increases from 1.07 to 1.32 when the number of users per group increases from $M = 2$ to $M = 3$. Consequently, the RS-based MIMO-NOMA proves to be a more robust solution in overloaded regimes.

**Figure 2b** demonstrates the impact of CSIT uncertainty on the MMF rate performance. The results indicate that the MMF rate performance of the conventional MIMO-NOMA system degrades more significantly when CSIT transitions from perfect to imperfect with $\eta = 0.5$. Therefore, the RS-based MIMO-NOMA system is more robust to CSIT uncertainty fluctuations. The gap between MMF rates of the RS-based

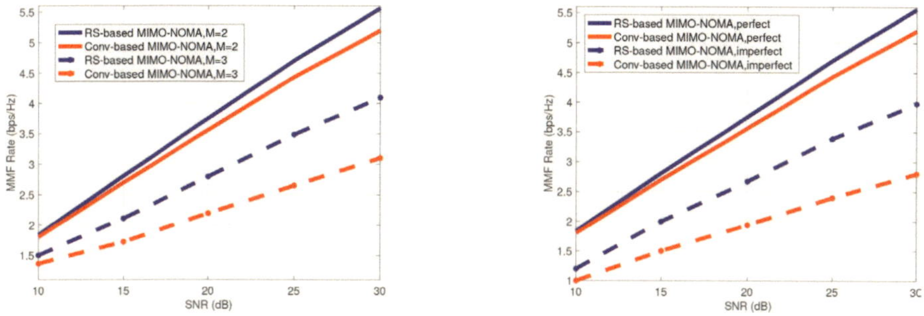


**Figure 2.**
*Comparison of achievable MMF rate performance for RS-based MIMO-NOMA and Conv-based MIMO-NOMA.*

MIMO-NOMA when CSIT changes from perfect to imperfect with $\eta = 0.5$ is 1.5880 bps/Hz. However, this gap is much higher in the conventional MIMO-NOMA system, amounting to 2.4 bps/Hz.

### 5.3 Sum-rate analysis results

This section investigate the performance of the proposed RS-based MIMO-NOMA scheme in terms of sum-rate. The objective is to maximize the overall system throughput by optimizing power allocation. The sum-rate performance is investigated under varying numbers of users per group and degrees of CSIT uncertainty.

**Figure 3** illustrates the sum-rate versus SNR comparison of RS-based MIMO-NOMA and Conv-based MIMO-NOMA. **Figure 3a** compares the sum-rate for different numbers of users per group, considering cases with two and three users per group. The results show that increasing the number of users per group decreases the sum-rate in both cases. Moreover, the gain of the RS-based MIMO-NOMA over the Conv-MIMO-NOMA is not considerably high, even when the number of users increases from $M = 2$ to $M = 3$.

**Figure 3b** explores the impact of CSIT uncertainty on the sum-rate performance. The results indicate that the sum-rate performance of the conventional MIMO-NOMA system experiences a more significant decline when CSIT transitions from perfect to imperfect with $\eta = 0.5$. Therefore, the RS-based MIMO-NOMA system exhibits greater robustness against CSIT uncertainty fluctuations. The gap between sum-rates of the RS-based MIMO-NOMA when CSIT changes from perfect to imperfect with

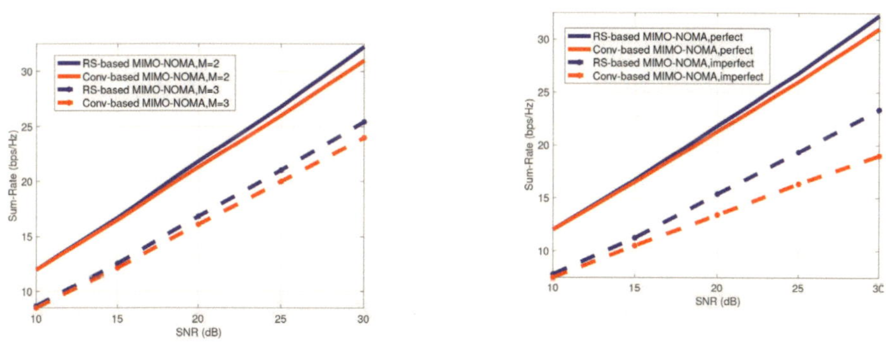


**Figure 3.**
*Achievable sum-rate performance comparison for RS-based MIMO-NOMA and Conv-based MIMO-NOMA.*

$\eta = 0.5$ is around 9 bps/Hz. In contrast, this gap is substantially larger in the conventional MIMO-NOMA system, amounting to 12 bps/Hz, roughly a 33% decline.

**Figures 2** and **3** illustrate that the proposed RS-based MIMO-NOMA scheme effectively exploits the rate-splitting technique to enhance its performance in overloaded scenarios and under imperfect CSIT, particularly when compared to the conventional MIMO-NOMA system. This improvement can be attributed to the RS-based MIMO-NOMA's ability to mitigate interbeam interference and efficiently allocate power among users, thus providing superior service to a larger number of users within each group even under imperfect CSIT. Overall, these results emphasize the advantages of adopting the RS-based MIMO-NOMA framework in practical network deployments, particularly in high-density and overloaded scenarios and under imperfect CSIT.

## 6. Conclusion

In this chapter, we have presented a novel scheme that combines the RS technique in MIMO systems with NOMA scheme for wireless communication systems, aiming to improve performance and capacity under imperfect CSIT and overloaded regime. The proposed scheme has considered a general and realistic scenario with both unicast and multicast users, focusing on increasing system throughput and optimizing precoding vectors for enhanced performance.

Furthermore, we have introduced a technique that transforms a non-convex optimization problem into a convex problem. By employing the WMMSE-rate relationship and an AO algorithm, the proposed technique successfully tackles the non-convex problem, allowing for the maximization of both the minimum rate and sum-rate of the system, particularly concentrating on the rate of the weakest user in each group under imperfect CSIT.

The comprehensive analysis provided in this chapter covers both tutorial background and novel ideas, offering valuable insights into the design and performance of future MIMO-NOMA systems that employ RS techniqusers. The findings demonstrate the potential of the proposed RS-based MIMO-NOMA scheme in addressing the challenges posed by imperfect CSIT and overloaded regimes in realistic scenarios with unicast and multicast users.

## Author details

Sareh Majidi Ivari*, Mohammad Reza Soleymani and Yousef R. Shayan
Concordia University, Montreal, Canada

*Address all correspondence to: sarehh.majidi@gmail.com

## References

[1] Cisco. Cisco Annual Internet Report (2018–2023) White Paper. Cisco; 2021. Available from: https://www.cisco.com/c/en/us/solutions/collateral/executive-perspectives/annual-internet-report/white-paper-c11-741490.html

[2] Ding Z, Adachi F, Poor HV. The application of MIMO to non-orthogonal multiple access. IEEE Transactions on Wireless Communications. 2016;**15**(1): 537-552. DOI: 10.1109/TWC.2015.2475746

[3] Huang Y, Zhang C, Wang J, Jing Y, Yang L, You X. Signal processing for MIMO-NOMA: Present and future challenges. IEEE Wireless Communications. 2018;**25**(2):32-38. DOI: 10.1109/MWC.2018.1700108

[4] Ali S, Hossain E, Kim DI. Non-orthogonal multiple access (NOMA) for downlink multiuser MIMO systems: User clustering, beamforming, and power allocation. IEEE Access. 2017;**5**: 565-577. DOI: 10.1109/ACCESS.2016.2646183

[5] Chen X, Zhang Z, Zhong C, Jia R, Ng DWK. Fully non-orthogonal communication for massive access. IEEE Transactions on Communications. 2018; **66**(4):1717-1731. DOI: 10.1109/TCOMM.2017.2779150

[6] Senel K, Cheng HV, Björnson E, Larsson EG. What role can NOMA play in massive MIMO? IEEE Journal of Selected Topics in Signal Processing. 2019;**13**(3):597-611. DOI: 10.1109/JSTSP.2019.2899252

[7] Nguyen N, Zeng M, Dobre OA, Poor HV. Securing massive MIMO-NOMA networks with ZF beamforming and artificial noise. In: 2019 IEEE Global Communications Conference (GLOBECOM). Waikoloa, HI, USA: IEEE GLOBCOM; 2019. pp. 1-6

[8] Min K, Kim T, Jung M. Performance analysis of multiuser massive MIMO with multi-antenna users: Asymptotic data rate and its application. ICT Express. 2023. DOI: 10.1016/j.icte.2023.01.003

[9] Björnson E, Sanguinetti L, Debbah M. Massive MIMO and small cells: Improving energy efficiency by optimal soft-cell coordination. International Journal of Wireless Information Networks. 2014;**21**(2):133-149

[10] Mao Y, Clerckx B, Li VO. Rate-splitting multiple access for downlink communication systems: Bridging, generalizing, and outperforming SDMA and NOMA. Journal on Wireless Communications and Networking. 2018; **2018**:133. DOI: 10.1186/s13638-018-1104-7

[11] Love D, Heath R, Lau V, Gesbert D, Rao B, Andrews M. An overview of limited feedback in wireless communication systems. IEEE Journal on Selected Areas in Communications. 2008;**26**(8):1341-1365

[12] Turan N, Fesl B, Koller M, Joham M, Utschick W. A versatile low-complexity feedback scheme for FDD systems via generative modeling. arXiv. 2023

[13] Sadeghi M, Björnson E, Larsson EG, Yuen C, Marzetta T. Joint unicast and multi-group multicast transmission in massive MIMO systems. IEEE Transactions on Wireless Communications. 2018;**17**(10): 6375-6388. DOI: 10.1109/TWC.2018.2854554

[14] Joudeh H, Clerckx B. RS for MISO wireless networks: A promising

PHY-layer strategy for LTE evolution. IEEE Communications Magazine. 2016; **54**(5):98-105

[15] Mao Y, Clerckx B. RS multiple access for downlink communication systems: Bridging, generalizing, and outperforming SDMA and NOMA. EURASIP Journal on Wireless Communications and Networking. 2018; **2018**(1):1-21

[16] Joudeh H, Clerckx B. Sum-rate maximization for linearly precoded downlink multiuser MISO systems with partial CSIT: A RS approach. IEEE Transactions on Communications. 2016; **64**(11):4847-4861

[17] Clerckx B, Kim I, Kim J, Zhang R, Poor HV. Is NOMA efficient in multi-antenna networks? A critical look at next generation multiple access Techniqusers. IEEE Communications Magazine. 2020; **58**(2):64-71

[18] Kim J, Kim I-M. Achievable rates of spatially coupled multiple-access channels with RS, superposition coding, and successive cancellation decoding. IEEE Transactions on Wireless Communications. 2018;**17**(10):6761-6775

[19] Lee B, Shin W. Max-min fairness precoder design for RS multiple access: Impact of imperfect channel knowledge. IEEE Transactions on Vehicular Technology. 2023;**72**(1):1355-1359. DOI: 10.1109/TVT.2022.3206808

[20] Joudeh H, Clerckx B. RS for max-min fair multigroup multicast beamforming in overloaded systems. IEEE Transactions on Wireless Communications. 2017;**16**(11): 7276-7289. DOI: 10.1109/TWC. 2017.2744629

[21] Yin L, Clerckx B. RS multiple access for multigroup multicast and multibeam satellite systems. IEEE Transactions on Communications. 2021;**69**(2):976-990. DOI: 10.1109/TCOMM.2020.3037596

[22] Yin L, Dizdar O, Clerckx B. RS multiple access for multigroup multicast cellular and satellite communications: PHY layer design and link-level simulations. In: 2021 IEEE International Conference on Communications Workshops (ICC Workshops); 2021 Jun 14–18; Montreal, Canada. IEEE; 2021. pp. 1-6. DOI: 10.1109/ICCWorkshops 50388.2021.9473795

[23] Zeng J, Lv T, Ni W, Liu RP, Beaulieu NC, Guo YJ. Ensuring max–min fairness of UL SIMO-NOMA: A RS approach. IEEE Transactions on Vehicular Technology. 2019;**68**(11): 11080-11093. DOI: 10.1109/TVT.2019. 2943511

[24] Joudeh H, Clerckx B. Robust transmission in downlink multiuser MISO systems: A rate-splitting approach. IEEE Transactions on Signal Processing. 2016;**64**(23):6227-6242

[25] Joudeh H, Clerckx B. Sum rate maximization for MU-MISO with partial CSIT using joint multicasting and broadcasting. In: Proc. IEEE Int. Conf. Commun. Londan, UK: IEEE International Conference on Communications (ICC); 2015. pp. 4733-4738. DOI: 10.1109/ ICC.2015.7249071

[26] Hao C, Wu Y, Clerckx B. Rate analysis of two-receiver MISO Broadcast Channel with finite rate feedback: A rate-splitting approach. IEEE Transactions on Communications. 2015; **63**(9):3232-3246. DOI: 10.1109/ TCOMM.2015.2453270

[27] Ivari SM, Caus M, Vazquez MA, Soleymani MR, Shayan YR, Perez-Neira AI. Precoding and scheduling in multibeam multicast NOMA based

satellite communication systems. In: 2021 IEEE Int. Conf. Commun. Workshops (ICC Workshops). Montreal, Canada: IEEE International Conference on Communications (ICC); 2021. pp. 1-6. DOI: 10.1109/ ICCWorkshops50388.2021.9473484

[28] Xiang Z, Tao M, Wang X. Massive MIMO multicasting in noncooperative cellular networks. IEEE Journal on Selected Areas in Communications. 2014;**32**(6):1180-1193. DOI: 10.1109/ JSAC.2014.2328144

Chapter 5

# Massive MIMO without CSI: When Non-Coherent Communication Meets Many Antennas

*Manuel José López Morales, Kun Chen-Hu and Ana García Armada*

**Abstract**

Under high-mobility scenarios, the traditional coherent demodulation schemes (CDS) have limited performance, because reference signals cannot effectively track the channel variations with an affordable overhead. As an alternative solution, non-coherent demodulation schemes (NCDS) based on differential modulation have been proposed. Even in the absence of reference signals, they are capable of outperforming the CDS with a reduced complexity. The literature on NCDS laid the theoretical foundations for simplified channel and signal models, often single-carrier and spatially uncorrelated flat-fading channels. This chapter explains the most recent results assuming orthogonal frequency division multiplexing (OFDM) signaling and realistic channel models.

**Keywords:** channel estimation, differential modulation, non-coherent, high-mobility, OFDM

## 1. Introduction

Massive multiple-input multiple-output (MIMO) [1] is a key technology for the advancement of wireless communications, especially in the evolution from the current fifth generation (5G) [2] to the forthcoming sixth generation (6G) [3–6] of mobile communication systems. Typically, the base station (BS) is equipped with a very large number of radiating elements, while the user equipment (UE) is only equipped with one single antenna or very few. Under this scenario, the BS can either simultaneously spatially multiplex several data streams to many UEs or enhance the quality of some links by exploiting spatial diversity. In order to fully exploit the benefits of MIMO technology, accurate channel state information (CSI) between the BS and the UEs is a must; otherwise, the performance is significantly degraded [7, 8].

Coherent demodulation scheme (CDS) is the typically chosen technique for exploiting massive MIMO systems. The acquisition of CSI is obtained by transmitting some reference signals or pilot symbols per antenna, which is known as pilot symbol-assisted modulation (PSAM) [9]. At the receiver, the CSI is estimated by typically using the Least-Squares criterion (LS) [10]. Finally, the pre/post-equalization matrices are computed in order to compensate for the effects of the channel by typically using the zero-forcing (ZF) or minimum mean squared error (MMSE) criteria [11]. However, the transmission of reference signals produces an excessive overhead in the system since these pilot symbols are mapped in the physical resources in the data frame. In order to alleviate this issue, time division duplexing (TDD) is the preferable choice since the channel reciprocity can be assumed, and hence, the CSI is only estimated in the uplink (UL) and reused in the downlink (DL) [12].

Nevertheless, acquiring accurate CSI without sacrificing the performance of the system is significantly limited and cannot be adopted in the new challenging scenarios considered in 6G, such as high-mobility communications and low-powered networks [8, 13]. On the one hand, CDS requires that the coherence time of the channel impulse response remains for long symbol periods, otherwise, a huge amount of reference signals must be transmitted to constantly track the fast channel variations, which is the typical case in autonomous vehicles, drone communications and satellite links. On the other hand, CDS requires links with a medium/high signal-to-noise ratio (SNR) in order to provide accurate enough CSI, otherwise the computed equalization matrices are not correct and degrade the performance of the system. To improve the quality of the CSI, the channel estimates must be obtained in several independent physical resources for the same UE and averaged out to reduce the noise and interference effects. Last but not least, in scenarios with many spatially multiplexing UEs, to avoid the pilot contamination produced among the UEs [14]. This results in a even larger training overhead, which will also be detrimental for the data efficiency.

Non-coherent demodulation scheme (NCDS) is an appealing alternative to be combined with massive MIMO since it can demodulate the transmitted information without the knowledge of CSI, with the same asymptotic performance as coherent schemes [8]. Thus, the huge amount of required reference signals in CDS is entirely avoided and the complexity of transceivers is also reduced. Many works in the literature showed that the NCDS detection can provide an acceptable performance in very fast time-varying scenarios [8, 13, 15–21], while the coherent scheme fails. Additionally, NCDS is flexible and can be integrated in an orthogonal frequency division multiplexing (OFDM) [22]. Compared to the CDS, its performance superiority in scenarios with stringent condition makes it a good candidate for future communication systems in high-speed scenarios.

Some works have targeted the UL scenario [17, 20], in which one single-antenna UE transmits the differential symbols, while the BS exploits the spatial diversity produced by large number of antennas. An NCDS scheme based on differential $M$-ary phase shift keying (DMPSK) constellations was exploited [17], allowing differential detection while leveraging the advantages of an increased number of receive antennas. Later, [20] combined the NCDS with the OFDM multi-carrier waveform, in order to combat the frequency-selective channel. The differential symbols are mapped in the two-dimensional (time and frequency) resource grid. In [19], the NCDS is combined with precoding based on beamforming, where assuming that a beam-management procedure is executed beforehand. Recently, a combination of CDS and NCDS is also explored [13] in order to take advantage of both techniques. To achieve this, a blind channel estimation is proposed utilizing reconstructed non-coherent data, which can be later used to perform UL filtering of coherent data resulting in a hybrid demodulation scheme (HDS). Additionally, Lopez-Morales and Garcia-Armada [15] also proposed using a multi-user precoding for the DL combined with DMPSK to avoid the use of pilot symbols.

An overview of NCDS combined with massive MIMO-OFDM under different scenarios is provided in this chapter. Section 2 explains the UL of the non-coherent massive MIMO based on DMPSK and blind channel estimation. Section 3 provides the two possibilities to perform the DL in the non-coherent massive MIMO based on DMSPK. Section 4 details the multi-user approach for the UL of the NC massive MIMO based on constellation multiplexing. Section 5 compares the CDS, NCDS and HDS schemes in different scenarios. Finally, Section 6 concludes the chapter and gives insights into future research lines.

## 2. Non-coherent massive MIMO in UL

Two wireless transceivers are considered in this scenario. One is a BS equipped with $V$ antennas, while the other is a UE equipped with a single antenna. The chosen waveform is the well-known OFDM, composed of $K$ subcarriers with a subcarrier spacing of $\Delta f$ Hz and a cyclic prefix (CP), whose length is measured in samples ($L_{CP}$), to mitigate the multi-path effects of the channel. A set of $N$ contiguous OFDM symbols is assumed to be transmitted in a burst. Note that multiple UEs can be multiplexed in either time or frequency dimensions thanks to the two-dimensional resource grid provided by the OFDM. Additionally, the UEs can be also mapped in the constellation domain, whose details are given in Section 4.

### 2.1 Fundamentals of differential encoding and decoding in OFDM

Typically, NCDS based on differential modulation is performed using the time domain scheme. This scheme is represented in **Figure 1a**, where the red arrows indicate the direction in which differential modulation and demodulation are performed. In this case, it occurs between resources that belong to the same frequency and contiguous symbols in the time domain. The differential encoding can be described as

$$\tilde{x}_{k,n} = \begin{cases} \tilde{r}_{k,n}, & n = 1 \\ \tilde{x}_{k,n-1}\tilde{s}_{k,n-1}, & 2 \leq n \leq N \end{cases}, \quad 1 \leq k \leq K, \tag{1}$$

where $\tilde{r}_{k,1}$ is the reference symbol transmitted by the UE at the $k$th subcarrier of the first OFDM symbol, while $\tilde{s}_{k,n}$ and $\tilde{x}_{k,n}$ are complex data and differential symbols, respectively, at the $k$th subcarrier and $n$th OFDM symbol transmitted by the UE. The data symbol $\tilde{s}_{k,n}$ needs to meet the condition

$$\tilde{s}_{k,n} \in \mathfrak{M}, \quad \mathbb{E}\left\{|\tilde{s}_{k,n}|^2\right\} = 1 \quad 1 \leq k \leq K, \quad 1 \leq n \leq N - 1, \tag{2}$$

where $\mathfrak{M}$ denotes the set of symbols of a PSK constellation due to the fact that the differential encoding can only transmit information in the phase component and its average energy is normalized to one. One drawback of implementing the mapping scheme in the time domain is the increased latency and memory consumption. This is because the scheme requires waiting for two complete OFDM symbols to be received in order to obtain $\tilde{s}_{k,n}$. In the time domain implementation, a differential decoding of two contiguous symbols is performed (as shown in **Figure 1a**). Furthermore, this

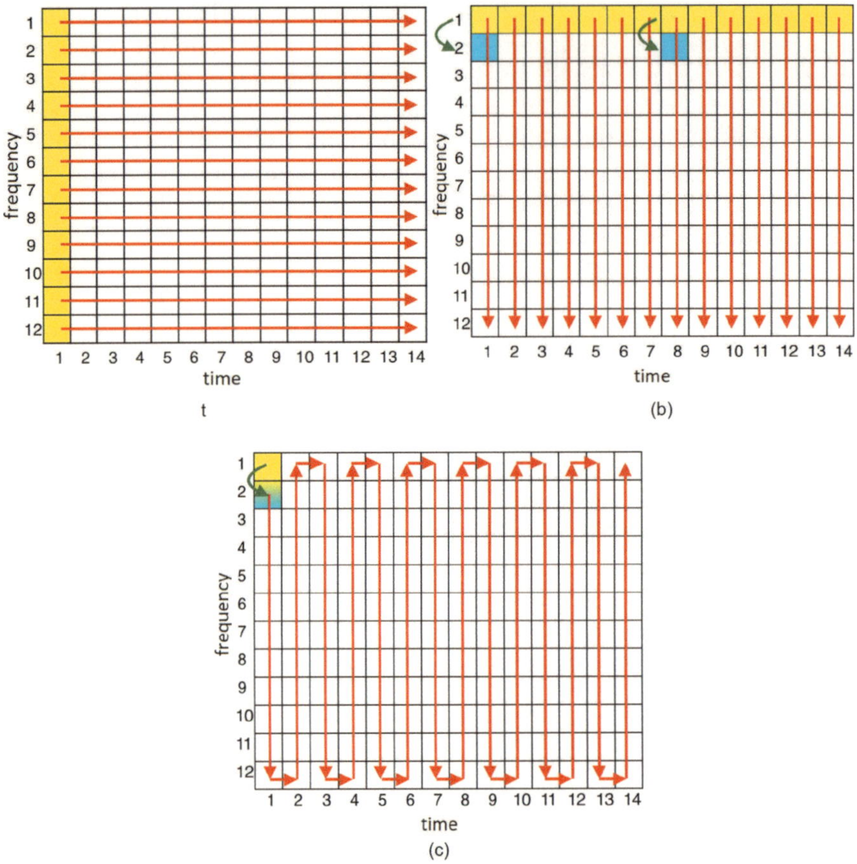


**Figure 1.**
*Differential modulation mapping schemes in an OFDM resource grid when $K = 12$, $N = 14$ and $\mathcal{I}_N = \{1, 8\}$. The yellow and blue boxes denote the reference symbols required by the differential modulation and phase difference estimation, respectively.*

implementation cannot be used when there is a high Doppler spread because two consecutive OFDM symbols may not experience similar channel responses.

Alternatively, the frequency domain scheme can be also used to implement the differential modulation technique, by exploiting the frequency dimension (as shown in **Figure 1b**). In this scheme, the differential symbols are mapped into contiguous frequency resources of the same OFDM symbol, according to [20] as

$$\tilde{x}_{k,n} = \begin{cases} \tilde{r}_{k,n}, & k = 1, \\ \tilde{x}_{k-1,n}\tilde{p}_{k,n}, & k = 2, n \in \mathcal{I}_N, \quad 1 \leq n \leq N \\ \tilde{x}_{k-1,n}\tilde{s}_{k-1,n}, & \text{otherwise} \end{cases} \tag{3}$$

where $\tilde{r}_{1,n}$ and $\tilde{p}_{2,n}$ are two reference symbols for different purposes. The set $\mathcal{I}_N$ contains the indexes that correspond to the OFDM symbols carrying $p_{2,n}$. As explained before, The first reference symbol is necessary for differential demodulation, as previously explained. The second type is required to estimate the phase difference between two subcarriers resulting from frequency-domain mapping, as detailed in [20]. This scheme has the advantage of reduced latency and robustness against high

Doppler spreads. It is reasonable to assume that contiguous subcarriers have similar channel responses due to the much larger number of subcarriers compared to the number of channel taps. However, the benefits come at the expense of an additional phase estimation and compensation procedure. Although this additional phase component is negligible for non-frequency-selective channels, it must be compensated for strongly frequency-selective channels. When diversity is employed, only one additional reference pilot is needed for all OFDM symbols within the coherence time ($p_{2,n}$), resulting in minimal overhead impact.

In [20], both time and frequency domain schemes are presented. However, if the number of allocated resources is reduced ($K\downarrow$ and/or $N\downarrow$), both schemes may result in significant overhead. For instance, in massive machine type communication (mMTC) scenarios, mechanical devices send short packets of only a few bytes. Adopting any of the two presented schemes implies sending a significant number of reference symbols. To address this issue, we propose a new mapping scheme called the mixed domain scheme (see **Figure 1c**). In this scheme, we first differentially encode the data symbols as

$$\tilde{x}_j = \begin{cases} \tilde{r}_j, & j = 1 \\ \tilde{x}_{j-1}\tilde{p}_j, & j = 2 \\ \tilde{x}_{j-1}\tilde{s}_{j-1}, & 3 \leq j \leq KN \end{cases}, \tag{4}$$

where $j$ denotes the resource index. Then, the differential symbols $\tilde{x}_j$ are allocated to the two-dimensional resource grid as

$$\tilde{x}_{k,n} = \tilde{x}_j | (k,n) = f(j), \quad 1 \leq j \leq KN, \tag{5}$$

where $f(\bullet)$ is the resource mapping policy function. **Figure 1c** shows a recommended example of a mapping policy function, where the dramatic reduction of reference signals can be observed. This policy mainly follows the frequency domain scheme, except for the edge subcarriers of the block, which follow a time domain scheme. This proposal cannot only significantly reduce the number of reference symbols, but it is also capable of taking all advantages of a frequency domain scheme. Moreover, in the case of time-varying channels, only those complex symbols placed at both edge subcarriers may suffer from an additional degradation.

To maintain conciseness and simplify notation, we adopt the frequency domain scheme for the remainder of this chapter. However, note that the techniques presented in the following sections can be applied to both time and mixed domain schemes without any modification.

Once, the differential symbols are obtained by using (3), the OFDM symbol can be obtained by performing an inverse discrete Fourier transform (IDFT) as

$$x_{m,n} = \frac{1}{\sqrt{K}} \sum_{k=1}^{K} \exp\left(j\frac{2\pi}{K}(k-1)(m-1)\right)\tilde{x}_{k,n}, \quad 1 \leq m \leq K, \quad 1 \leq n \leq N. \tag{6}$$

Then, a CP, whose length is given by $L_{CP}$ is appended to each OFDM symbol **s** in order to absorb the multi-path effect.

At the receiver, the CP is discarded from the received signal, and hence, the linear convolution between the multi-tap channel and transmitted data symbols is converted to a circular one. Hence, the received signal at the $v$-th antenna at the BS is given by

$$y_{m,n,v} = \sum_{\tau=1}^{L_{CH}} h_{\tau,n,v} x_{\mathrm{mod}(m-\tau,K),n} + w_{m,n,v}, 1 \le m \le K, 1 \le n \le N, 1 \le v \le V, \tag{7}$$

where $w_{m,n}$ is the additive white Gaussian noise (AWGN) at $m$-th sample in the $n$-OFDM symbol, and it is distributed as $\mathcal{CN}(0, \sigma_w^2)$. Following [7], the channel coefficients suffer from time variability and an autoregressive model approximates the temporally correlated fading channel coefficients of subcarrier $k$ at time instant $n$ as

$$h_{\tau,n',v} = \alpha_d h_{\tau,n,v} + w'_{\tau,n',v}, \quad \alpha_d = J_0\left(2\pi d f_D\left(\frac{K + L_{CP}}{K\Delta f}\right)\right) < 1, \tag{8}$$

where $n'$ refers to a time instant in the future with respect to $n$ ($d = |n' - n|$ time difference in OFDM symbols), $\alpha_d$ is the temporal correlation parameter, $J_0(\cdot)$ denotes the zero-th order Bessel function of the first kind and $f_D$ represents the maximum Doppler spread experienced by the transmitted signal, also in Hertz. Similar to CDS, NCDS requires that the channel impulse response should be quasi-static during, at least, one OFDM symbol, otherwise inter-symbol and inter-carrier interferences (ISI and ICI, respectively) will appear. Consequently, the length of the OFDM symbols ($K \downarrow$) should be reduced as the Doppler effect is higher ($f_D \uparrow$).

Then a discrete Fourier transform is performed to obtain the received symbols in the frequency domain as

$$\tilde{y}_{k,n,v} = \frac{1}{\sqrt{K}} \sum_{m=1}^{K} \exp\left(-j\frac{2\pi}{K}(n-1)(m-1)\right) y_{m,n,v}, \tag{9}$$

where $1 \le m \le K, 1 \le n \le N, 1 \le v \le V$ and the received signal in the frequency domain can be modeled as

$$\tilde{y}_{k,n,v} = \tilde{h}_{k,n,v}\tilde{x}_{k,n} + \tilde{w}_{k,n,v} \quad 1 \le k \le K, \quad 1 \le n \le N, \quad 1 \le v \le V, \tag{10}$$

where $\tilde{h}_{k,n,v}$ and $\tilde{w}_{k,n,v}$ is the channel frequency response and noise in the frequency domain, respectively, in the $k$th subcarrier and $n$th OFDM symbol at $v$th antenna.

Later, a differential demodulation is performed in the frequency domain to undo (3) as

$$\tilde{z}_{k,n} = \frac{1}{V}\sum_{v=1}^{V} \tilde{y}^*_{k-1,n,v}\tilde{y}_{k,n,v} = \sum_{i=1}^{4} T_{k,n,v,i}, \quad 2 \le k \le k-1, \quad 1 \le n \le N, \tag{11}$$

$$T_{k,n,v,1} = \frac{1}{V}\sum_{v=1}^{V} \tilde{w}_{k-1,n,v}\tilde{w}_{k,n,v}, \quad T_{k,n,v,2} = \frac{1}{V}\sum_{v=1}^{V} \tilde{h}_{k-1,n,v}\tilde{x}_{k-1,n}\tilde{w}_{k,n,v}, \tag{12}$$

$$T_{k,n,v,3} = \frac{1}{V}\sum_{v=1}^{V} \tilde{w}_{k-1,n,v}\tilde{h}_{k,n,v}\tilde{x}_{k,n}, \quad T_{k,n,v,4} = \frac{1}{V}\sum_{v=1}^{V} \tilde{h}_{k-1,n,v}\tilde{h}_{k,n,v}\tilde{x}_{k-1,n}\tilde{x}_{k,n}, \tag{13}$$

where $z_{k,n}$ is the decision variable and $T_{k,n,v,i}$, $1 \le i \le 4$ denotes each term out of four produced by differential demodulation. Note that the first three terms correspond to noise and interference terms, while the last one is the desired data term.

Making use of the Law of Large Numbers, when the number of the antennas tends to infinity ($V \to \infty$), the fourth terms can be simplified as

$$\mathbb{E}\left\{|T_{k,n,v,1}|^2\right\} = \mathbb{E}\left\{|T_{k,n,v,2}|^2\right\} = \mathbb{E}\left\{|T_{k,n,v,3}|^2\right\} = 0, \tag{14}$$

$$\mathbb{E}\left\{|T_{k,n,v,4}|^2\right\} = \rho_f \exp(j\theta_f) \times \left\{\begin{array}{ll} \tilde{p}_{k,n}, & k = 2, n \in \mathcal{I}_N \\ \tilde{s}_{k-1,n}, & 3 \le k \le K \end{array}\right\}, \tag{15}$$

where $2 \le k \le K$, $1 \le n \le N$, the first three terms, which correspond to the interference and noise terms, vanished since the channel frequency response, noise and data symbols are independent random variables to each other, while the fourth term remains. Note that the pilot and data symbols in the fourth term are scaled by the correlation between two contiguous channel frequency responses at subcarriers $k-1$ and $k$, whose modulus and phase are given by $\rho_f$ and $\theta_f$, respectively. This scaling factor is producing a common phase rotation to the received symbols $z_{k,n}$, which can be easily estimated and equalized by transmitting a pilot symbol ($\tilde{p}_{k,n}$) before performing the symbol decision.

If the number of antennas ($V$) is not large enough, the three terms given in (14) are not zero. Hence, the received signal is polluted by noise and self-interference. The performance measured in signal-to-noise and interference ratio (SINR) for the multi-user case is given in Section 4, which corresponds to the generalization of the single-user case.

The performance given by (11)–(13) assumed an ideal case, where hardware impairments are not considered. However, it is well-known that OFDM combined with the traditional CDS is very sensitive to phase noise (PN) [23, 24]. The effect of this PN is due to the instabilities of the local oscillators, which are typically modeled according to a classical Wiener random walk process. Its negative effect not only will degrade the received symbols, but it will also add a common phase error. According to 5G [6], the phase-tracking reference signal (PT-RS) is proposed to be added in order to estimate and equalize this phase error, and hence, the overhead of the system is further increased. On the other hand, according to [25], when NCDS is combined with OFDM it does not require any additional PN estimation and equalization since it is inherently robust to these effects thanks to the use of the differential modulation, and no additional reference signal is required.

### 2.2 Blind channel estimation based on differential detection

As it has been explained in the previous subsection, the non-coherent massive MIMO is capable of obtaining the transmitted data in the UL without the CSI and post-equalization. However an interesting question arises, could we estimate an accurate enough CSI given the non-coherently detected symbols? In the end, these non-coherently detected data symbols can be seen as a new type of reference signals, which can be utilized in CDS for channel estimation and equalization, without rising the overhead since the non-coherent data symbols convey data information.

Assuming that accurate CSI can be successfully obtained by using the NCDS, these estimates can be exploited in two ways. On the one hand, the estimates can be used to compute the precoding matrices and used in the DL in TDD mode [21], and hence, the overhead generated by transmitted reference signals in the UL is avoided, as will be shown in Section 3.2. On the other hand, CDS and NCDS can be merged in the UL, namely to produce a HDS, where the traditional pilot symbols transmitted in CDS are replaced by non-coherent data symbols. The latter can be jointly used for data transmission, channel estimation and the computation of post-coding matrices. Consequently, the efficiency of the UL transmission is increased [13] (**Figure 2**).

The steps for the blind channel estimation based on NCDS can be summarized as follows:

1. Firstly, the symbol decision is performed over $\tilde{z}_{k,n}$ as

$$\hat{\hat{s}}_{k,n} = \ddot{s}_j | j = \arg\min_j \left\{ \left|\tilde{z}_{k,n} - \ddot{s}_j\right| \right\}, \quad \ddot{s}_j \in \mathfrak{M}, \quad 1 \leq j \leq |\mathfrak{M}|, \tag{16}$$

   where $2 \leq k \leq K$, $1 \leq n \leq N$, $\hat{\hat{s}}_{k,n}$ are the decided symbols at $k$th subcarrier in $n$th OFDM symbol, $\ddot{s}_j$ corresponds to the $j$th symbol of the constellation $\mathfrak{M}$ whose number of elements is given by $|\mathfrak{M}|$.

2. Then, the differential data sequence is reconstructed ($\hat{\hat{x}}_{k,n}$) by using the frequency domain scheme, given in (3), and replacing the transmitted symbols ($\tilde{s}_{k,n}$) with the decided ones ($\hat{\hat{s}}_{k,n}$).

3. Finally, the channel is estimated for each subcarrier and OFDM symbol ($\hat{\hat{h}}_{k,n,v}$) by utilizing the reconstructed differential data sequence $\hat{\hat{x}}_{k,n}$ as a pilot symbol with any estimation technique. For instance, a LS criterion [10] can be used as

$$\hat{\hat{h}}_{k,n,v} = \hat{\hat{x}}_{k,n}^{-1} \tilde{y}_{k,n,v}. \tag{17}$$

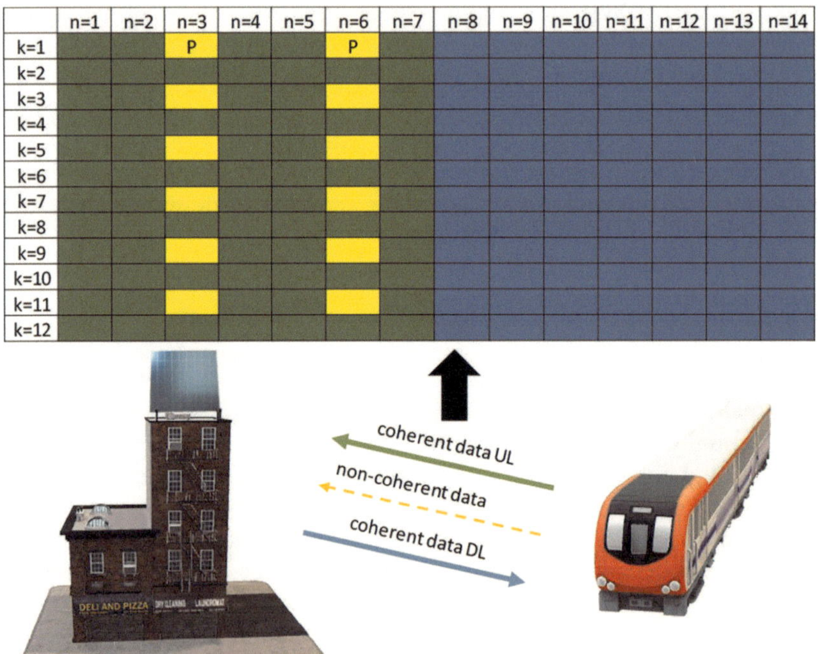


**Figure 2.**
*Example of a unit block for a proposed HDS scheme.*

Note that an additional error term in the channel estimation, with respect to the classical PSAM [9], is produced by a possible mismatch between transmitted data symbols $\tilde{x}_{k,n}$ and reconstructed differential symbols $\hat{\tilde{x}}_{k,n}$, whose error was characterized in [13, 21]. The estimated channel at $k$th subcarrier will be used in another subcarrier index $k'$, such that $k \neq k'$. Hence, the channel estimation error is composed of two independent components ([13], Eq. (24)) as shown below

$$e_d^2 = \mathbb{E}\left\{\left|\hat{\tilde{h}}_{k,n,v} - \tilde{h}_{k,n,v}\right|^2\right\} = \sigma_{x,d}^2 + \sigma_b^2 = 2\left(1 - \alpha_d \delta_u^{n,k}\right) + \sigma_w^2, \tag{18}$$

where $\sigma_{x,d}^2$ is the channel estimation error that comes from compensation and estimation in different time instants with a possible mismatch between transmitted and reconstructed differential symbol. The term $\delta_u^{n,k}$ is computed as

$$\delta_{k,n} = \mathbb{E}\left\{\cos\left(\angle(\tilde{x}_{k,n}) - \angle\left(\hat{\tilde{x}}_{k,n}\right)\right)\right\} \approx \frac{1 - P_{k,n} - \left(1 - P_{k,n,u}\right)^N}{(N-1)P_{k,n}}, \tag{19}$$

where $P_{k,n}$ is the error probability for the UL of each user. To find the details of the derivations the interested reader is referred to [15].

The MSE of channel estimation, as given in (18), shows that when either $\alpha_d$ or $\delta_{k,n}$ is zero, the channel error estimation is highest, while both need to be 1 to prevent any increase in the channel estimation error compared to the PSAM. Various MSE curves are displayed for different values of $\alpha_d$ and SNR (**Figure 3**).

To ensure that the channel is properly estimated in a certain time-frequency resource, some error-detecting code (such as a cyclic redundancy code) can be added to a data stream of non-coherent data. With this, and performing the channel estimation with reconstructed data that we are sure is correct, the channel estimation error will be the same as that of the PSAM.

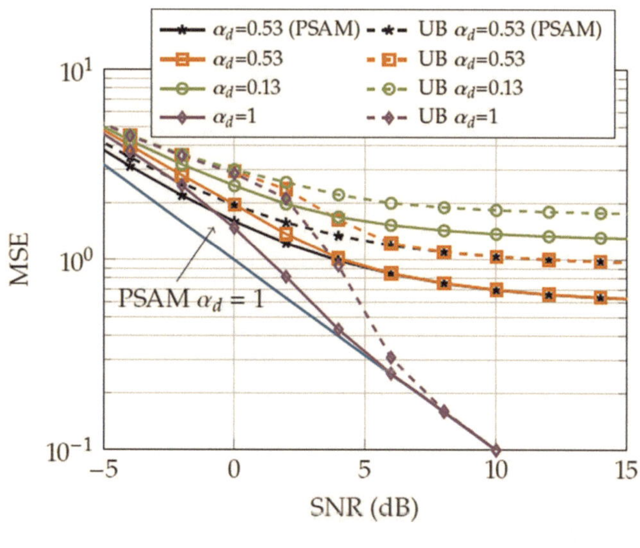


**Figure 3.**
*MSE of channel estimation for $M_{UL} = 16$ and $R = 100$. The continuous line shows the result obtained from the Monte Carlo simulation, while the dashed line represents the theoretical upper bound. The blue line corresponds to the PSAM method without considering channel time variability, which represents the best-case scenario.*

## 3. Non-coherent massive MIMO in DL

In this scenario, one multi-antenna BS simultaneously serves $U$ UEs in the DL. It is assumed that the parameters of the OFDM system are the same as described for the UL.

### 3.1 Non-coherent massive MIMO in FDD mode

In FDD, the multi-path channel coefficients between the UL and DL are fully uncorrelated, and the channel reciprocity property cannot be assumed as in TDD. Consequently, the massive number of antennas at the BS used for transmission can only be exploited in spatial diversity mode since the channel estimates of the $V$ antennas per user in the DL are not available. However, the exploitation of the diversity from the transmitter without knowledge of the channel is still a challenge, due to the fact that techniques based on block codes [26] failed to exploit a large number of antennas at the transmitter, since their complexity is proportional to the number of antennas. Even though the mapping schemes proposed for the UL are still valid, a few more ingredients are needed to make NCDS suitable for the DL, detailed in the following subsections.

#### *3.1.1 Precoding based on beamforming or codebook selection*

The NCDS can be combined with the precoding technique based on beamforming or codebook selection at the expense of using some (reduced) channel knowledge. At the BS, it is assumed that either the angular position or the best codebook index of the UE of interest is available, which is obtained through a beam-management procedure. Given this additional information, the data is sent over a non-coherently processed link. For the sake of conciseness, beamforming is the chosen technique for the rest of the document. Note that the detailed procedures for the beamforming in the following sections can be easily adapted for the codebook selection scheme.

The combination of NCDS with a practical beamforming technique based on knowing the angular position of the UE is proposed in [19]. The beam-management procedure defined in 5G [6] is suggested to be performed as a first step. This procedure is responsible for accurately determining the angle of the spatial clusters of the propagation channel contributing to the signal of each UE, by transmitting some reference signals. These reference signals are the synchronization signals (SS) and channel state information-reference signals (CSI-RS). The former is used when a UE would like to enter the system for the first time, while the latter is exploited for updating the angular position of an existing UE in the system. Note that, this beam-management procedure must be executed, at least, once per channel coherence time in order to constantly update the estimated angular positions of the current and new UEs.

At the transmitter, the BS transmits the data stream to all the UEs by using beamforming as

$$\tilde{x}_{k,n,v} = \sum_{u=1}^{U} \tilde{b}_{k,n,v,u} \tilde{x}_{k,n,u}, \quad 1 \le v \le V, \quad 1 \le k \le K, \quad 1 \le n \le N \tag{20}$$

where $\tilde{x}_{k,n,v}$ and $\tilde{b}_{n,v}$ are the precoded data symbol and the precoding coefficient, respectively, for the $v$th antenna and $u$th UE of the BS placed at the $k$th subcarrier and

$n$th OFDM symbol. This precoding coefficient is obtained according to either the estimated angular position or the best codebook index for each UE, and thus, it is in charge of focusing the energy in the obtained specific direction. In this way, the energy received by the UE is enhanced since its path loss is compensated. Similarly, precoding can be used in the UL for the BS to receive the signal from this spatial direction.

#### *3.1.2 Diversity in the frequency domain*

In order to enhance SINR gain for a good performance of NCDS [17], averaging in dimensions other than space, such as time or frequency, is proposed in [19]. Since the number of antennas at the UE is usually limited, this additional source of diversity may be particularly necessary to multiplex several UEs or enable critical services. The use of the frequency dimension is explained in [20], where each OFDM symbol can be processed independently, providing the advantage of easy extension to averaging in time (processing multiple consecutive OFDM symbols) or space (increasing the number of receive antennas of the UE when feasible).

To exploit frequency diversity, the same differential complex symbol is transmitted in multiple frequency resources. After performing the differential encoding for the $u$th UE, the $Q$ differential symbols are replicated at the transmitter as

$$\tilde{x}_{k,n,u} = \tilde{x}_{q,n,u} | q = \mathrm{mod}(k-1, Q) + 1, \quad K = Q \times F, \quad 1 \le k \le K, \quad 1 \le n \le N, \quad (21)$$

where $F$ is the frequency repetition/averaging factor.

The non-coherent detection at the receiver exploits the frequency diversity, where the received data in the subcarriers that carry the same transmitted data are averaged as

$$\tilde{z}_{q,n,u} = \frac{1}{F}\sum_{k=0}^{F-1} y^*_{q+kQ-1,n,u} y_{q+kQ,n,u}, \quad 2 \le q \le Q, \quad 1 \le n \le N, \quad 1 \le u \le U. \quad (22)$$

With this scheme there is a trade-off between overhead and robustness. According to [19], even though the frequency diversity add an additional overhead, it still outperforms the CDS in terms of throughput for some particular scenarios with high mobility.

### **3.2 Non-coherent massive MIMO in TDD mode**

As was explained in Section 2.2, the channel could be blindly estimated utilizing the reconstructed data in the UL of a non-coherent massive MIMO scheme. Therefore, once the channel is available, it can be used for precoding in the DL transmission to spatially separate the users. To avoid the use of demodulation pilots and thus avoid any pilot signal in the TDD time slot, it is preferred to use a DMPSK also in the DL signals. The use of demodulation pilots is needed in the DL of any coherent scheme to compensate for inefficiencies in the precoder, which can be caused by an erroneous channel estimation, by the use of a simple and not so powerful precoder (such as the MRT) and by the fact that the power in transmission is limited by the RF circuitry, which may cause that some precoders are not realizable. By using a DMPSK in the DL, the transmitted signals will be much more robust against errors in amplitude and phase, compared to the QAM constellations.

To improve clarity and conciseness, we will be using matrix notation throughout this document. Boldface uppercase letters will represent matrices, boldface lowercase letters will represent vectors and normal letters will represent scalar quantities. Specifically, $[\mathbf{A}]_{m,n}$ refers to the element in the $m$th row and $n$th column of matrix $\mathbf{A}$, and $[\mathbf{a}]_n$ represents the $n$th element of vector $\mathbf{a}$.

In the DL, the symbols of all the users are stacked in $\tilde{\mathbf{x}}_{k,n}$ of size $(U \times 1)$ for each time instant $n$ and subcarrier $k$ and are precoded before transmission using the precoding matrix $\tilde{\mathbf{B}}_{k,n} = (\tilde{\mathbf{H}}_{k,n})^H = \left[\tilde{\mathbf{b}}_{k,n,1}, \cdots, \tilde{\mathbf{b}}_{k,n,U}\right]$ for maximum ratio transmission (MRT). The channel for each user is defined as $\tilde{\mathbf{h}}_{k,n,u} = \left[\tilde{h}_{k,n,1,u}, \cdots, \tilde{h}_{k,n,V,u}\right]^T$. The DL channel is composed as $\tilde{\mathbf{H}}_{k,n} = \left[\tilde{\mathbf{h}}_{k,n,1}, \cdots, \tilde{\mathbf{h}}_{k,n,U}\right]^T$, where the DL channels of all users are stacked. Thus, in the DL the received signal is

$$\tilde{\mathbf{y}}_{k,n} = \tilde{\mathbf{H}}_{k,n}\tilde{\mathbf{B}}_{k,n}\tilde{\mathbf{x}}_{k,n} + \tilde{\nu}_{k,n}, \tag{23}$$

where the noise vector $\tilde{\nu}_n^k$ is a $U \times 1$ vector where each element represents the noise at the receiver of user $u$ and is distributed as $\tilde{\nu}_{k,n,u} \sim \mathcal{CN}(0, \sigma_u^2)$. In the case of applying MRT in the DL of the BS, the matrix in (23) can be separated into the desired user and the rest of the users. Therefore, we can rewrite (23) as follows

$$\tilde{y}_{k,n,u} = \tilde{\mathbf{h}}_{k,n,u}\tilde{\mathbf{b}}_{k,n,u}\tilde{\mathbf{x}}_{k,n} + \sum_{u' \neq u} \tilde{\mathbf{h}}_{k,n,u'}\tilde{\mathbf{b}}_{k,n,u'}\tilde{\mathbf{x}}_{k,n} + \tilde{\nu}_{k,n}. \tag{24}$$

To analyze the effect of imperfect channel estimation for the proposed scheme in the next Section of the DL transmission, we assume the following definition [27], $\hat{\mathbf{H}}_{k,n} = \sqrt{1-e_d^2}\tilde{\mathbf{H}}_{k,n} + \tilde{\mathbf{H}}_{k,e}$, where $\tilde{\mathbf{H}}_{k,e} \sim \mathcal{CN}(\mathbf{0}, e_d^2\mathbf{I})$ is an error component which is uncorrelated with $\mathbf{H}_{k,n}$. By performing some straightforward manipulations which can be found in [15], the distribution of $\tilde{y}_{k,n,u}$ (for $x_{k,n,u} = 1$, without loss of generality[1]) is

$$\Re\left\{\tilde{y}_{k,n,u}\right\} \sim R\sqrt{1-e_d^2} + \mathcal{N}\left(0, \frac{R(U - e_d^2 + 1) + \sigma_u^2}{2}\right) = \mu_\Re + \mathcal{N}(0, \sigma_\Re^2) \tag{25}$$

$$\Im\left\{\tilde{y}_{k,n,u}\right\} \sim \mathcal{N}\left(0, \frac{R(U + e_d^2 - 1) + \sigma_u^2}{2}\right) = \mathcal{N}(0, \sigma_\Im^2). \tag{26}$$

The differential decoding performed in reception for the received signal at each user as $\tilde{z}_{k,n,u} = \tilde{y}^*_{k,n-1,u}\tilde{y}_{k,n,u}$ results in the product of complex normally distributed variables, where in order to find the distribution of the received symbol, we have to consider the product of two complex variables. Applying again some straightforward manipulations which can be found in [15], we have

$$\Re\{\tilde{z}_{k,n,u}\} \sim \mathcal{N}\left(\mu_\Re^2, 2\mu_\Re^2\sigma_\Re^2 + \sigma_\Re^4 + \sigma_\Im^4\right), \quad \Im\{\tilde{z}_{k,n,u}\} \sim \mathcal{N}\left(0, 2\sigma_\Im^2\left(\mu_\Re^2 + \sigma_\Re^2\right)\right), \tag{27}$$

so the SER for the DL of user $u$ is computed using ([13], Appendix A).

---

[1] The error is computed for $\left[\tilde{\mathbf{x}}_n^k\right]_u = 1$ for simplicity but is the same for the rest of the symbols.

## 4. Multi-user non-coherent massive MIMO based on DMPSK

In the previous sections, only a single UE is mapped in each time/frequency resource of the OFDM for the non-coherent massive MIMO system based on DMPSK. Hence, the case of multiple UEs is presented in this Section, where its access strategy is based on a mapping the different UEs in the constellation domain. Each UE transmits its individual constellation and they superimpose in the receiver, resulting in a joint-constellation. Since there is no CSI available, a joint decision must be made. Therefore, ensuring a bijective relation between the individual constellations and the joint-constellation is important, resulting in a crucial constellation design problem to increase the multi-user performance. For this, in this Section, the system model of the multiple UE is briefly introduced first, which shows that joint-constellation distribution depends on the individual one, hindering classical design strategies to be utilized. Then, two design approaches that are based on utilizing artificial intelligence are described, followed by a proposal of some multi-user constellations.

### 4.1 System model

The constellation design for the simultaneous transmission of multiple UEs can be applied in UL or DL. For the sake of simplicity and without loss of generalization, UL is considered. The UEs transmit to the BS concurrently using the non-coherent scheme described in Section 2.1. During the $n$th OFDM symbol, the transmitted bits by the $u$th UE are arranged in a vector $\mathbf{b}_{n,u}$ having a dimension of $(Nb, u \times 1)$. Here, $N_{b,u}$ denotes the number of bits for user $u$. The vector $\mathbf{b}n, u$ is then transformed into a complex symbol $\tilde{s}_{k,n,u}$, given by

$$\tilde{s}_{k,n,u} = g_B(\varpi_u, \mathbf{b}_{n,u}) \in \mathfrak{M}_u, \quad 1 \le k \le K-1, \quad 1 \le n \le N, \quad 1 \le u \le U, \tag{28}$$

$$\mathfrak{M}_u = \{c_{u,1}, \ldots, c_{u,M_u}\}, \quad M_u = |\mathfrak{M}_u| = 2^{N_{b,u}}, \quad c_i^u \in \mathbb{C}, |c_i^u| = 1, c_i^u \ne c_{i'}^u \forall i \ne i', \tag{29}$$

where the $g_B(\cdot)$ is the bit mapping function, $\mathfrak{M}_u$ denotes the individual constellation set for the $u$th UE (constrained to constant modulus to facilitate the use of the differential modulation) and $\varpi_u$ of size $(M_u \times 1)$ denotes the bit mapping policy for the $u$th UE which satisfies that $[\varpi_u]_i \in \{1, \ldots, M_u\}, 1 \le i \le M_u, [\varpi]_i \ne [\varpi]_{i'}, \forall i \ne i'$. We define $\Pi = \left[\varpi_1^T \ \cdots \ \varpi_U^T\right]^T$ a vector of size $\left(\sum_{u=1}^{U} M_u \times 1\right)$ that contains the bit mapping policies of all UEs. The complex symbols of each UE are differentially encoded and mapped in the OFDM symbol as described in (3) and transmitted to the wireless channel using an OFDM system.

At the BS, the received signal at $k$th subcarrier in the $n$th OFDM symbol can be described as

$$\tilde{\mathbf{y}}_{k,n} = \mathbf{H}_{k,n}\beta\tilde{\mathbf{x}}_{k,n} + \tilde{\mathbf{w}}_{k,n}, \quad \beta = \mathrm{diag}\left(\left[\sqrt{\beta_1}, \cdots, \sqrt{\beta_U}\right]\right), \tag{30}$$

where $\tilde{\mathbf{x}}_{k,n} = \left[\tilde{x}_{k,n,1}, \cdots, \tilde{x}_{k,n,U}\right]^T$, $\tilde{\mathbf{w}}_{k,n} = \left[\tilde{w}_{k,n,1}, \cdots, \tilde{w}_{k,n,U}\right]^T$, and $\beta_u$ represents the ratio of the received average power of the $u$th UE, with $1 \le \beta_u \le \beta_{\max}$. This ratio is directly proportional to the combination of the large-scale channel effects and the power control employed by each user. The design of constellations takes into account the impact of varying $\beta_u$ values on the performance of each user. To prevent significant performance differences between users, a maximum value of $\beta_{\max}$ is considered.

Again, the phase difference of two consecutive symbols received at each antenna is non-coherently detected as

$$\tilde{z}_{k,n} = \frac{\left(\tilde{\mathbf{y}}_{k,n-1}\right)^H \tilde{\mathbf{y}}_{k,n}}{R} = \frac{1}{R}\left(\tilde{\mathbf{x}}_{k,n-1}\right)^H \beta\left(\mathbf{H}^{n-1}\right)^H \mathbf{H}\beta\tilde{\mathbf{x}}_{k,n} + \frac{1}{R}\left(\tilde{\mathbf{x}}_{k,n-1}\right)^H \beta\left(\mathbf{H}^{n-1}\right)^H \tilde{\mathbf{w}}_{k,n} + \frac{1}{R}\left(\tilde{\mathbf{w}}_{k,n-1}\right)^H \mathbf{H}_{k,n}\beta\tilde{\mathbf{x}}_{k,n} + \frac{1}{R}\left(\tilde{\mathbf{w}}_{k,n-1}\right)^H \tilde{\mathbf{w}}_{k,n}, \tag{31}$$

which is a generalization of (11) to multiple UEs mapped in the constellation domain. For a very large number of antennas, using the asymptotic property of massive SIMO, by making use of the Law of Large Numbers, assuming that $\mathbf{H}_{k,n-1} \approx \mathbf{H}_{k,n}$, we know that $\frac{1}{R}\left(\mathbf{H}_{k,n-1}\right)^H \mathbf{H}_{k,n} \overset{R\to\infty}{\rightarrow} \mathbf{I}_U$, and thus

$$z_{k,n} \overset{R\to\infty}{\rightarrow} \varsigma_{k,n} = \sum_{u=1}^{U} \beta_u s_{k,n,u} \in \mathfrak{M}, \quad M = |\mathfrak{M}| = \prod_u M_u, \tag{32}$$

where the joint-symbol $\varsigma_{k,n}$ is the result of superimposing the symbols sent by the users, where $\mathfrak{M}$ represents the joint-constellation set. **Figure 4** illustrates the joint-constellation set formed by two specific individual constellations, which are designed using the proposed methods. We define $\mathbf{b}_{i,u}$ as a $\left(N_b^u \times 1\right)$ vector containing the bits for the $u$th UE and the $i$th joint-symbol according to the mapping $\Pi$. Furthermore, we define $\mathbf{b}_i = \left[\mathbf{b}_{i,1}^T; \cdots; \mathbf{b}_{i,U}^T\right]^T$ as a $\left(\sum_{u=1}^{U} N_b^u \times 1\right)$ vector containing all the $\mathbf{b}_{i,u}$ vectors for the $i$th joint-symbol of all UEs. The terms of (31) are independent, and their distribution is shown in [18]. Therefore, the conditional PDF of $z_{k,n}$ given the transmitted symbols of each UE can be analytically obtained as a convolution of the PDF of each of the terms. Assuming equiprobable joint-constellation elements, the decision of $\varsigma_{k,n}$ while receiving $z_{k,n}$ can be done using (32) and maximum likelihood detection as

$$\hat{\varsigma}_{k,n,ML} = \arg\max_{\varsigma_{k,n}} \left\{ f\left(z_{k,n} | \varsigma_{k,n}\right) \right\} \in \mathfrak{M}. \tag{33}$$

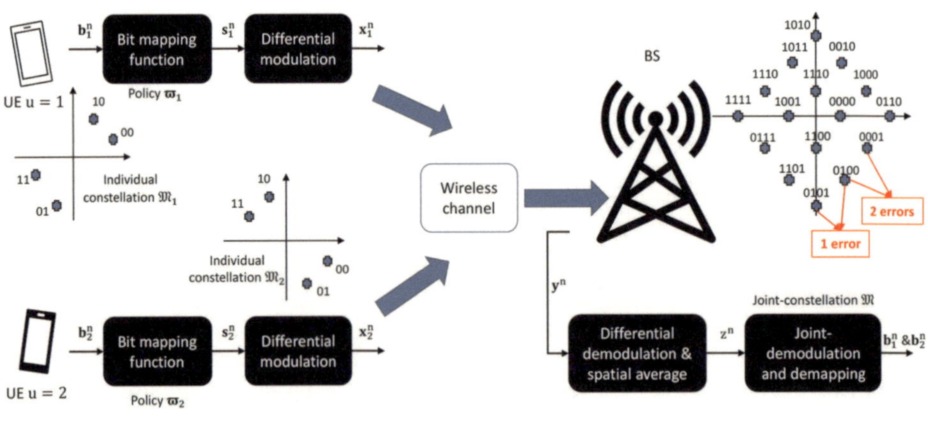


**Figure 4.**
*Block diagram that illustrates the NC scheme in the UL for the specific scenario of $U = 2$, where $\beta_1 = \beta_2 = 1$. The diagram also shows two distinct cases of individual constellations, namely $\mathfrak{M}1$ and $\mathfrak{M}2$. These individual constellations are designed using the proposed methods to generate a QAM joint-constellation denoted as $\mathfrak{M}$.*

Based on the previous analysis, in order to minimize interference among the different elements of the joint-constellation and reduce the symbol error rate (SER) or bit error rate (BER), it is necessary to place them strategically. However, this results in a significant increase in the complexity of the constellation design, as the probability density function (PDF) varies for each joint-symbol depending on the individual constellations. Additionally, even if an optimal joint-constellation is identified, the individual constant modulus constellations must be capable of generating that joint-constellation while also fulfilling individual requirements, which may not be feasible.

One of the most relevant parameters to produce high performance in terms of SER/BER is enlarging the minimum distance between the elements in the joint-constellation. For comparison purposes, it is normalized as $\hat{d}_{\min} = d_{\min}/\sqrt{\sum_{u=1}^{U}\beta_u^2}$. The value of this distance for the typically used constellations [16, 17] is 0.39 for Type A, 0.6325 for Type B, 0.4142 for equally error protection (EEP) and 0.6325 for the Monte Carlo Optimization (MCO). Type A exhibits an exponential reduction in distance as the number of users and/or constellation sizes increase. Type B, on the other hand, is limited to DQPSK and requires specific average receive powers. The normalized minimum distance (NMD) is crucial to performance, as demonstrated in [17], and a larger NMD results in better performance. However, as the number of users $U$ and/or constellation sizes $M_u$ increase, the NMD of the joint-constellation decreases, leading to a decrease in performance. Regular M-QAM joint-constellations maximize the NMD, which can be calculated as $((M-1)/6)^{-1/2}$. Therefore, the minimum distance of any joint-constellation must satisfy $0 < \hat{d}_{\min} \leq ((M-1)/6)^{-1/2}$, with $M$ calculated using (32). Moreover, the distribution of the received symbols around the theoretical values in the joint-constellation depends on the individual constellations chosen by each UE. If the phases of the individual constellation elements that make up the joint-constellation element are similar, the interference power projected on its direction is larger, and vice versa. The interference shapes of the joint-constellation elements are dependent on the individual constellations, and minimizing the effect of interference by altering the joint-constellation shape requires the use of different individual constellations, resulting in a recursive problem in the design process. Additionally, EEP suffers from distance reduction in the inner circle, which is inherent to the constellation definition structure and can even result in a distance of 0 in certain configurations. Consequently, the constellation design problem is mathematically intractable and cannot be solved using classical constellation design techniques.

### 4.2 Multi-user constellation design approaches for NC massive MIMO

Since the constellation design for the multi-user NC massive MIMO scenario implies solving a non-tractable optimization problem, two main approaches have been exploited in the literature, such as the "guess and try" approach and the artificial intelligence techniques specially designed for solving non-convex optimization problems. In the case of multi-user constellations, [16, 17] proposed a small set of suboptimal constellations for the NC based on DMPSK, namely Type A, Type B and EEP. Type A was designed to separate users over sub-quadrants, Type B involved separating elements through power control of the users and EEP placed the constellation elements of each user with a certain phase shift relative to the others. In this sense, these constellations are suboptimal since they do not maximize the probabilistic minimum distance in the joint-constellation and do not focus on any bit mapping policy, which is also critical to minimize the BER. Recently, [18] defined an optimization problem to find the individual constellations and the bit mapping policies that give a proper joint-constellation in terms of BER performance. This is the first constellation design proposal for NC massive MIMO multi-user constellations that is based on evolutionary computation algorithms (a subfield of artificial intelligence techniques) to solve a mathematically intractable problem.

The optimization problem of finding the best individual constellations that result in an optimal joint-constellation and bit mapping policy is mathematically intractable and thus we utilize evolutionary computation algorithms [28] to solve them. We propose using the MCO, where no assumptions on the joint-constellation shape are considered and the bit-mapping policy is co-designed together with the joint-constellation shape. MCO defines a single optimization problem capable of providing the individual constellations and the bit mapping policy of all UEs at once. It is based on the Monte Carlo method to numerically evaluate the performance in terms of BER of the candidates at each iteration. The MCO optimization problem is expressed as

$$\begin{aligned} \min_{\tilde{\mathbf{c}}_u, \beta} \quad & \alpha_1 \sum_{u=1}^{U} [\varepsilon]_u + \alpha_2 \sum_{u=1}^{U} \beta_u, \quad \text{where} \quad \varepsilon = g_M(\sigma_w^2, R, \Pi, \beta, \hat{\mathbf{c}}, N_s, N_r) \\ \text{s.t.} \quad & \left| [\tilde{\mathbf{c}}_u]_{i_u} \right|^2 = 1, \quad 0 \le \angle\left( [\tilde{\mathbf{c}}_u]_{i_u} \right) < 2\pi, \quad u = 1, \cdots, U; \; i_u = 1, \cdots, M_u; \\ & 1 \le \beta_u \le \beta_{\max}, \quad [\hat{\mathbf{c}}] = [\tilde{\mathbf{c}}_1, \cdots, \tilde{\mathbf{c}}_U]^T, \quad \alpha_1 + \alpha_2 = 1, \quad \varpi_u \in \mathfrak{B}_u, \end{aligned} \tag{34}$$

where $\varepsilon$ is a vector of size $(U \times 1)$ that contains the BER of each UE and $g_M(\cdot)$ denotes a function to obtain this BER for a particular set of system parameters. These system parameters are $\Pi$ which is a bit mapping policy for the individual constellations, $N_r$ and $N_s$ are the number of iterations and the number of symbols of the Monte Carlo simulation. This optimization problem is non-convex and NP-hard, so we propose solving it again by using numerical methods based on EC [28]. **Figure 5** provides a block diagram of the implementation of MCO, where $N_G$ is the number of generations and $N_P$ is the population size of the EC algorithm. The interested reader is referred to [18] for more explanations of the MCO.

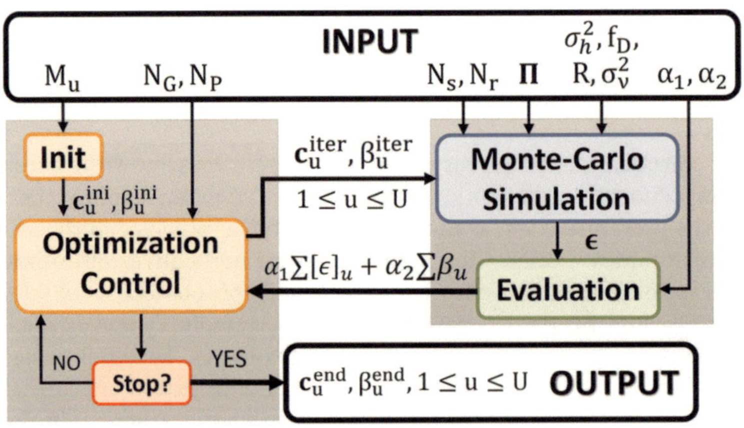


**Figure 5.**
*Block diagram of the MCO.*

### 4.3 Proposed multi-user constellations

We provide a set of optimized constellations in ([18], Table II). While each constellation has been determined for a certain $R$ and $\rho$, it can be used for any values in a realistic range. To read the table, for each scenario, there are $U$ vectors of the form $\Phi = \left[\Phi_1^u \Phi_2^u \cdots \Phi_{M_u}^u\right]$, where $\Phi_{m_u}^u$ is the phase in radians for the constellation element $m_u$ of user $u$ ($1 \leq m_u \leq M_u$, $1 \leq u \leq U$, where $M_u$ is the constellation size of user $u$). A constellation element $m_u$ of user $u$ can be found as $s_{m_u}^u = \exp j\Phi_{m_u}^u$. The mapping of element $m_u$ is obtained with a decimal to the binary conversion of $m_u - 1$.

## 5. Comparison among coherent, non-coherent and hybrid schemes in massive MIMO

As mentioned before, CDS and NCDS have their benefits and limitations since CDS is suitable for slowly varying and high SNR scenarios, while NCDS is recommendable in the opposite scenarios. Comparatively, CDS can provide high throughput to many users while the NCDS can provide a lower throughput for fewer users, but working in scenarios where the CDS would fail. Consequently, HDS is also proposed in [13], where it is capable of trading-off both CDS and NCDS in order to get the benefits of each scheme, at the expense of a little increment in the channel estimation error. Here we provide a comparison in terms of throughput between the HDS and the CDS for different time and frequency variability. Specifically, we show the percentage improvement in the throughput of the HDS with respect to the CDS for the different required number of pilots in each dimension time ($N_p$) and frequency ($K_p$) for 14 OFDM symbols and 12 subcarrier frequencies (**Table 1**).

In **Figure 6a**, a comparison between the coherent (CDS), non-coherent (NCDS), superimposed training (ST, [29]) and hybrid scheme (HDS, [13]) is shown. It can be seen that the HDS outperforms all the other alternatives in fast-varying channels for all SNR ranges. Additionally, we compare the performance of the coherent and the NC massive MIMO for the DL approach including spatial multiplexing proposed in [15]. This approach blindly estimates the channel using reconstructed differential data in the uplink. We can see that the proposed scheme (N) works better than the coherent scheme (C) in case the coherence time $n_c$ is smaller than 2 times the TDD slot duration. In scenarios where the coherence time is 1.5 times the DL slot duration, even with channel prediction, the coherent scheme performs worse than the proposed scheme. This can be seen in curves C,6,∞,cP and N,6,∞. The reason for this is that the proposed scheme is much more robust than the coherent scheme in these situations.

| $N_p \parallel K_p$ | 1 | 2 | 3 | 4 | 6 | 12 |
|---|---|---|---|---|---|---|
| **1** | 0% | 0.5% | 0.9% | 1.4% | 2.3% | 5.3% |
| **2** | 0% | 0.9% | 1.9% | 2.8% | 4.8% | 11.5% |
| **4** | 0% | 1.9% | 3.8% | 5.9% | 10.4% | 27.5% |
| **7** | 0% | 3.4% | 7.1% | 11.2% | 20.8% | 68.7% |
| **14** | 0% | 7.5% | 16.7% | 28.1% | 62.5% | ∞ |

**Table 1.**
*Percentage improvement of the throughput for the HDS with respect to the CDS.*

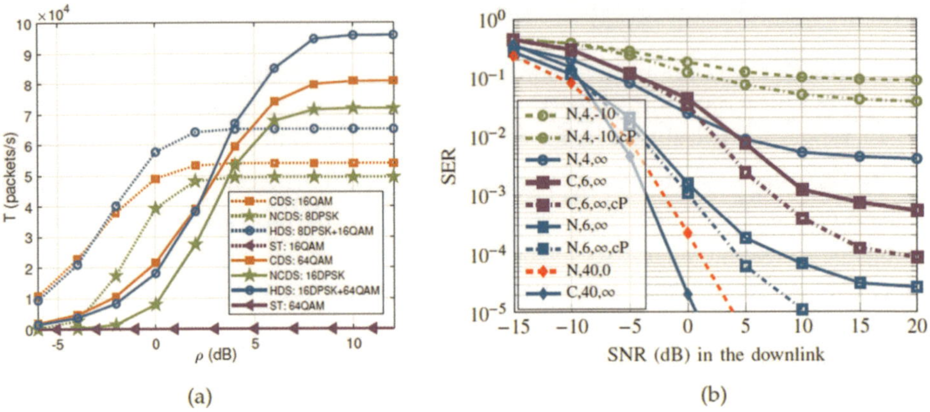


**Figure 6.**
*Throughput comparison of CDS, HDS, ST and NCDS for different constellation sizes, $R = 64$, $K_p = 6$ and $N_p = 7$ (left) and (right) SER comparison between classical (C, dashed) and proposed (N, continuous) schemes in the DL, labeled from left to right with the legend written as "technique (N,C), $n_c$ (4,6,40) coherence time, SNR (dB) uplink for channel estimation" for $R = 100$ antennas, $\tau_d = 4$ DL time slot, $M_{DL} = 4$ DL constellation size and 2 users. cP refers to the inclusion of channel prediction. (a) Throughput comparison of CDS and NCDS and (b) SER comparison between C and NC.*

We now consider a multi-path time-varying channel and an implementation with OFDM modulation according to the 5G new radio numerology. To obtain these results, the coherence time is calculated as $T_c = 0.15 f_D^{-1}$, where $f_D$ is the maximum Doppler frequency. We also consider that the duration of an OFDM symbol is the inverse of the separation between subcarriers $T_s = 1/\Delta_f$. In [13], the coherent scheme employs channel estimation based on zero-forcing with PSAM. The results, which are shown in **Figure 6**, are based on multi-path channels with a delay spread ($\sigma_\tau < 1\ \mu s$), resulting in a minimum coherence bandwidth of $B_c \approx 1/(5\sigma_\tau) = 200$ kHz. In the NC scheme, differential encoding is performed over the frequency domain [19], and 4 out of 14 OFDM symbols are dedicated to reference signals for each slot, following the 5G

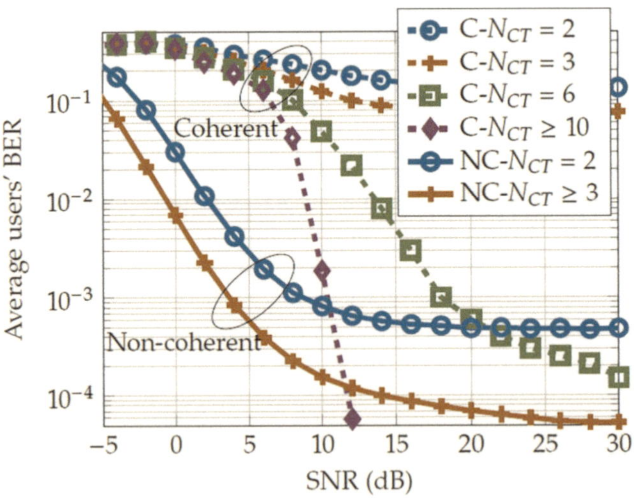


**Figure 7.**
*Non-coherent ($M_u = [4\ \ 4]$ and $\beta_u = [1\ \ 1]$) ([18], Table II) vs. coherent scheme (2 users with regular QPSK) for R=128, for different $N_{CT}$.*

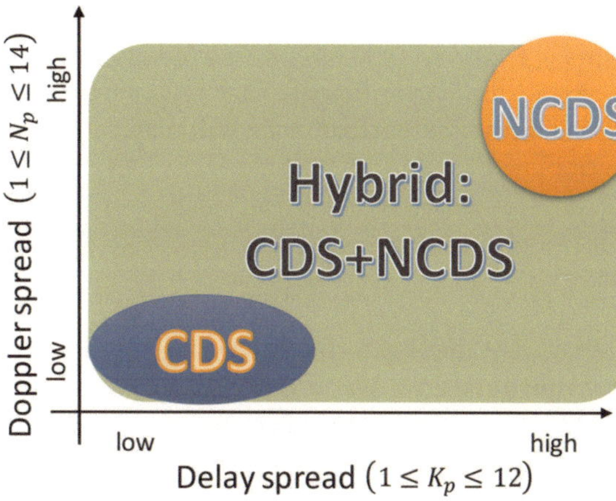


**Figure 8.**
*CDS, NCDS or HDS depending on channel variability as in **Figure** 2. Image taken from [13].*

standard. Due to channel estimation overhead, the SNR ($\rho$) for the coherent scheme is penalized as $10\rho/14$. The NC outperforms the coherent scheme for high $\rho$, except for $N_{CT} \geq 10$. Moreover, for all $\rho$ values, the NC outperforms the coherent scheme when $N_{CT} \leq 5$. In addition, even for large $N_{CT}$, the NC outperforms the coherent counterpart in the low $\rho$ regime (**Figures** 7 and **8**).

## 6. Conclusions

This chapter has provided a review of non-coherent massive MIMO based on DMPSK, which leverages the advantage of using an huge number of antennas in the BS either by not using requiring or by obtaining this CSI without transmitting any reference signals. In the case of UL, three different mapping schemes have been proposed for the OFDM. Additionally, a blind channel estimation using reconstructed differentially encoded data has been also proposed. In the case of DL, two proposals are given, one for FDD and the other for TDD. The first one corresponds to a precoding based on either beamforming or codebook selection, while the second one accounts for a precoding based on the channel estimated in the UL. Additionally, we have indicated how the multi-user version of the NC massive MIMO based on DMPSK can be implemented via constellation design. Lastly, a comparison of the coherent, non-coherent and hybrid schemes in terms of performance is provided to demonstrate that the NC alternative is better for the scenarios with a high variability in time and/or frequency, with a low SNR and with many users.

Moreover, it has been observed that the performance of NCDS is highly dependent on the spatial separation of the multiplexed UEs, whether in terms of constellation or space. Hence, scheduling algorithms that optimize a specific performance metric while considering this factor are crucial. While NCDS outperforms CDS in dynamic channel scenarios with moderate SNR and a large number of users, it becomes less advantageous in quasi-static channels, high SNR, or a small number of users. Therefore, hybrid schemes that combine both paradigms, such as the one proposed in [13], are recommended for such scenarios.

Furthermore, the integration of sensing with communication is one of the main goals of 6G mobile communications [3]. In these systems, efficient CSI exploitation under various scenarios will be crucial, and hence, the use of non-coherent techniques to create hybrid systems is expected to be an interesting alternative to increase overall system efficiency. In conclusion, we anticipate that this review of NCDS characteristics, implementation feasibility and performance will inspire new research and advancements in this field.

## Acknowledgements

This work has received funding from the European Union Horizon 2020 research and innovative programme under the Marie skÅ,odowska-Curie ETN TeamUp5G, grant agreement No. 813391 and the Spanish National project IRENE-EARTH (PID2020-115323RB-C33/AEI/10.13039/501100011033).

## Conflict of interest

The authors declare no conflict of interest.

## Abbreviations

| | |
|---|---|
| 5G | fifth generation |
| 6G | sixth generation |
| AWGN | additive white Gaussian noise |
| BER | bit error rate |
| BS | base station |
| CP | cyclic prefix |
| CSI | channel state information |
| CSI-RS | channel state information-reference signals |
| DL | downlink |
| DMPSK | differential $M$-ary phase shift keying |
| DSP | digital signal processing |
| EEP | equally error protection |
| ETN | Educational and Training Network |
| HDS | hybrid demodulation scheme |
| ISI | inter-symbol interference |
| IDFT | inverse discrete Fourier transform |
| ICI | inter-carrier interference |
| IoT | Internet of Things |
| LS | least squares |
| MCO | Monte Carlo Optimization |
| MIMO | multiple-input multiple-output |
| MMSE | minimum mean squared error |
| mMTC | massive machine type communications |
| MRT | maximum ratio transmission |
| NCDS | non-coherent demodulation scheme |
| OFDM | orthogonal frequency division multiplexing |
| PSAM | pilot symbol assisted modulation |
| SER | symbol error rate |

| | |
|---|---|
| SINR | signal-to-noise and interference ratio |
| SNR | signal-to-noise ratio |
| SS | synchronization signals |
| ST | superimposed training |
| TDD | time division duplexing |
| UAV | unmanned aerial vehicles |
| UC3M | University Carlos III de Madrid |
| UE | user equipment |
| UL | uplink |
| ZF | zero forcing |

## Author details

Manuel José López Morales*, Kun Chen-Hu and Ana García Armada
Universidad Carlos III de Madrid, Leganés, Madrid, Spain

*Address all correspondence to: mjlopez@tsc.uc3m.es

## References

[1] Lu L, Li GY, Swindlehurst AL, Ashikhmin A, Zhang R. An overview of massive MIMO: Benefits and challenges. IEEE Journal on Selected Topics in Signal Processing. 2014;**8**(5):742-758

[2] Evolved Universal Terrestrial Radio Access (E-UTRA); Physical channels and modulation, 3GPP Std. 36.211, 2017

[3] Letaief KB, Shi Y, Lu J, Lu J. Edge artificial intelligence for 6g: Vision, enabling technologies, and applications. IEEE Journal on Selected Areas in Communications. 2022;**40**(1):5-36

[4] Masaracchia A, Sharma V, Canberk B, Dobre OA, Duong TQ. Digital twin for 6G: Taxonomy, research challenges, and the road ahead. IEEE Open Journal of the Communications Society. 2022;**3**:2137-2150

[5] Nguyen DC, Ding M, Pathirana PN, Seneviratne A, Li J, Niyato D, et al. 6G internet of things: A comprehensive survey. IEEE Internet of Things Journal. 2022;**9**(1):359-383

[6] 5G-NR; Physical channels and modulation (Release 17), 3GPP Std. 38.211, 2022

[7] Truong KT, Heath RW. Effects of channel aging in massive MIMO systems. Journal of the Communications Network. 2013;**15**(4):338-351

[8] Chowdhury M, Manolakos A, Goldsmith AJ. Coherent versus noncoherent massive SIMO systems: Which has better performance? In: 2015 IEEE International Conference on Communications (ICC), London, UK. 2015. pp. 1691-1696. DOI: 10.1109/ICC.2015.7248568

[9] Cavers JK. An analysis of pilot symbol assisted modulation for Rayleigh fading channels (mobile radio). IEEE Transactions on Vehicular Technology. 1991;**40**(4):686-693

[10] Lin J. Least-squares channel estimation for mobile OFDM communication on time-varying frequency-selective fading channels. IEEE Transactions on Vehicular Technology. 2008;**57**(6):3538-3550

[11] Hedayat A, Nosratinia A. Outage and diversity of linear receivers in flat-fading MIMO channels. IEEE Transactions on Signal Processing. 2007;**55**(12):5868-5873

[12] De Figueiredo FAP, Cardoso FACM, Moerman I, Fraidenraich G. Channel estimation for massive MIMO TDD systems assuming pilot contamination and frequency selective fading. IEEE Access. 2017;**5**:17733-17741

[13] Lopez-Morales MJ, Chen-Hu K, Garcia-Armada A. Differential data-aided channel estimation for up-link massive SIMO-OFDM. IEEE Open Journal of the Communications Society. 2020;**1**:976-989

[14] Elijah O, Leow CY, Rahman TA, Nunoo S, Iliya SZ. A comprehensive survey of pilot contamination in massive mimo—5g system. IEEE Communications Surveys & Tutorials. 2015;**18**(2):905-923

[15] Lopez-Morales MJ, Garcia-Armada A. Channel estimation and prediction in a pilot-less massive mimo tdd using non-coherent dmpsk. IEEE Access. 2022;**10**: 112327-112341

[16] Baeza VM, Armada AG, Zhang W, El-Hajjar M, Hanzo L. A noncoherent multiuser large-scale SIMO system

relying on M-ary DPSK and BICM-ID. IEEE Transactions on Vehicular Technology. 2018;**67**(2):1809-1814

[17] Armada AG, Hanzo L. A non-coherent multi-user large scale SIMO system relaying on M-ary DPSK. In: 2015 IEEE International Conference on Communications (ICC), London, UK. 2015. pp. 2517-2522. DOI: 10.1109/ICC.2015.7248703

[18] Lopez-Morales MJ, Chen-Hu K, Garcia-Armada A, Dobre OA. Constellation design for multiuser non-coherent massive simo based on dmpsk modulation. IEEE Transactions on Communications. 2022;**70**(12):8181-8195

[19] Chen-Hu K, Liu Y, Armada AG. Non-coherent massive MIMO-OFDM down-link based on differential modulation. In: IEEE Transactions on Vehicular Technology. Vol. 69, No. 10. Oct 2020. pp. 11281-11294. DOI: 10.1109/TVT.2020.3008913

[20] Chen-Hu K, Armada AG. Non-coherent multiuser massive MIMO-OFDM with differential modulation. In: ICC 2019 - 2019 IEEE International Conference on Communications (ICC), Shanghai, China. 2019. pp. 1-6. DOI: 10.1109/ICC.2019.8761447

[21] Morales MJL, Chen-Hu K, Armada G. "Pilot-less massive MIMO TDD system with blind channel estimation using non-coherent DMPSK," in 2022 IEEE Global Communications Conference (GLOBECOM). 2022. pp. 1-6

[22] Hwang T, Yang C, Wu G, Li S, Li GY. OFDM and its wireless applications: A survey. IEEE Transactions on Vehicular Technology. 2009;**58**(4):1673-1694

[23] Corvaja R, Armada AG. Phase noise degradation in massive MIMO downlink with zero-forcing and maximum ratio transmission precoding. IEEE Transactions on Vehicular Technology. 2016;**65**(10):8052-8059

[24] Ghozlan H, Kramer G. Models and information rates for wiener phase noise channels. IEEE Transactions on Information Theory. 2017;**63**(4):2376-2393

[25] Chen-Hu K, Liu Y, Armada AG. Non-coherent massive MIMO-OFDM for communications in high mobility scenarios. ITU Journal on Future and Evolving Technologies. 2020;**1**(1):13-24

[26] Cabrejas J, Roger S, Calabuig D, Fouad YMM, Gohary RH, Monserrat JF, et al. Non-coherent open-loop MIMO communications over temporally-correlated channels. IEEE Access. 2016;**4**:6161-6170

[27] Mi D, Dianati M, Zhang L, Muhaidat S, Tafazolli R. Massive mimo performance with imperfect channel reciprocity and channel estimation error. IEEE Transactions on Communications. 2017;**65**(9):3734-3749

[28] Sloss AN, Gustafson S. 2019 Evolutionary algorithms review. arXiv preprint arXiv:1906.08870. 2019

[29] Estrada-Jiménez JC, Chen-Hu K, García MJF, García Armada A. "Power allocation and capacity analysis for FBMC-OQAM with superimposed training." IEEE Access. vol. 7. 2019. pp. 46968-46976

Chapter 6

# Spatial Multiplexing for MIMO/Massive MIMO

*Haonan Wang and Ang Li*

## Abstract

In this chapter, we will discuss how to achieve spatial multiplexing in multiple-input multiple-output (MIMO) communications through precoding design, for both traditional small-scale MIMO systems and massive MIMO systems. The mathematical description for MIMO communications will first be introduced, based on which we discuss both block-level precoding and the emerging symbol-level precoding techniques. We begin with simple and closed-form block-level precoders such as maximum ratio transmission (MRT), zero-forcing (ZF), and regularized ZF (RZF), followed by the classic symbol-level precoding schemes such as Tomlinson-Harashima precoder (THP) and vector perturbation (VP) precoder. Subsequently, we introduce optimization-based precoding solutions, including power minimization, SINR balancing, symbol-level interference exploitation, etc. We extend our discussion to massive MIMO systems and particularly focus on precoding designs for hardware-efficient massive MIMO systems, such as hybrid analog-digital precoding, low-bit precoding, nonlinearity-aware precoding, etc.

**Keywords:** MIMO, massive MIMO, spatial multiplexing, precoding, beamforming

## 1. Introduction

In recent years, the demand for high-speed wireless communication has grown exponentially, driven by the proliferation of smart devices, the Internet of Things (IoT), and the increasing need for reliable and efficient data transmission [1]. To meet these demands, multiple-input multiple-output (MIMO) technology has emerged as a promising solution, offering significant improvements in spectral efficiency, capacity, and reliability. In this chapter, we will explore the concept of spatial multiplexing in MIMO communications, focusing on precoding design for both traditional small-scale MIMO systems and massive MIMO systems.

MIMO communication systems employ multiple antennas at both the transmitter and receiver ends to exploit the spatial domain, enabling the simultaneous transmission of multiple data streams over the same frequency band [2]. This spatial multiplexing capability is the key factor in achieving the high data rates and improved link reliability that MIMO systems offer. Precoding is a crucial technique in MIMO communications, as it allows the transmitter to pre-process the signals before transmission, effectively mitigating inter-stream interference and optimizing the received signal quality. We will begin our discussion with a mathematical description of MIMO communications, providing a solid foundation for understanding the principles and techniques involved in precoding design. Based on this mathematical framework, we will dive deep into both block-level precoding and the emerging symbol-level precoding technique.

Block-level precoding techniques, such as maximum ratio transmission (MRT), zero-forcing (ZF), and regularized ZF (RZF), offer simple and closed-form solutions for mitigating inter-stream interference. These methods have been widely adopted in small-scale MIMO systems due to their ease of implementation and relatively low computational complexity. We will also discuss classic symbol-level precoding schemes, including the Tomlinson-Harashima precoder (THP) and vector perturbation (VP) precoder, which offer improved performance by exploiting the inherent structure of the transmitted symbols. As we move beyond these basic precoding techniques, we will introduce optimization-based precoding solutions that aim to further enhance the performance of MIMO systems. These approaches include power minimization, SINR balancing, and symbol-level interference exploitation, among others. By optimizing various performance metrics, these advanced precoding techniques can achieve significant gains in spectral efficiency and link reliability.

In the latter part of the chapter, we will extend our discussion to massive MIMO systems, which employ a large number of antennas at the transmitter and receiver to achieve even greater spatial multiplexing gains. While the basic principles of precoding design remain applicable to massive MIMO systems, the increased scale and complexity of these systems introduce new challenges and opportunities for precoding optimization. In particular, we will focus on precoding designs for hardware-efficient massive MIMO systems, such as hybrid analog-digital precoding, low-bit precoding, and nonlinearity-aware precoding. These techniques aim to address the practical limitations of massive MIMO systems, including hardware constraints, power consumption, and implementation complexity, while still achieving desired performance gains.

In conclusion, this chapter will provide a comprehensive overview of spatial multiplexing in MIMO communications, with a focus on precoding design for both small-scale and massive MIMO systems. By exploring a wide range of precoding techniques, from simple closed-form solutions to advanced optimization-based approaches, we aim to offer the reader a deep understanding of the principles and methods involved in achieving high-performance MIMO communications.

## 2. Body of the manuscript

In Section 3, we will provide an introduction to the MIMO communication system, which will include a mathematical description of the MIMO system, performance metrics of MIMO communications, and emerging massive MIMO techniques. In Section 4, we will explain traditional precoding design, which will include preliminaries on precoding and classical precoding schemes. Subsequently, in Section 5, we will discuss optimization-based precoding to demonstrate the use of convex optimization in precoding design. Finally, in recognition of the wide application of massive MIMO, Section 6 will introduce hardware-efficient precoding as a means of achieving a favorable balance between communication performance and power consumption.

## 3. MIMO communication systems

Due to the increasing demand for higher data rates and reliability for wireless networks, MIMO techniques have appeared and received extensive research attention. To support spatial multiplexing, parallel data streams can be transmitted simultaneously with multiple antennas deployed at the BS. To improve reliability, space-time coding techniques can be employed by sending copies of the same information across the antenna array. In this section, we present an overview of the fundamental concepts of multi-antenna technology, which serves as a foundation for the subsequent discussion on precoding. Given that spatial multiplexing is the primary focus of this chapter, our attention is primarily directed toward multi-user multi-input single-output (MU-MISO) systems.

### 3.1 Mathematical description for MIMO communications

In a wireless multi-user MISO (MU-MISO) system, as depicted in **Figure 1**, the data symbol vector is denoted as **s**, and one BS with $N_t$ antennas transmits wireless signals to $K$ single-antenna receivers. Mathematically, the signal vector at the receiver can be expressed as.where $h_{i,j}$ denotes the complex channel gain between the $i$-th receiver and the $j$-th transmit antenna, $x_j$ denotes the transmit signal on the $j$-th transmit antenna, $y_i$ denotes the received signal of the $j$-th receiver, and $n_i$ denotes the additive Gaussian noise corresponding to the $i$-th receiver. Based on that, the $k$-th user's received signal can be expressed as

$$\begin{bmatrix} y_1 \\ y_2 \\ \vdots \\ y_K \end{bmatrix} = \begin{bmatrix} h_{1,1} & h_{1,2} & \cdots & h_{1,N_t} \\ h_{2,1} & h_{2,2} & \cdots & h_{2,N_t} \\ \vdots & \vdots & \ddots & \vdots \\ h_{K,1} & h_{K,2} & \cdots & h_{K,N_t} \end{bmatrix} \begin{bmatrix} x_1 \\ x_2 \\ \vdots \\ x_{N_t} \end{bmatrix} + \begin{bmatrix} n_1 \\ n_2 \\ \vdots \\ n_K \end{bmatrix}, \tag{1}$$

$$y_k = \mathbf{h}_k^{\mathrm{T}}\mathbf{x} + n_k, \tag{2}$$

where $y_k$ denotes the $k$-th user's received signal, $\mathbf{h}_k \in \mathbb{C}_{N_t\times 1}$ denotes the $k$-th user's channel vector, $\mathbf{x} \in \mathbb{C}_{N_t\times 1}$ denotes the transmit signal vector, and $n_k$ denotes the additive noise vector which follows the complex Gaussian distribution $\mathbb{CN}(0, \sigma_k^2\mathbf{I})$ with the zero mean and $\sigma_k^2$ noise power. The combining process is eliminated at the receiver side, for the single-antenna configuration. Based on (2), the transmission process in MU-MISO can be reorganized into a matrix form, as shown below:

$$\mathbf{y} = \mathbf{Hx} + \mathbf{n}, \tag{3}$$

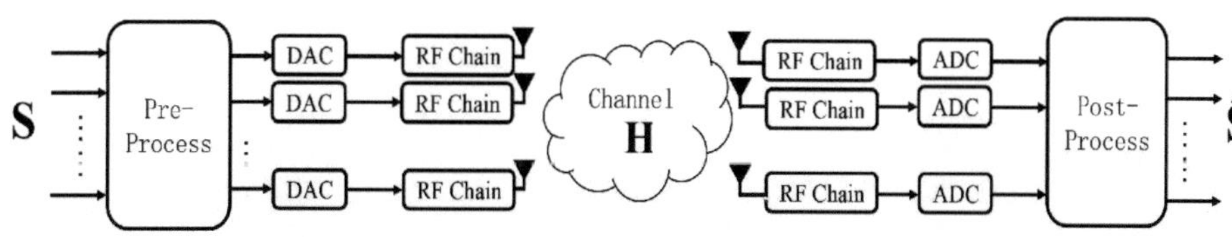


**Figure 1.**
*A block diagram of MU-MISO systems.*

with $\mathbf{y} = [y_1, y_2, \dots, y_K]^{\mathrm{T}}, \mathbf{H} = [\mathbf{h}_1, \mathbf{h}_2, \dots, \mathbf{h}_K]^{\mathrm{T}}$, and $\mathbf{n} = [n_1, n_2, \dots, n_K]^{\mathrm{T}}$.

To mitigate the detrimental impact of channel fading, the transmitter performs precoding on the symbol vector to obtain the transmitted signal, expressed as $\mathbf{x} = \mathbf{Ws}$. Precoding is achieved using a matrix $\mathbf{W} \in \mathbb{C}_{N_t \times K}$. The design of the precoding matrix $\mathbf{W}$ is the crucial signal processing procedure in MIMO downlink transmission, as it enables each receiver to achieve a received signal $y_k$ that closely approximates the original symbol $s_k$.

## 3.2 Performance metrics for MIMO communications

In order to measure the communication performance of MIMO systems, bit error rate (BER) and channel capacity are the two performance metrics that are usually employed, as explained below.

### *3.2.1 BER*

Bit Error Rate (BER) refers to the proportion of erroneously transmitted bits to the total number of transmitted bits during the transmission process and is the most commonly used performance metric to evaluate the reliability of digital communication systems. Its mathematical definition can be given as

$$P_b = \frac{N_e}{N_b}, \tag{4}$$

where $N_e$ denotes the erroneous transmitted bits, and $N_b$ denotes the total transmitted bits.

### *3.2.2 Channel capacity*

The channel capacity represents the maximum rate of information transmission that can be sustained by a communication system when the bit error rate approaches zero. Its mathematical definition is given as the maximum mutual information between the input and output signals of the channel, which represents the extent to which the received signal preserves information about the transmitted signal after the channel. More specifically, the channel capacity is determined by identifying the input distribution that maximizes the mutual information, subject to the constraints of the channel's physical properties and the power limitations of the system. Therefore, it serves as a fundamental limit on the data transmission rate and is a crucial performance metric for evaluating the effectiveness of communication systems. The definition of channel capacity can be expressed as

$$\mathrm{C} = \max \mathrm{I}\,(\text{input}; \text{output}), \tag{5}$$

where C denotes the channel capacity, and $\mathrm{I}\,(x; y)$ denotes the mutual information between $x$ and $y$. For SISO systems, when both the transmitter and receiver have perfect Channel State Information (CSI), the channel capacity can be obtained as

$$\mathrm{C} = B \log_2 (1 + \gamma), \tag{6}$$

where $B$ denotes the system bandwidth, and $\gamma$ denotes the receive SNR. The physical interpretation of (8) has been discussed in ref. [2].

In the context of MIMO systems, it is feasible to decompose the channel into a sum of multiple SISO channels via singular value decomposition (SVD) [2]. Subsequently, utilizing "water-filling" power allocation strategy [2], it is possible to harness the full potential of the system and achieve channel capacity. In an ideal scenario where both the transmitter and receiver possess perfect CSI, the channel capacity of an $N_r \times N_t$ MIMO channel can be captured precisely using the following equation:

$$C = \log_2 \det\left(\mathbf{I}_{N_r} + \frac{\rho}{N_r}\mathbf{H}\mathbf{H}^{\mathrm{H}}\right), \tag{7}$$

where $\rho$ denotes the transmit SNR.

### 3.3 Massive MIMO

As mobile communication technologies continue to evolve, wireless network capacity and communication quality have become increasingly critical. Traditional wireless communication systems face limitations that prevent them from satisfying the modern industry's demands for high-speed, high-capacity, and high-quality communication. Massive MIMO technology has emerged as a promising solution to these challenges.

Massive MIMO is an extension of conventional MIMO technology [3, 4]. In contrast to the typical tens-of-antenna configuration in traditional MIMO systems for signal transmission and reception, Massive MIMO employs significantly more antennas, for example, hundreds or even thousands of antennas.

Massive MIMO technology enjoys wide applications in various fields of wireless communications, such as 5G and IoT [5]. It has several notable features: channel hardening, favorable propagation, power concentration, capacity enhancement, interference reduction, and spectral efficiency improvement. In particular, channel hardening refers to the property that as the antenna array size increases, the relative fluctuations of channel coefficients decrease [5]. Although randomness still exists, its impact on communication approximates that of non-fading channels. Favorable propagation is a phenomenon in which the channels of different users become nearly orthogonal in the spatial domain as the number of antennas at the base station increases significantly. This leads to a substantial reduction in inter-user interference and further improved spectral efficiency, making massive MIMO a promising technology for future wireless communication systems. Power concentration refers to Massive MIMO's ability to focus transmitted power more efficiently through finer beamforming techniques, especially for millimeter-wave communication where channel gain drops off precipitously with distance [6]. Capacity enhancement is achieved by processing more data streams than traditional MIMO systems, leading to improved network capacity. Interference reduction is accomplished through spatial multiplexing and beamforming, which minimize inter-signal interference and enhance signal quality and reliability. Last, spectral efficiency improvement results from more efficient utilization of bandwidth resources, which enhances data transmission speeds.

However, Massive MIMO technology still faces certain challenges in engineering applications, such as high power consumption [7] and hardware costs. To be more specific, traditional MIMO systems equip each antenna with radio frequency (RF) chains and high-resolution digital-to-analog converters (DACs), causing significant power loss when the antenna array is large. In such a scenario, the advanced signal processing mechanisms required to handle a large number of antennas for signal transmission and reception are generally more complex, necessitating much more energy consumption than traditional wireless communication systems. From this perspective, hardware-efficient precoding techniques hold significant research value and promising application prospects.

## 4. Traditional precoding

In this section, we will introduce traditional precoding to discuss its working mechanism and design principle. Preliminaries will be first introduced, as the basis of further discussion. Based on that, we mainly introduce the linear block-level precoding schemes with closed-form solutions, including MRT, ZF, and RZF. After that, the traditional non-linear symbol-level precoding will be discussed, including THP and VP.

### 4.1 Preliminaries on precoding

First, we will introduce the preliminaries of the precoding process in the downlink MIMO system, as the basis of further discussion.

Without loss of generality, we mainly consider a downlink MU-MISO system, where $K$ single-antenna users are served by a common base station with $N_t$ transmit antennas at the same time. Considering that users are generally separated spatially, based on CSI, the BS needs to employ signal processing techniques before transmission such that the destructive effect of channel fading and inter-user interference can be eliminated as much as possible. This is the initial motivation for precoding. Mathematically, the precoding process can be expressed as

$$\mathbf{x} = \sum_{k=1}^{K} \mathbf{w}_k s_k = \mathbf{W}\mathbf{s}, \tag{8}$$

where $\mathbf{w}_k \in \mathbb{C}_{N_t \times 1}$ denotes the $k$-th user's precoding vector and $s_k$ is the $k$-th user's data symbol, which is drawn from a specific modulation constellation. Based on that, with the general precoding matrix $\mathbf{W} = [\mathbf{w}_1, \mathbf{w}_2, \ldots, \mathbf{w}_K] \in \mathbb{C}_{N_t \times K}$ and date symbol vector $\mathbf{s} = [s_1, s_2, \ldots, s_K]^{\mathrm{T}} \in \mathbb{C}_{K \times 1}$, the received signal for the $k$-th user can be expressed as

$$y_k = \mathbf{h}_k^{\mathrm{T}} \mathbf{x} + n_k = \mathbf{h}_k^{\mathrm{T}} \mathbf{W}\mathbf{s} + n_k, \tag{9}$$

where $y_k$ is the received signal for the $k$-th user, $\mathbf{h}_k \in \mathbb{C}_{N_t \times 1}$ is the complex channel vector between the BS and the $k$-th user, and $n_k \sim \mathbb{CN}(0, \sigma^2)$ is the additive Gaussian noise with zero mean and $\sigma^2$ noise power. Based on that, the transmission process can be given as

$$\mathbf{y} = \mathbf{H}\mathbf{W}\mathbf{s} + \mathbf{n}, \tag{10}$$

where $\mathbf{y} \in \mathbb{C}_{K\times 1}$ denotes the received signal vector, $\mathbf{H} \in \mathbb{C}_{K\times N_t}$ denotes the channel matrix, and $\mathbf{n} \in \mathbb{C}_{K\times 1}$ denotes the additive noise vector.

In traditional communication systems, the presence of interference can significantly degrade the quality of the received signal. This is particularly true in multi-user systems, where signals for different users are superimposed over the spatial channel. In such scenarios, the transmitted signals from different users can interfere with each other, leading to reduced signal quality at the receiver.

The insight of precoding is to design the precoding matrix **W** such that the received signal **y** can approach the data symbol vector **s** as much as possible. In the following subsections, we will introduce linear closed-form block-level precoding, which is a classical type of precoding.

### 4.2 Linear closed-form precoding

The classical linear block-level precoding schemes have been widely used in practical engineering systems since they can ensure satisfactory communication performance with low computational complexity. In this subsection, we will mainly discuss the specific linear closed-form precoding, including MRT, ZF, and RZF, to show the principle of precoding design and the physical mechanism of the precoding effect.

Specifically, the precoding matrix of **MRT** can be given as [4].

$$\mathbf{W}_{\mathrm{MRT}} = \frac{1}{f_{\mathrm{MRT}}} \cdot \mathbf{H}^{\mathrm{H}} = \sqrt{\frac{P_0}{\mathrm{tr}\{\mathbf{H}\mathbf{H}^{\mathrm{H}}\}}}\mathbf{H}^{\mathrm{H}}, \tag{11}$$

where $f_{\mathrm{MRT}} = \sqrt{\frac{\mathrm{tr}\{\mathbf{H}\mathbf{H}^{\mathrm{H}}\}}{P_0}}$ denotes the normalization factor to ensure the satisfaction of the transmit power constraint, and $P_0$ denotes the total transmit power. Considering that MRT can maximize the signal gain at the intended user, its performance is promising in noise-limited scenarios (low SNR regimes or large-scale MIMO scenarios), while its performance is limited in interference-limited scenarios.

**Zero-Forcing (ZF)** precoding is another classical precoding method that has been extensively used in practical applications [8]. By employing a Moore-Penrose inverse of the channel matrix **H** as the precoding matrix, ZF precoding can create an ideal environment where each user's effective channel is orthogonal with each other. Based on that, inter-user interference can be eliminated as much as possible. The ZF precoding matrix can be expressed as

$$\mathbf{W}_{\mathrm{ZF}} = \frac{1}{f_{\mathrm{ZF}}} \cdot \mathbf{H}^{\mathrm{H}}\left(\mathbf{H}\mathbf{H}^{\mathrm{H}}\right)^{-1} = \sqrt{\frac{P_0}{\mathrm{tr}\left\{\left(\mathbf{H}\mathbf{H}^{\mathrm{H}}\right)^{-1}\right\}}}\mathbf{H}^{\mathrm{H}}\left(\mathbf{H}\mathbf{H}^{\mathrm{H}}\right)^{-1}, N_t \geq K, \tag{12}$$

where $f_{\mathrm{ZF}} = \sqrt{\frac{\mathrm{tr}\left\{\left(\mathbf{H}\mathbf{H}^{\mathrm{H}}\right)^{-1}\right\}}{P_0}}$ denotes the normalization factor for ZF precoding. ZF precoding is shown to achieve improved performance over MRT in the high SNR regime. The main idea of ZF precoding is to create orthogonal effective channels among all the users to fully eliminate inter-user interference. For its low computational complexity, ZF precoding has been widely used in practical engineering systems. However, the noise amplification effect limits its performance, especially in low SNR regions, which has been improved by RZF precoding.

By introducing a regularization factor to handle the noise amplification effect, the **RZF** precoding can further improve the performance of ZF precoding [9]. The RZF precoding matrix can be given by

$$\begin{aligned}\mathbf{W}_{\mathrm{RZF}} &= \frac{1}{f_{\mathrm{RZF}}} \cdot \mathbf{H}^H\left(\mathbf{H}\mathbf{H}^H + \alpha \cdot \mathbf{I}\right)^{-1} \\ &= \sqrt{\frac{P_0}{\operatorname{tr}\left\{\left(\mathbf{H}\mathbf{H}^H + \alpha \cdot \mathbf{I}\right)^{-1}\mathbf{H}\mathbf{H}^H\left(\mathbf{H}\mathbf{H}^H + \alpha \cdot \mathbf{I}\right)^{-1}\right\}}}\mathbf{H}^H\left(\mathbf{H}\mathbf{H}^H + \alpha \cdot \mathbf{I}\right)^{-1},\end{aligned} \tag{13}$$

where $f_{\mathrm{RZF}} = \sqrt{\frac{\operatorname{tr}\left\{\left(\mathbf{H}\mathbf{H}^H+\alpha\cdot\mathbf{I}\right)^{-1}\mathbf{H}\mathbf{H}^H\left(\mathbf{H}\mathbf{H}^H+\alpha\cdot\mathbf{I}\right)^{-1}\right\}}{P_0}}$ denotes the normalization factor for RZF precoding, and $\alpha$ denotes the regularization factor whose optimal value is $\alpha^* = K\sigma^2$.

## 4.3 Non-linear symbol-level precoding

Compared with linear precoding, non-linear precoding can achieve better performance by employing more sophisticated precoding techniques, at the cost of relatively high computational complexity. Generally speaking, based on CSI and the data symbol, non-linear precoding manipulates signal at the symbol level, which leads to a better communication performance but higher processing complexity. The transmitted signal of non-linear precoding is no longer a linearly weighted combination of symbol vectors. In this subsection, we will introduce classical non-linear precoding schemes to show their working mechanism.

**Dirty Paper Coding (DPC)** is able to reduce the destructive effect of inter-user interference and further achieve channel capacity in MIMO systems [10]. However, assuming perfect CSI and that interference information can be obtained at the transmitter, the capacity-achieving DPC requires an infinite-length coding and a high-complexity searching algorithm, which limits its application in practical systems.

Considering the high complexity of DPC, **Tomlinson-Harashima Precoding (THP)** has been proposed as an alternating near-capacity scheme whose computational complexity is relatively acceptable in practice. The basic idea of THP is to pre-distort the symbols before they are transmitted over the communication channel [11]. This pre-distortion is achieved by adding a feedback loop to the transmitting system, which modifies the symbols based on the previous symbols that have been transmitted. The feedback loop effectively cancels out the distortion introduced by the communication channel, leading to a higher quality and more reliable signal at the receiver. **Figure 2** shows the architecture of the THP precoding system.

Specifically, THP first decomposes the channel matrix into

$$\mathbf{H} = \mathbf{L}\mathbf{F}^{\mathrm{H}}, \tag{14}$$

with a lower-triangle matrix $\mathbf{L}$ and a unitary matrix $\mathbf{F}$. Based on that, the transmitted signal vector $\mathbf{x}$ for THP can be further expressed as

$$\mathbf{x}_{\mathrm{THP}} = \mathbf{F}\tilde{\mathbf{x}}_{\mathrm{THP}}, \tag{15}$$

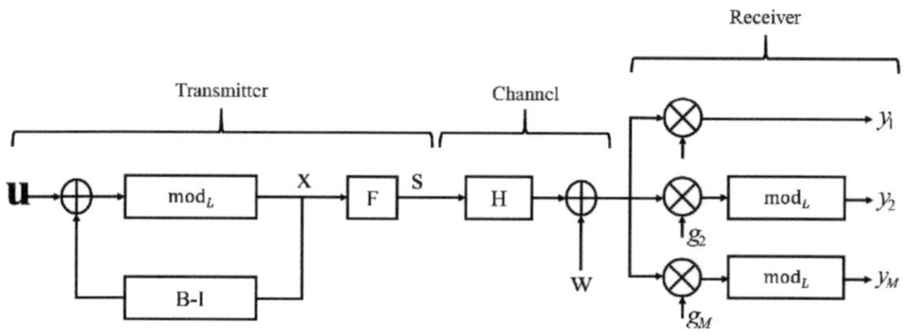


**Figure 2.**
*The geometrical representation of THP.*

where $\tilde{\mathbf{x}}$ can be obtained by

$$[\tilde{\mathbf{x}}_{\text{THP}}]_k = \text{mod}_\tau \left\{ s_k - \sum_{l=1}^{k-1} [\mathbf{B}]_{k,l} \, [\tilde{\mathbf{x}}_{\text{THP}}]_l \right\}, \forall k \in \{1, 2, \cdots, K\}. \tag{16}$$

$\text{mod}_\tau \{x\}$ denotes a complex modulo function, given by

$$\text{mod}_\tau\{x\} = \left( \Re(x) - \tau \cdot \lfloor \frac{\Re(x) + \tau/2}{\tau} \rfloor \right) + j \left( \Im(x) - \tau \cdot \lfloor \frac{\Im(x) + \tau/2}{\tau} \rfloor \right), \tag{17}$$

where $\tau$ denotes the modulo basis and $\lfloor \cdot \rfloor$ denotes the floor approximating function. Based on the analysis above, the effective THP channel can be expressed as

$$\mathbf{B} = \mathbf{GHF}, \tag{18}$$

where $\mathbf{G}$ is a diagonal matrix that contains the complex scaling gain corresponding to each user, which is actually the inverse of the corresponding diagonal entry in $\mathbf{L}$, i.e.,

$$g_k = [\mathbf{G}]_{k,k} = \frac{1}{[\mathbf{L}]_{k,k}}. \tag{19}$$

At the receiver side, the scaling compensation operation and the modulo operation are also required prior to the demodulation.

Considering that the performance of ZF precoding is mainly limited by its noise amplification effect, the **Vector- Perturbation (VP)** precoding [12] has been proposed as an improvement [12]. Based on the ZF precoding, VP precoding introduces a perturbation vector to the symbol vector, resulting in a transmitted signal that aligns better with the main eigenvector direction of the channel inverse matrix. This reduces the noise amplification factor and further lowers the noise amplification effect of ZF. Therefore, compared to ZF, VP can achieve significant performance gains. To be more specific, the VP precoding process can be expressed as

$$\mathbf{x}_{\text{VP}} = \frac{1}{f_{\text{VP}}} \cdot \mathbf{H}^{\text{H}} \left( \mathbf{H}\mathbf{H}^{\text{H}} \right)^{-1} (\mathbf{s} + \tau \cdot \mathbf{l}), \tag{20}$$

where $\tau = 2|c|_{\max} + \Delta$ denotes the modulo basis corresponding to the modulation level, $|c|_{\max}$ denotes the modulus value of the maximum amplitude modulation constellation point, and $\Delta$ is the minimum distance among the constellation points. $\mathbf{l} \in \mathbb{CZ}^{K\times 1}$ denotes the complex integer perturbation vector, given as

$$\mathbf{l} = \underset{\mathbf{l}\in \mathbb{CZ}^{K\times 1}}{\arg\min} \left\| \mathbf{H}^{\mathrm{H}} \left(\mathbf{H}\mathbf{H}^{\mathrm{H}}\right)^{-1} (\mathbf{s} + \tau \cdot \mathbf{l}) \right\|_2^2, \tag{21}$$

which can be obtained by the sphere decoder. Based on that, the normalization factor of VP precoding can be obtained by

$$y_k = \frac{1}{f_{\mathrm{VP}}} \cdot \mathbf{h}_k \mathbf{x}_{\mathrm{VP}} + n_k = \frac{1}{f_{\mathrm{VP}}} (s_k + \tau l_k) + n_k, \tag{22}$$

where $l_k$ denotes the $k$-th element of the perturbation vector $\mathbf{l}$. In order to eliminate the perturbation component $\tau l_k$ at the receiver side, the receiver needs to accomplish the module operation after the power compensation, as shown below:

$$\begin{aligned} r_k &= \mathrm{mod}_\tau\{f_{\mathrm{VP}} y_k\} \\ &= \mathrm{mod}\tau\left\{s_k + \tau l_k + f_{\mathrm{VP}n_k}\right\} \\ &= s_k + f_{\mathrm{VP}} \hat{n}_k, \end{aligned} \tag{23}$$

where $\hat{n}_k$ denotes the effective noise of the $k$-th user.

## 5. Optimization-based precoding

With the deepening of research on precoding technology, an increasing number of mathematical tools, such as convex optimization, have been introduced into the precoding design process to improve precoding performance as much as possible. In addition, optimization-based precoding can flexibly serve various communication targets, and therefore has a wide range of applications in practical engineering systems.

### 5.1 Block-level precoding

#### *5.1.1 Preliminary*

Based on the analysis above, due to the linear relationship between the transmitted signal vector $\mathbf{x}$, the symbol vector $\mathbf{s}$, and the precoding matrix $\mathbf{W}$, the transmitted signal $\mathbf{x}$ can be regarded as a linear weighted combination of the precoding matrix $\mathbf{W}$, where the weighting coefficients are given by the symbol vector $\mathbf{s}$. Therefore, the wireless transmission process of (7) and (8) can be reformulated in the following form:

$$y_k = \mathbf{h}_k \sum_{i=1}^{K} \mathbf{w}_i s_i + n_k = \mathbf{h}_k \mathbf{w}_k s_k + \mathbf{h}_k \sum_{i\neq k}^{K} \mathbf{w}_i s_i + n_k, \tag{24}$$

where the first component denotes the expected received signal of the $k$-th user, the second component denotes the interference, and the third component denotes the additive noise. Based on that, the received SINR of the $k$-th user can be given as

$$\gamma_k = \frac{|\mathbf{h}_k \mathbf{w}_k|^2}{\sum_{i\neq k}^{K} |\mathbf{h}_k \mathbf{w}_i|^2 + \sigma^2}. \tag{25}$$

Based on the analysis above, there are two main schemes for optimization-based block-level precoding, as discussed in the following.

### *5.1.2 Power minimization (PM) scheme*

Power minimization precoding, also known as minimum power beamforming[1], is a technique used to minimize the total transmitted power subject to a set of quality of service (QoS) constraints. The goal of this technique is to transmit the signal with the minimum possible power while ensuring that the received signal quality meets the desired level. This technique is particularly useful in situations where power consumption is a critical issue or in large-scale MIMO systems where the number of antennas is much larger than the number of users.

The PM design problem can be formulated as below [13]:

$$\begin{aligned} \mathcal{P}_1 : \min_{\mathbf{w}_i} & \sum_{i=1}^{K} \|\mathbf{w}_i\|_F^2 \\ \text{s.t.} \quad & \frac{|\mathbf{h}_k \mathbf{w}_k|^2}{\sum_{i\neq k}^{K} |\mathbf{h}_k \mathbf{w}_i|^2 + \sigma^2} \geq \Gamma_k, \forall k \in \{1, 2, \cdots, K\} \end{aligned} \tag{26}$$

where $\Gamma_k$ denotes the SINR threshold for the $k$-th user. It is proved that $\mathcal{P}_1$ is convex which can be solved via convex optimization algorithms efficiently. In addition to conventional convex optimization algorithms, literature has revealed an uplink-downlink duality in ref. [14], which has led to the development of an efficient iterative algorithm for solving downlink precoding optimization. Meanwhile, after transforming PM optimization into a semi-definite programming (SDP) problem, the semi-definite relaxation (SDR) approach [15–17] can be used to design the precoding matrix efficiently.

### *5.1.3 SINR balancing (SB) scheme*

SINR balancing precoding is a technique used to balance the signal-to--interference-plus-noise ratio (SINR) across all users in a multi-user system. The goal of this technique is to allocate the transmit power among the users such that each user experiences an equal SINR. This technique is particularly useful in situations where there are multiple users with different channel conditions, as it ensures that each user receives an equal quality of service. To be more specific, the SB design problem can be formulated as below [18]:

[1] It is noted that in this chapter the term 'beamforming' and 'precoding' are interchangeable.

$$\begin{aligned} \mathcal{P}_2: \quad & \max_{\mathbf{w}_i} \min_k \gamma_k \\ \text{s.t.} \quad & \gamma_k = \frac{|\mathbf{h}_k \mathbf{w}_k|^2}{\sum_{i \neq k}^{K} |\mathbf{h}_k \mathbf{w}_i|^2 + \sigma^2}, \forall k \in \{1, 2, \cdots, K\} \\ & \sum_{i=1}^{K} \|\mathbf{w}_i\|_F^2 \leq P_0 \end{aligned} \tag{27}$$

where $P_0$ is the maximum transmit power. Unlike the PM design problem, $\mathcal{P}_2$ is non-convex, which brings difficulties to the optimal precoding design. However, SB precoding can be efficiently designed through the bisection search method in ref. [16], or via an iterative algorithm in [14].

## 5.2 Symbol-level precoding

Block-level precoding is a precoding design based on CSI and is generally independent of the transmitted symbols. These algorithms tend to eliminate inter-user interference. In recent years, symbol-level precoding has received increasing attention [19]. Compared with block-level precoding, symbol-level precoding accomplishes precoding design based on both CSI and transmitted symbols, which gives it the ability to manipulate interference vectors more wisely compared with block-level precoding. With symbol-level precoding, the system can manage and utilize inter-user interference, which offers an additional power gain to improve system performance. In this subsection, we first introduce the concept of constructive interference (CI) to reveal the main idea of interference exploitation and then discuss the design problem of symbol-level precoding in different scenarios.

### *5.2.1 Concept for interference exploitation*

Interference is commonly considered a factor that limits performance in wireless communication systems. It arises due to the superimposition of transmit signals for different users in the wireless channel during multi-user transmission. Precoding strategies capitalize on the availability of CSI at the base station, along with data symbol information, to predict interference before transmission. Information theory analysis reveals that known interference will not affect the broadcast channel's capacity when CSI is available at the transmitter. However, most existing linear precoding schemes aim to eliminate, avoid or limit interference, and operate on a block level. Recent studies suggest that constructive interference (CI) precoding via Symbol-Level Precoding (SLP) can control both the power and direction of interfering signals, allowing interference to contribute to error-less signal detection and improve system performance [20]. Interference exploitation techniques are most useful in systems where interference can be predicted. In this subsection, we will give an illustrative example to demonstrate the division of instantaneous interference into CI and destructive interference (DI) [20].

Let us consider a scenario where the desired symbol $u$ is from a nominal BPSK constellation, with the assumption that $u = 1$. We use $i$ to denote the interfering signal and discuss two cases: (i) $i > 0$ and (ii) $i < 0$.

In the first case, when $i > 0$, as shown in **Figure 3(a)**, the received signal can be expressed as $\tilde{y} = h_u u + \tilde{h}_i i + n = \tilde{r} + n$, where $\tilde{r}$ represents the received signal excluding noise, and $n$ denotes the additive noise at the receiver side. **Figure 3(a)** shows that $\text{Proj}_{\text{O}\tilde{\text{E}}}(\tilde{r}) > \text{Proj}_{\text{O}\tilde{\text{E}}}(h_u u)$, which means that the interference has pushed $r$ further away

*DOI: http://dx.doi.org/10.5772/intechopen.112041*

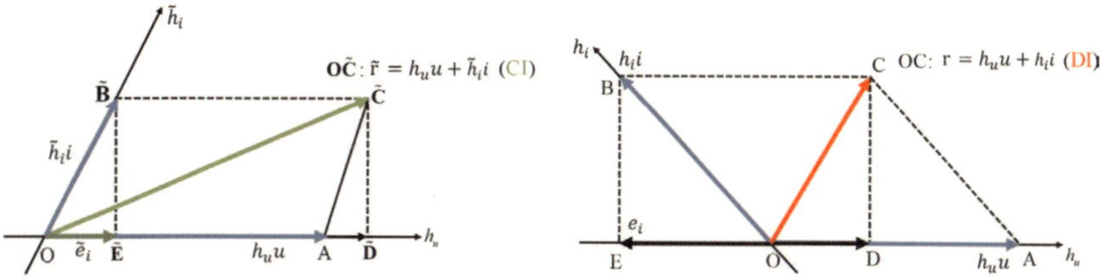


(a) CI (b) DI

**Figure 3.**
*The geometrical representation of CI and DI.*

from the detection threshold of BPSK when compared to the original data symbol $u$. Here $\text{Proj}_{\mathbf{d}}(\mathbf{x})$ denotes the projection of vector $\mathbf{x}$ on the direction of $\mathbf{d}$. In this situation, the interfering signal is actually constructive and contributes to the useful signal power. Given a fixed noise power, $\tilde{y} = \tilde{r} + n$ is more likely to be detected correctly than the interference-free case $y' = h_u u + n$. Thus, we can expect improved performance.

On the other hand, in the second case, when $i < 0$, as shown in **Figure 3(b)**, the interfering signal causes the received signal $r$ to move closer to the detection threshold. In this case, the interfering signal reduces the useful signal power and is therefore destructive. The noiseless received signal $r = h_u u + h_i i$ is more susceptible to noise than $r' = u$ in this scenario.

In summary, symbol-level precoding offers more precise interference management and control, with the added benefit of improved performance through beneficial interference. This makes it a better communication performance option compared to traditional block-level precoding. Next, we will introduce the design principles of symbol-level precoding by discussing classical CI-SLP precoding methods.

### *5.2.2 Phase rotation metric*

As depicted in **Figure 4**, CI-SLP is a technique that manipulates inter-user interference to ensure that the noise-free receive signal falls within the constructive region.

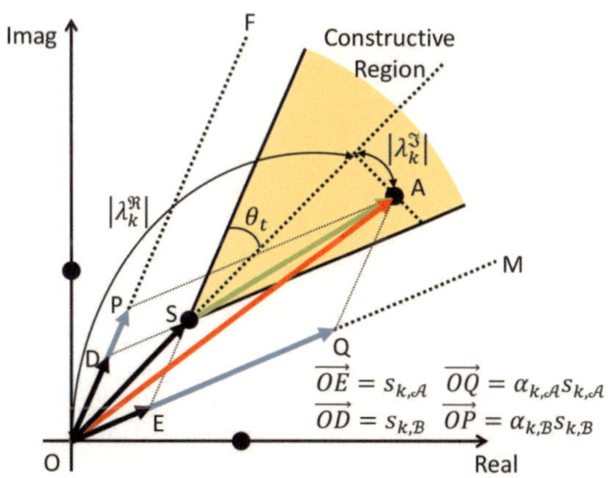


**Figure 4.**
*CI-SLP, 'phase-rotation' metric, 8-PSK.*

The SLP matrix **W** is designed to maximize the distance between the worst user's constructive region and the detection threshold, thereby improving the transmission performance. Masouros [21] first proposed the "phase rotation" metric for PSK modulated systems. Based on this metric, the noise-free receive signal can be expressed as follows [22]:

$$\vec{OA} = \mathbf{h}_k^{\mathrm{T}}\mathbf{W}\mathbf{s} = \lambda_k s_k. \tag{28}$$

The constructive factor $\lambda_k$ quantifies the constructive effect of interference exploitation for that user. Based on this factor, the constructive region can be described as follows:

$$\begin{aligned} &\theta_{AB} \leq \theta_t \Rightarrow \tan\theta_{AB} \leq \tan\theta_t \\ &\Rightarrow \frac{\left|j \cdot \lambda_k^{\mathcal{I}} s_k\right|}{\left|\left[\lambda_k^{\mathcal{R}} - \sqrt{\Gamma_k \sigma^2}\right] s_k\right|} \leq \tan\theta_t \\ &\Rightarrow \left[\lambda_k^{\mathcal{R}} - \sqrt{\Gamma_k \sigma^2}\right] \tan\theta_t \geq \left|\lambda_k^{\mathcal{I}}\right| \end{aligned} \tag{29}$$

According to the transmit power minimization criterion, the CI-SLP design problem is shown below

$$\begin{aligned} \mathcal{P}_3 :\ &\min_{\mathbf{W}} \|\mathbf{W}\mathbf{s}\|_{\mathrm{F}}^2 \\ &\text{s.t.} \mathbf{h}_k \mathbf{W}\mathbf{s} = \lambda_k s_k, \forall k \in \{1, 2, \cdots, K\} \\ &\left[\lambda_k^{\mathcal{R}} - \sqrt{\Gamma_k \sigma^2}\right] \tan\theta_t \geq \left|\Gamma_k^{\mathcal{I}}\right|, \forall k \in \{1, 2, \cdots, K\}, \end{aligned} \tag{30}$$

where $\Gamma_k^{\mathcal{I}}$ denotes the Quality of Serves (QoS) threshold of the $k$-th user.

The convexity of $\mathcal{P}_3$ can be proven, similar to the traditional PM problem, enabling the use of several convex optimization algorithms to solve this problem conveniently. Similarly, the CI-SLP design problem based on the SB criterion can be formulated as

$$\begin{aligned} \mathcal{P}_4 :\ \ & \max{}_{\mathbf{W},t} t \\ \text{s.t.}\ \ & \mathbf{h}_k \mathbf{W}\mathbf{s} = \lambda_k s_k, \forall k \in \{1, 2, \cdots, K\} \\ & \left[\lambda_k^{\mathcal{R}} - t\right] \tan\theta_t \geq \left|\lambda_k^{\mathcal{I}}\right|, \forall k \in \{1, 2, \cdots, K\} \\ & \|\mathbf{W}\mathbf{s}\|_{\mathrm{F}}^2 \leq P_0. \end{aligned} \tag{31}$$

It is worth noting that the convexity of the equation shown above can also be proven, which distinguishes it from the traditional SB problem and renders it more mathematically tractable.

### *5.2.3 Symbol scaling metric*

In QAM modulation, the interference exploitation is conditional, unlike PSK modulation. The constellation signal points of QAM modulation can be classified into four groups based on their interference exploitation characteristics, as shown in **Figure 5**. Group A' represents signal points that do not exploit any interference, while Group B′ and Group C′ represent signal points that exploit interference in the real and

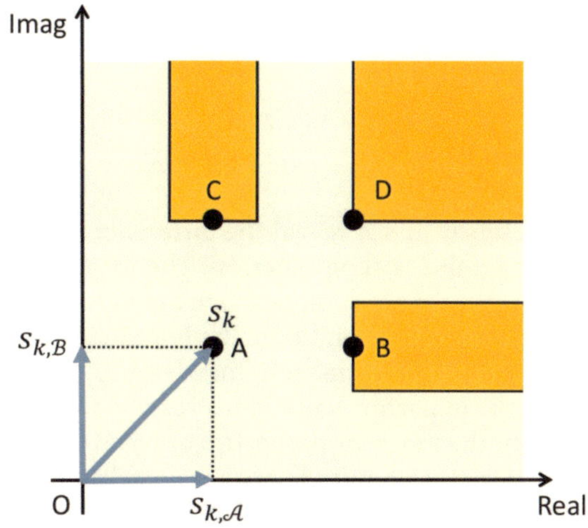


**Figure 5.**
*CI-SLP, 'symbol-scaling' metric, 16-QAM.*

imaginary parts, respectively. Group D′ represents signal points that exploit interference in both the real and imaginary parts, resulting in full interference exploitation.

The interference exploitation procedure via the "symbol-scaling" [23] metric and decomposition of the noiseless receive signal of the $k$-th user can be described as follows:

$$\mathbf{h}_k^{\mathrm{T}}\mathbf{W}\mathbf{s} = \boldsymbol{\alpha}_k^{\mathrm{T}}\mathbf{s}_k, \tag{32}$$

where

$$\boldsymbol{\alpha}_k = \left[\alpha_k^{\mathcal{A}}, \alpha_k^{\mathcal{B}}\right]^{\mathrm{T}}, \mathbf{s}_k = \left[s_k^{\mathcal{A}}, s_k^{\mathcal{B}}\right]^{\mathrm{T}} \tag{33}$$

with

$$s_k^{\mathcal{A}} = \Re(s_k), s_k^{\mathcal{B}} = (s_k), \ \ k = 1, 2, \ldots, K. \tag{34}$$

Based on that, the CI-SLP design problem in QAM-modulated systems can be described as follows

$$\begin{aligned}
\mathcal{P}_5: \ \max_{\mathbf{W}, \boldsymbol{\Omega}_k, t} \quad & t \\
\text{s.t.} \quad & \mathbf{h}_k^T\mathbf{W}\mathbf{s} = \boldsymbol{\alpha}_k^{\mathrm{T}}\mathbf{s}_k, \forall k \in \mathcal{K} \\
& t \leq \alpha_m^{\mathcal{O}}, \forall \alpha_m^{\mathcal{O}} \in \mathcal{O} \\
& t = \alpha_n^{\mathcal{I}}, \forall \alpha_n^{\mathcal{I}} \in \mathcal{I} \\
& \|\mathbf{W}\mathbf{s}\|_2^2 \leq p_0.
\end{aligned} \tag{35}$$

The set $\mathcal{O}$ comprises the indices of successful interference exploitation corresponding to the real part of the symbol in group B′, the imaginary part of the symbol in group C′, and both the real and imaginary parts of the symbol in group D′. Conversely, the set $\mathcal{I}$ comprises the indices of unsuccessful interference exploitation corresponding to the imaginary part of the symbol in group B′, the real part of the symbol in group C′, and both the real and imaginary parts of the symbol in group A'. It follows that $\mathcal{O}$ and $\mathcal{I}$ satisfy the following relationship:

$$\begin{aligned} \mathcal{O} \cup \mathcal{I} = \mathcal{K}, \mathcal{O} \cap \mathcal{I} = \varnothing, \\ \text{card}\{\mathcal{O}\} + \text{card}\{\mathcal{I}\} = 2K. \end{aligned} \tag{36}$$

The definitions of the sets $\mathcal{O}$ and $\mathcal{I}$ reveal the difference between the phase rotation criterion and the symbol scaling criterion. The former exploits interference unconditionally, i.e., all constellation points participate in interference exploitation, while the latter exploits interference conditionally. For QAM modulation systems, the inner constellation points do not participate in interference exploitation, and beneficial interference only results in performance gains for the outer constellation points. This difference arises from the inherent properties of QAM and PSK modulation schemes. In PSK modulation, the amplitude of the constellation points does not carry any information, and therefore, any constellation point can be exploited for interference without adversely affecting the detection of other constellation points. However, for the inner constellation points in QAM modulation, interference vectors that push the noiseless receive signal points in any direction will adversely affect the error decision of other constellation points. It is worth noting that these two design criteria only differ in their description of the interference exploitation process and are essentially equivalent. Li et al. [23] has proven that under PSK modulation, the symbol scaling criterion and the phase rotation criterion are equivalent, as depicted in **Figure 4**, where the symbol-scaling metric is also applicable. Therefore, the symbol scaling criterion is more universal in this sense.

## 6. Hardware-efficient precoding

The use of technologies such as General Artificial Intelligence (AI), has led to a surge in users' demand for mobile data traffic. One way to address this issue is to utilize massive MIMO systems, which employ a large number of antennas at the base station to improve data rate and link reliability. This approach allows signals to be dynamically adjusted in both horizontal and vertical directions, reducing interference between small areas and enabling more accurate pointing toward specific users. However, directly applying Massive MIMO technology to traditional communication system architectures can result in new problems [3]. To be more specific, traditional MIMO systems equip each antenna with RF chains and high-resolution DACs, causing significant power loss when the antenna array is large. To solve this issue, there are three general approaches: reducing the number of RF chains, lowering the resolution of the DACs, or employing power-efficient nonlinear power amplifiers. However, these hardware-efficient architectures introduce new challenges to precoding designs, which will be explained in more detail in the following.

### 6.1 Hybrid analog-digital (HAD) precoding

Fully-digital precoders can be used in traditional sub-6 GHz bands, but for millimeter wave (mmWave) communications, the cost and power consumption of hardware components make this approach impractical. To solve this issue, researchers have developed the hybrid analog-digital structure, which provides a promising trade-off between the cost, complexity, and capacity of the mmWave network. This structure reduces hardware complexity and power consumption by reducing the total number of RF chains. Specifically, the mmWave transceivers first process data streams with a low-dimension digital precoder, followed by high-dimension analog precoding using low-cost phase shifters, switches [24], or lens [25]. While the performance of the hybrid precoder is usually inferior to that of a fully-digital precoder, it offers a cost-efficient and energy-efficient solution for mmWave communication.

In an MU-MIMO system illustrated in **Figure 6**, $N_t$ transmit antennas are utilized by the BS to serve $K$ single-antenna users simultaneously. The transmitter has $N_{\mathrm{RF}}^t$ RF chains, where $N_{\mathrm{RF}}^t \ll N_t$. In this subsection, we use phase shifter-based hybrid architecture as an illustrative example, without loss of generality.

Based on that, the transmit symbol vector $\mathbf{x}$ can be expressed as

$$\mathbf{x} = \mathbf{F}_{\mathrm{RF}}\mathbf{F}_{\mathrm{BB}}\mathbf{s}, \tag{37}$$

where $\mathbf{F}_{\mathrm{RF}} \in \mathbb{C}_{N_{\mathrm{RF}}^t \times N_t}$ denotes the hybrid precoding matrix, $\mathbf{F}_{\mathrm{BB}} \in \mathbb{C}_{K \times N_{\mathrm{RF}}^t}$ denotes the digital baseband precoding matrix, and $\mathbf{s} \in \mathbb{C}_{K\times 1}$ denotes the data symbol vector with $\mathbb{E}\{\mathbf{s}\mathbf{s}^{\mathrm{H}}\} = \frac{1}{K}\mathbf{I}_K$, respectively. Considering that the hybrid precoding matrix is the mathematical description of phase shifters, we have the constant-module constraint for the hybrid precoder, as shown below:

$$|\mathbf{F}_{\mathrm{RF}}(i,j)| = 1, \quad 1 \le i \le N_{\mathrm{RF}}^t, \quad 1 \le j \le N_t. \tag{38}$$

Meanwhile, the power constraint at the transmit side can be expressed as

$$\|\mathbf{F}_{\mathrm{BB}}\mathbf{F}_{\mathrm{RF}}\|_{\mathrm{F}}^2 = P_0, \tag{39}$$

where $P_0$ is the maximum transmit power.

Based on that, the $k$-th user's received signal can be expressed as

$$y_k = \mathbf{h}_k^{\mathrm{H}}\mathbf{F}_{\mathrm{RF}}\mathbf{F}_{\mathrm{BB}}\mathbf{s} + n_k, \tag{40}$$

where $\mathbf{h}_k \in \mathbb{C}_{N_t \times 1}$ denotes the complex channel matrix for the $k$-th user, and $n_k \sim \mathcal{CN}(0, \sigma_k^2)$ denotes the additive Gaussian noise vector for the $k$-th user with the zero-mean and $\sigma_k^2$ noise power.

Aimed at maximizing the spectral efficiency, a common HAD precoding design problem can be formulated as [26].

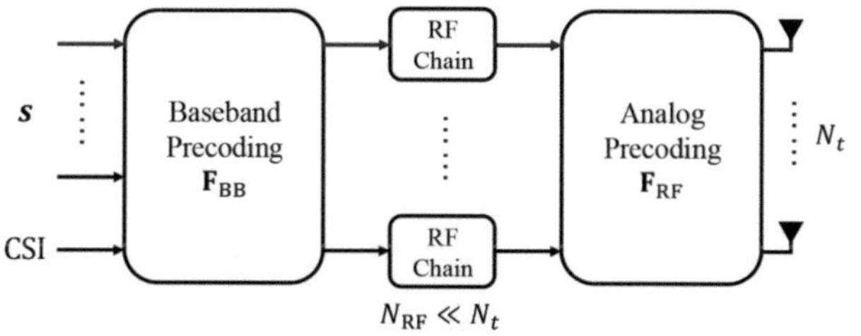


**Figure 6.**
*The HAD MIMO system.*

$$\mathcal{P}_6: \max_{\mathbf{F}_{\mathrm{RF}}, \mathbf{f}_k^{\mathrm{BB}}} \sum_{k=1}^{K} \log_2\left(1+\frac{\left|\mathbf{h}_k^{\mathrm{H}}\mathbf{F}_{\mathrm{RF}}\mathbf{f}_k^{\mathrm{BB}}\right|^2}{\sum_{i\neq k}\left|\mathbf{h}_k^{\mathrm{H}}\mathbf{F}_{\mathrm{RF}}\mathbf{f}_i^{\mathrm{BB}}\right|^2+\sigma_k^2}\right)$$
$$\text{s.t. } \mathbf{F}_{\mathrm{RF}} \in \mathcal{F}, \forall 1 \le k \le K,$$
$$\left\|\mathbf{F}_{\mathrm{RF}}\left[\mathbf{f}_1^{\mathrm{BB}}, \mathbf{f}_2^{\mathrm{BB}}, \dots, \mathbf{f}_K^{\mathrm{BB}}\right]\right\|_F^2 = P_0, \tag{41}$$

where $\mathcal{F}$ denotes the available region of $\mathbf{F}_{\mathrm{RF}}$, as defined below:

$$\mathcal{F} = \left\{\mathbf{F}_{\mathrm{RF}} \middle| \left|\mathbf{F}_{\mathrm{RF}}(i,j)\right| = 1,\ 1 \le i \le N_{\mathrm{RF}}^t,\ 1 \le j \le N_t\right\}. \tag{42}$$

The non-convexity of $\mathcal{P}_6$ is due to the constant-module constraint of $\mathbf{F}_{\mathrm{RF}}$, making it difficult to solve. To address this issue, a two-stage hybrid precoding algorithm was proposed in ref. [27] where the analog precoder maximizes the effective channel gain and the digital precoder mitigates multi-user interference based on the ZF principle. In ref. [28], it was demonstrated that hybrid precoding can achieve any fully-digital precoding when the number of RF chains is twice the number of data streams, and a near-optimal hybrid precoding design was proposed for single-user and multi-user transmissions with fewer RF chains. Reference [29] focused specifically on partially-connected structures in multi-user scenarios and proposed hybrid precoding designs based on successive interference cancelation (SIC). This approach decomposes the total spectral efficiency optimization problem into a series of sub-rate optimization problems that can be solved efficiently using the power iteration algorithm. Other works on hybrid precoding include low-complexity designs based on MRT [30], virtual path selection [31], and SVD [32].

## 6.2 Low-bit precoding

Using low-resolution DACs instead of high-resolution DACs in massive MIMO architecture can be an effective way to reduce the power consumption of BS. This approach reduces the power consumption per RF chain, as depicted in **Figure** 7, instead of reducing the number of RF chains like in the hybrid architecture.

High-resolution DACs are required for each transmit signal to avoid signal distortion, but they consume significant power due to their linear relationship with bandwidth and exponential relationship with resolution [33]. Large-scale antenna arrays, with hundreds of antenna elements, require a significantly large number of DACs, posing practical challenges. To address this issue, low-resolution DACs, particularly 1-bit DACs, can substantially simplify hardware and reduce the corresponding power

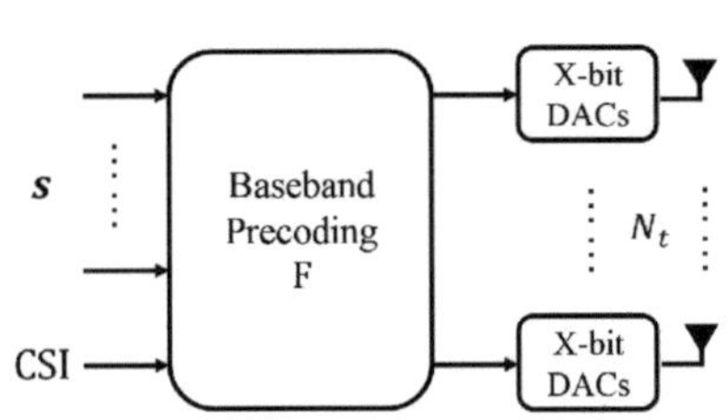


**Figure 7.**
*The architecture of low-bit MIMO system.*

consumption at the BS. Furthermore, 1-bit DACs generate CE signals, which facilitate the use of power-efficient amplifiers, further reducing hardware complexity. The common low-bit precoding design problem can be formulated as [34].

$$\begin{aligned} \mathcal{P}_7: & \min_{\mathbf{x}} \|\mathbf{s} - \beta_{\text{DAC}} \cdot \mathbf{Hx}\|_2^2 + K\beta_{\text{DAC}}^2 \sigma^2 \\ & \text{s.t.} \mathbf{x} \in \mathcal{X}_{\text{DAC}} \\ & \beta_{\text{DAC}} > 0. \end{aligned} \tag{43}$$

The optimization problem $\mathcal{P}_7$ seeks to minimize the MSE between transmitted and received symbols using low-resolution DACs. For 1-bit DACs, the set of output signals is denoted as $\mathcal{X}_{\text{DAC}} = \left\{ \pm\sqrt{\frac{P_0}{2N_t}} \pm \sqrt{\frac{P_0}{2N_t}} \cdot j \right\}$. In ref. [35], a non-linear precoding method based on a biconvex relaxation framework achieved promising performance with a low computational cost. Its corresponding VLSI design architectures were illustrated in refs. [36]. Alternatively, Jacobsson et al. [37] proposed several 1-bit precoding schemes based on SDR, sphere encoding, and squared $l_\infty$-norm relaxation, while Landau and de Lamare [38] described a 1-bit precoding method based on the branch-and-bound framework that can theoretically achieve optimal performance. Other downlink precoding designs for low-resolution DACs include SER minimization in refs. [39, 40] and alternating minimization in ref. [34]. Nonlinear precoding designs tend to outperform linear methods when low-resolution DACs are used at the transmitter. For example, CI-based symbol-level precoding design has been discussed in low-resolution DACs systems [41–43]. Several efficient solutions [43–45] have been proposed for the NP-hard optimization problem, both for 1-bit and few-bit DACs systems.

### 6.3 Nonlinearity-aware precoding

In a massive multiple-input-multiple-output (MIMO) system, the integration of power-efficient nonlinear power amplifiers (PAs) can reduce the power consumption of each RF chain, similar to the architecture of low-bit digital-to-analog converters. Consequently, this leads to an improved energy efficiency of the system. However, in traditional multi-antenna systems, the limited linear region of nonlinear PAs causes significant signal distortions when transmitting signals with high peak-to-average power ratios (PAPRs). This consequently negatively impacts system performance.

To resolve the issue of PAPR, traditional research falls into two categories: (a) constant envelope precoding (CEP) schemes that maintain signal power at a constant value, commonly known as SLP schemes; and (b) frame-level precoding matrix optimization aimed at reducing the PAPR of the transmit signal. CEP eliminates the performance loss introduced by nonlinear PAs by limiting the amplitude of the transmit signal to a constant value, while the low-PAPR precoding relaxes the strict CE constraint by allowing the maximum PAPR to a certain value. In recent years, there has been a growing body of literature that explores the precoding design based on the knowledge of the nonlinear response characteristics of PAs. This approach represents a departure from the traditional emphasis solely on reducing the peak-to-average power ratio (PAPR) of transmitted signals. To be more specific, nonlinearity-aware precoding utilizes a clipping function to model the response characteristics of nonlinear PAs and developed a precoder that can resist both interference and PA nonlinearity by describing the modeled response characteristics [46].

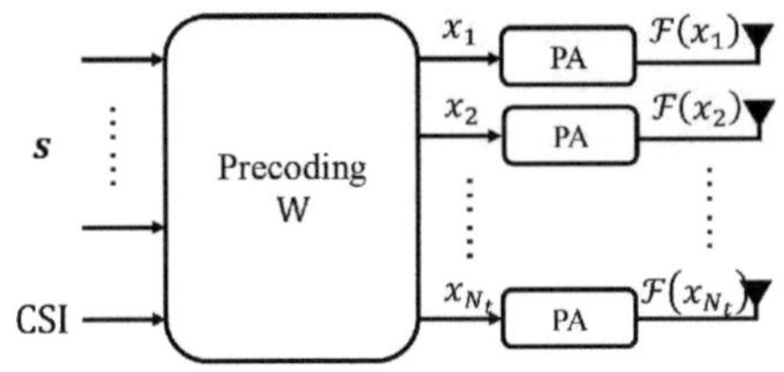


**Figure 8.**
*The nonlinearity-aware precoding system.*

The nonlinearity-aware precoding system can be shown in **Figure 8**. Considering a multi-user MISO system, the $k$-th user's received signal can be expressed as

$$y_k = \mathbf{h}_k^T \mathscr{F}(\mathbf{Ws}) + n_k, \tag{44}$$

where $\mathscr{F}(\cdot) : \mathbb{C} \to \mathbb{C}$ is the nonlinearity function that delineates the input-output response properties of nonlinear power amplifiers [47]. Based on that, the nonlinearity-aware precoding design problem aimed at maximizing the sum rate can be expressed as

$$\begin{aligned} \mathcal{P}_8 : \max_{\mathbf{W} \in \mathbb{C}^{K \times N_t}} \quad & R_{\text{sum}}(\mathbf{W}) \\ \text{s.t.} \quad & \mathbb{F}\left[\|\phi(\mathbf{Ws})\|^2\right] = P_{\text{t}}, \end{aligned} \tag{45}$$

where $P_t$ denotes the maximum transmit power constraint. The problem has been addressed through the introduction of a distortion-aware beamforming (DAB) algorithm as proposed by [48]. This method adopts an iterative approach to optimize data rate while minimizing the effect of distortions. In addition, several other precoding strategies have been developed with a focus on accounting for nonlinearity in the system. Specifically, Aghdam et al. [49] studied a precoding scheme that incorporates power amplifier effects in massive MU-MIMO downlink systems and put forth a robust algorithm to mitigate interference and nonlinearity resulting from power amplifiers. Moreover, Zayani et al. [50] presented a power control mechanism and a precoding scheme for SU-MISO communication systems that utilize nonlinear power amplifiers at the base station. The proposed method maximizes the received SINR while utilizing an iterative precoding algorithm. Finally, Jee et al. [51] optimized both precoding and power allocation strategies jointly to maximize the achievable sum rate of MU-MIMO systems.

## 7. Conclusions

In this chapter, we have provided a comprehensive overview of precoding design for achieving spatial multiplexing in MIMO communications.

We began in Section 3 by introducing the fundamental concepts of MIMO systems, including the mathematical description of MIMO communications, performance metrics, and the increasingly important and widely used massive MIMO technology in 5G. These concepts laid a solid foundation for the subsequent discussions on the precoding design.

In Section 4, we discussed traditional precoding design methods, including closed-form linear block-level precoding techniques such as MRT, ZF, and RZF, as well as traditional nonlinear symbol-level precoding techniques such as THP and VP. Through these algorithms, we introduced the basic principles and guidelines of precoding design.

In Section 5, we discussed more complex precoding design methods based on convex optimization, including power minimization, SINR balancing, and the emerging CI-SLP precoding. These methods provide more flexibility and adaptability in precoding design and can achieve better performance in practical communication systems.

In Section 6, we focused on the hardware-efficient precoding design for massive MIMO systems in 5G. We discussed hybrid analog-digital precoding, low-bit precoding, and nonlinearity-aware precoding, which are essential for reducing power consumption and computational complexity while maintaining high communication performance.

Overall, this chapter highlights the importance of efficient precoding design for achieving efficient and reliable wireless transmission. Precoding design is a critical component of MIMO technology, and it requires a careful balance between communication performance, power consumption, and computational complexity. The discussions in this chapter provide a comprehensive understanding of the various precoding techniques that can be employed to achieve spatial multiplexing in MIMO communications and underscore the significance of efficient precoding design for realizing the full potential of MIMO technology in wireless communication systems.

## Nomenclature

| | |
|---|---|
| SISO | single-input single-output |
| MISO | multi-input single-output |
| MIMO | multi-input multi-output |
| MRT | maximum ratio transmission |
| ZF | zero-forcing |
| RZF | regularized zero-forcing |
| THP | Tomlinson-Harashima precoding |
| VP | vector perturbation |
| IoT | internet of things |
| PM | power minimization precoding |
| SNR | signal-to-noise ratio |
| SINR | signal-to-interference-and-noise ratio |
| SB | SINR balancing precoding |
| IE | interference exploitation |
| CI | constructive interference |
| DI | destructive interference |
| BLP | block-level precoding |
| SLP | symbol-level precoding |
| HAD | hybrid analog-digital precoding |
| CEP | constant envelope precoding |

**Author details**

Haonan Wang* and Ang Li*
Xi'an Jiaotong University, Xi'an, Shaanxi, China

*Address all correspondence to: whn8215858@stu.xjtu.edu.cn and ang.li.2020@xjtu.edu.cn

## References

[1] Andrews JG, Buzzi S, Choi W, et al. What will 5G Be? IEEE Journal on Selected Areas in Communications. 2014;**32**(6):1065-1082. DOI: 10.1109/JSAC.2014.2328098

[2] Tse D, Viswanath P. Fundamentals of Wireless Communication. Cambridge, UK: Cambridge University Press; 2005

[3] Swindlehurst AL, Ayanoglu E, Heydari P, et al. Millimeter-wave massive MIMO: The next wireless revolution? IEEE Communications Magazine. 2014;**52**(9):56-62. DOI: 10.1109/MCOM.2014.6894453

[4] Joham M, Utschick W, Nossek JA. Linear transmit processing in MIMO communications systems. IEEE Transactions on Signal Processing. 2005; **53**(8):2700-2712. DOI: 10.1109/TSP.2005.850331

[5] Lu L, Li GY, Swindlehurst AL, et al. An overview of massive MIMO: Benefits and challenges. IEEE Journal of Selected Topics in Signal Processing. 2014;**8**(5): 742-758. DOI: 10.1109/JSTSP.2014.2317671

[6] Hochwald BM, Marzetta TL, Tarokh V. Multiple-Antenna Channel hardening and its implications for rate feedback and scheduling. IEEE Transactions on Information Theory. 2004;**50**(9):1893-1909. DOI: 10.1109/TIT.2004.833345

[7] Heath RW, Gonzalez-Prelcic N, Rangan S, et al. An overview of signal processing techniques for Millimeter wave MIMO systems. IEEE Journal of Selected Topics in Signal Processing. 2016;**10**(3):436-453. DOI: 10.1109/JSTSP.2016.2523924

[8] Jiang Y, Varanasi MK, Li J. Performance analysis of ZF and MMSE equalizers for MIMO systems: An In-depth study of the high SNR regime. IEEE Transactions on Information Theory. 2011;**57**(4):2008-2026. DOI: 10.1109/TIT.2011.2112070

[9] Peel CB, Hochwald BM, Swindlehurst AL. A vector-perturbation technique for near-capacity multiantenna multiuser communication-part I: Channel inversion and regularization. IEEE Transactions on Communications. 2005;**53**(1):195-202. DOI: 10.1109/TCOMM.2004.840638

[10] Costa M. Writing on dirty paper. IEEE Transactions on Information Theory. 1983;**29**(3):439-441. DOI: 10.1109/TIT.1983.1056659

[11] Garcia-Rodriguez A, Masouros C. Power-efficient Tomlinson-Harashima precoding for the downlink of multi-user MISO systems. IEEE Transactions on Communications. 2014;**62**(6): 1884-1896. DOI: 10.1109/TCOMM.2014.2317189

[12] Hochwald BM, Peel CB, Swindlehurst AL. A vector-perturbation technique for near-capacity multiantenna multiuser communication-part II: Perturbation. IEEE Transactions on Communications. 2005;**53**(3): 537-544. DOI: 10.1109/TCOMM.2005.843995

[13] Bengtsson M, Ottersten B. Optimum and suboptimum transmit beamforming. In: Handbook of Antennas in Wireless Communications. Boca Raton, Florida, US: CRC Press; 2018. pp. 18-1-18-33. DOI: 10.1201/9781315220031-18

[14] Schubert M, Boche H. Solution of the multiuser downlink beamforming problem with individual SINR

constraints. IEEE Transactions on Vehicular Technology. 2004;**53**(1):18-28. DOI: 10.1109/TVT.2003.819629

[15] Huang Y, Palomar DP. Rank-constrained separable semidefinite programming with applications to optimal beamforming. IEEE Transactions on Signal Processing. 2009; **58**(2):664-678. DOI: 10.1109/TSP.2009.2031732

[16] Huang Y, Palomar DP. A dual perspective on separable semidefinite programming with applications to optimal downlink beamforming. IEEE Transactions on Signal Processing. 2010; **58**(8):4254-4271. DOI: 10.1109/TSP.2010.2049570

[17] Law KL, Wen X, Vu MT, et al. General rank multiuser downlink beamforming with shaping constraints using real-valued OSTBC. IEEE Transactions on Signal Processing. 2015; **63**(21):5758-5771. DOI: 10.1109/TSP.2015.2455516

[18] Wiesel A, Eldar YC, Shamai S. Linear precoding via conic optimization for fixed MIMO receivers. IEEE Transactions on Signal Processing. 2005; **54**(1):161-176. DOI: 10.1109/TSP.2005.861073

[19] Li A, Spano D, Krivochiza J, et al. A tutorial on interference exploitation via symbol-level precoding: Overview, state-of-the-art and future directions. IEEE Communications Surveys & Tutorials. 2020;**22**(2):796-839. DOI: 10.1109/COMST.2020.2980570

[20] Masouros C, Ratnarajah T, Sellathurai M, et al. Known interference in the cellular downlink: A performance limiting factor or a source of green signal power? IEEE Communications Magazine. 2013;**51**(10):162-171. DOI: 10.1109/MCOM.2013.6619580

[21] Masouros C. Correlation rotation linear precoding for MIMO broadcast communications. IEEE Transactions on Signal Processing. 2010;**59**(1):252-262. DOI: 10.1109/TSP.2010.2088395

[22] Li A, Masouros C. Interference exploitation precoding made practical: Optimal closed-form solutions for PSK modulations. IEEE Transactions on Wireless Communications. 2018;**17**(11): 7661-7676. DOI: 10.1109/TWC.2018.2869382

[23] Li A, Masouros C, Vucetic B, et al. Interference exploitation precoding for multi-level modulations: Closed-form solutions. IEEE Transactions on Communications. 2020;**69**(1):291-308. DOI: 10.1109/TCOMM.2020.3031616

[24] Méndez-Rial R, Rusu C, González-Prelcic N, et al. Hybrid MIMO architectures for Millimeter wave communications: Phase shifters or switches? IEEE Access. 2016;**4**:247-267. DOI: 10.1109/ACCESS.2015.2514261

[25] Brady J, Behdad N, Sayeed AM. Beamspace MIMO for Millimeter-wave communications: System architecture, Modeling, analysis, and measurements. IEEE Transactions on Antennas and Propagation. 2013;**61**(7):3814-3827. DOI: 10.1109/TAP.2013.2254442

[26] El Ayach O, Rajagopal S, Abu-Surra S, et al. Spatially sparse precoding in Millimeter wave MIMO systems. IEEE Transactions on Wireless Communications. 2014;**13**(3):1499-1513. DOI: 10.1109/TWC.2014.011714.130846

[27] Alkhateeb A, Leus G, Heath RW. Limited feedback hybrid precoding for multi-user Millimeter wave systems. IEEE Transactions on Wireless Communications. 2015;**14**(11): 6481-6494. DOI: 10.1109/TWC.2015.2455980

[28] Sohrabi F, Yu W. Hybrid digital and Analog beamforming Design for Large-Scale Antenna Arrays. IEEE Journal of Selected Topics in Signal Processing. 2016;**10**(3):501-513. DOI: 10.1109/JSTSP.2016.2520912

[29] Gao X, Dai L, Han S, et al. Energy-efficient hybrid Analog and digital precoding for mmWave MIMO systems with large antenna arrays. IEEE Journal on Selected Areas in Communications. 2016;**34**(4):998-1009. DOI: 10.1109/JSAC.2016.2549418

[30] Liang L, Xu W, Dong X. Low-complexity hybrid precoding in massive multiuser MIMO systems. IEEE Wireless Communications Letters. 2014;**3**(6):653-656. DOI: 10.1109/LWC.2014.2363831

[31] Li A, Masouros C. Hybrid Analog-digital Millimeter-wave MU-MIMO transmission with virtual path selection. IEEE Communications Letters. 2016;**21**(2):438-441. DOI: 10.1109/LCOMM.2016.2621741

[32] Li A, Masouros C. Hybrid precoding and combining Design for Millimeter-Wave Multi-User MIMO based on SVD. In: 2017 IEEE International Conference on Communications (ICC'17), 21-25 May 2017; Paris. New York: IEEE; 2017. pp. 1-6. DOI: 10.1109/ICC.2017.7996970

[33] Allen PE, Dobkin R, Holberg DR. CMOS Analog Circuit Design. Amsterdam: Elsevier; 2011

[34] Chen JC. Alternating minimization algorithms for one-bit precoding in massive multiuser MIMO systems. IEEE Transactions on Vehicular Technology. 2018;**67**(8):7394-7406. DOI: 10.1109/TVT.2018.2836335

[35] Castaneda O, Goldstein T, Studer C. POKEMON: A non-linear beamforming algorithm for 1-bit massive MIMO. In: 2017 IEEE International Conference on Acoustics, Speech and Signal Processing (ICASSP'17), 5-9 March, 2017; New Orleans, LA, USA. New York: IEEE; 2017. pp. 3464-3468. DOI: 10.1109/ICASSP.2017.7952800

[36] Castañeda O, Jacobsson S, Durisi G, et al. 1-bit massive MU-MIMO precoding in VLSI. IEEE Journal on Emerging and Selected Topics in Circuits and Systems. 2017;7(4):508-522. DOI: 10.1109/JETCAS.2017.2772191

[37] Jacobsson S, Durisi G, Coldrey M, et al. Quantized precoding for massive MU-MIMO. IEEE Transactions on Communications. 2017;**65**(11):4670-4684. DOI: 10.1109/TCOMM.2017.2723000

[38] Landau LTN, de Lamare RC. Branch-and-bound precoding for multiuser MIMO systems with 1-bit quantization. IEEE Wireless Communications Letters. 2017;**6**(6):770-773. DOI: 10.1109/LWC.2017.2740386

[39] Swindlehurst A, Saxena A, Mezghani A, et al. Minimum probability-of-error perturbation precoding for the one-bit massive MIMO downlink. In: 2017 IEEE International Conference on Acoustics, Speech and Signal Processing (ICASSP'17), 5-9 March, 2017; New Orleans, LA, USA. New York: IEEE; 2017. pp. 6483-6487. DOI: 10.1109/ICASSP.2017.7953405

[40] Shao M, Li Q, Ma WK. One-bit massive MIMO precoding via a minimum symbol-error probability design. In: 2018 IEEE International Conference on Acoustics, Speech and Signal Processing (ICASSP'18), 15-20 April, 2018; Calgary, AB, Canada. New York: IEEE; 2018. pp. 3579-3583. DOI: 10.1109/ICASSP.2018.8461980

[41] Jedda H, Mezghani A, Nossek JA, et al. Massive MIMO downlink 1-bit precoding with linear programming for PSK Signaling. In: 2017 IEEE 18th International Workshop on Signal Processing Advances in Wireless Communications (SPAWC'17), 3-6 July, 2017; Sapporo, Japan. New York: IEEE; 2017. pp. 1-5. DOI: 10.1109/SPAWC.2017.8227757

[42] Li A, Masouros C, Swindlehurst AL. 1-bit massive MIMO downlink based on constructive interference. In: 2018 26th European Signal Processing Conference (EUSIPCO'18), 3-7 September, 2018; Rome, Italy. New York: IEEE; 2018. pp. 927-931. DOI: 10.23919/EUSIPCO.2018.8553556

[43] Li A, Liu F, Masouros C, et al. Interference exploitation 1-bit massive MIMO precoding: A partial branch-and-bound solution with near-optimal performance. IEEE Transactions on Wireless Communications. 2020;**19**(5):3474-3489. DOI: 10.1109/TWC.2020.2973987

[44] Tsinos CG, Kalantari A, Chatzinotas S, et al. Symbol-level precoding with low resolution DACs for large-scale Array MU-MIMO systems. In: 2018 IEEE 19th International Workshop on Signal Processing Advances in Wireless Communications (SPAWC'18), 25-28 June, 2018; Kalamata, Greece. New York: IEEE; 2018. pp. 1-5. DOI: 10.1109/SPAWC.2018.8445995

[45] Bertsekas DP. Nonlinear programming. Journal of the Operational Research Society. 1997;**48**(3):334-334. DOI: 10.1057/palgrave.jors.2600425

[46] Sohrabi F, Liu YF, Yu W. One-bit precoding and constellation range design for massive MIMO with QAM signaling. IEEE Journal of Selected Topics in Signal Processing. 2018;**12**(3):557-570. DOI: 10.1109/JSTSP.2018.2823267

[47] Jedda H, Mezghani A, Swindlehurst AL, et al. Precoding under instantaneous per-antenna peak power constraint. In: 2017 25th European Signal Processing Conference (EUSIPCO'17), 28 August-2 September, 2017; Kos, Greece. New York: IEEE; 2017. pp. 863-867. DOI: 10.23919/EUSIPCO.2017.8081330

[48] Ding L, Zhou GT. Effects of even-order nonlinear terms on Predistortion linearization. In: Proceedings of 2002 IEEE 10th Digital Signal Processing Workshop, 2002 and the 2nd Signal Processing Education Workshop, 16 October 2002; Pine Mountain, GA, USA. New York: IEEE; 2002. pp. 1-6. DOI: 10.1109/DSPWS.2002.1231064

[49] Aghdam SR, Jacobsson S, Eriksson T. Distortion-aware linear precoding for Millimeter-wave multiuser MISO downlink. In: 2019 IEEE International Conference on Communications Workshops (ICC Workshops'19), 20-24 May 2019; Shanghai. New York: IEEE. pp. 1-6. DOI: 10.1109/ICCW.2019.8757031

[50] Zayani R, Shaïek H, Roviras D. Efficient precoding for massive MIMO downlink under PA nonlinearities. IEEE Communications Letters. 2019;**23**(9):1611-1615. DOI: 10.1109/LCOMM.2019.2924001

[51] Jee J, Kwon G, Park H. Precoding design and power control for SINR maximization of MISO system with nonlinear power amplifiers. IEEE Transactions on Vehicular Technology. 2020;**69**(11):14019-14024. DOI: 10.1109/TVT.2020.3026752

Chapter 7

# Deep Learning for MIMO Communications

*Yunlong Cai, Qiyu Hu, Guangyi Zhang and Kai Kang*


## Abstract

Recently, deep learning (DL) is becoming a key feature of next-generation multiple-input multiple-output (MIMO) transceiver design with learning and inference capabilities embedded in the network, which achieves greatly enhanced system performance. Popular topics include end-to-end (E2E) learning for transceiver design, deep reinforcement learning (DRL) for communications, and model-driven deep unfolding techniques. In particular, E2E learning treats the communication system design as an E2E data reconstruction task that seeks to jointly optimize transceiver components, so that encoding and decoding are fostered by the learned weights of deep neural networks (DNN). E2E learning can be employed to solve various problems in MIMO communications, such as channel state information (CSI) feedback, beamforming, signal detection, and channel estimation. Moreover, DRL has been widely applied to solve high-dimensional non-convex optimization problems in designing the transceiver. However, these DNNs generally suffer from an inability to be interpreted or generalized, and they often lack performance guarantees. To overcome such drawbacks, substantial researches have proposed to unfold the iterations of an iterative optimization algorithm into a layer-wise structure analogous to a DNN. Inspired by the great potential of these DL methods, it is important to investigate AI-empowered transceivers for future MIMO systems.


**Keywords:** deep learning, deep unfolding, beamforming, transceiver design, channel estimation

## 1. Introduction

In physical layer communications, the transceiver design is a core technology in multiple-input multiple-output (MIMO) systems, as shown in **Figure 1**. Iterative optimization algorithms for the transceiver design have achieved satisfactory system performance, but they generally require a large number of iterations and have the high-complexity computation, which makes it difficult to be deployed in practical systems. Recently, the deep learning (DL) method, as a primary technique in artificial intelligence, has received great attention in wireless communications, especially in physical layer communications. DL methods employ the deep neural networks (DNNs) and treat the algorithm as a "black-box". Compared to conventional optimization algorithms, DL methods can approximate high-complexity operations with

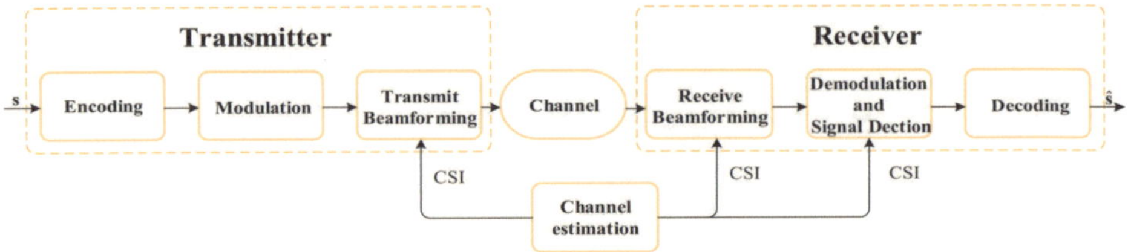


**Figure 1.**
*The architecture of transceiver in physical layer communications.*

lower computational complexity. These DNNs are usually data-driven models, which rely on a large number of data for training. However, it is difficult to obtain training samples in practical communication systems and these data-driven DNNs suffer from poor interpretability and generalization ability. In contrast, model-driven methods exploit known physical mechanisms and domain knowledge. Thus, they require less training samples and it makes the DNNs explainable. Some studies unfolded the iterative optimization algorithms into layer-wise networks with introduced trainable parameters, which reduce the iteration numbers and improve the system performance. This chapter discusses the application of DL-based approaches in physical layer communications, which includes parts of channel estimation and feedback, beamforming, detection, channel decoding and end-to-end learning. Each part will be introduced with data-driven and model-driven approaches.

### 1.1 Channel estimation and feedback

In massive MIMO systems, the base station (BS) relies on accurate channel state information (CSI) to achieve potential gains from multiple antennas. However, the large number of antennas brings challenges and huge overhead for channel estimation and feedback, where many DL-based methods have been proposed to exploit the features of CSI and reduce the overhead [1–7]. In [2], the authors exploited the channel sparsity in the angle domain and proposed a DNN for channel estimation and direction-of-arrival (DoA) estimation. The proposed DL method can learn the spatial structures of channels and achieve better performance than conventional methods. As for channel feedback, the CsiNet has been developed in ref. [3] for channel compression, feedback, and reconstruction. It employs the structure of an autoencoder, where an encoder and a decoder are designed for channel compression and construction, respectively. Compared to the traditional compressive sensing (CS) algorithm, the CsiNet improves the CSI recovery quality and compression ratio.

Model-driven DL approaches have also been applied for channel estimation and feedback [5–7]. In ref. [5], a learned denoising-based approximate message passing (LDAMP) network has been proposed for beamspace millimeter-wave (mmWave) MIMO channel estimation, where the convolutional denoising NN is merged into the AMP channel estimation algorithm. In addition, the authors of ref. [6] proposed a dynamic deep-unfolding neural networks (NNs) with adaptive depth for channel estimation, where the layers of NN vary from different inputs. As shown in **Figure 2 (a)**, $\mathcal{F}(\cdot)$ denotes each layer of the NN and a function $\phi$ is defined to control the depth of the NN. When the output of the function $\tau > \varepsilon$, the NN stops to output results. To estimate CSI, the sparse Bayesian learning (SBL) algorithm is unfolded into a layered network with introduced trainable parameters in the framework. In particular, some priori parameters which are difficult to determine in the SBL algorithm are set as

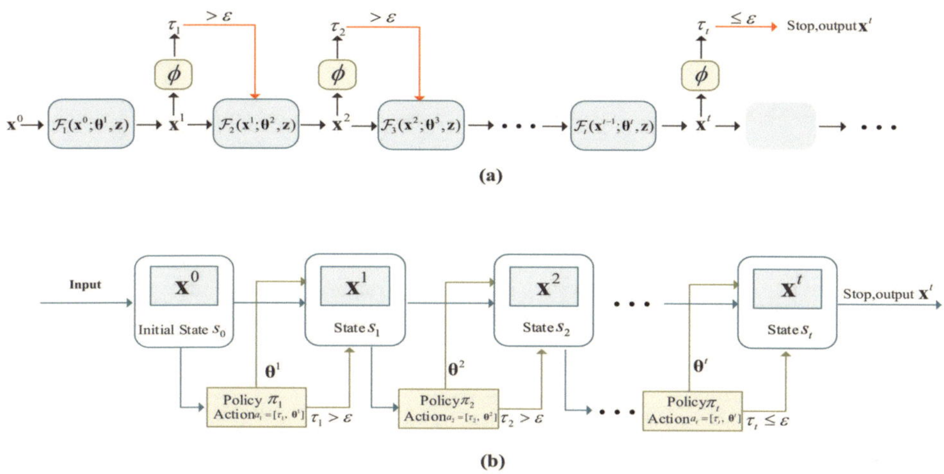


**Figure 2.**
*DDPG-driven deep-unfolding framework with adaptive depth.*

trainable parameters. The other trainable parameters are introduced to approximate the operations with high computational complexity. Besides, to avoid gradient explosion, the trainable parameters are updated by deep deterministic policy gradient (DDPG), rather than updated by the stochastic gradient descent (SGD) algorithm directly. As shown in **Figure 2(b)**, the state, action, and state transition of DDPG correspond to the optimization variables, trainable parameters, and architecture of NN, respectively. As for channel feedback, the authors of ref. [7] proposed a model-driven multiple-measurement-vectors learned approximate message passing (MMV-LAMP) network for channel estimation and feedback in frequency division duplex (FDD) systems, which reduces the pilot feedback overhead.

## 1.2 Beamforming

In massive MIMO systems, beamforming has been a key technique to improve the spectrum efficiency and achieve spatial multiplexing gains. Traditional beamforming algorithms require a large number of iterations and high-complexity computations, which impedes their application in practical systems, especially when the number of antennas is large. Thus, many DL-based approaches for beamforming design have been proposed [8–14]. In ref. [8], the authors proposed a DNN-enabled massive MIMO framework for effective hybrid beamforming. Compared to conventional schemes, the proposed framework achieves better performance with lower computational complexity. Besides, a DL-based joint channel feedback and beamforming approach has been designed in ref. [9]. In addition, deep reinforcement learning (DRL) based beamforming algorithms have also been developed in ref. [10, 11]. The authors of ref. [10] proposed a DRL hybrid beamforming scheme to improve the coverage range of THz communications in the reconfigurable intelligent surfaces (RIS) assisted system.

Apart from the aforementioned data-driven NNs, researchers developed the model-driven methods where iterative beamforming algorithms are unfolded into networks [12–14]. In ref. [12], the authors proposed an iterative algorithm-induced deep-unfolding neural network (IAIDNN) for digital beamforming shown in

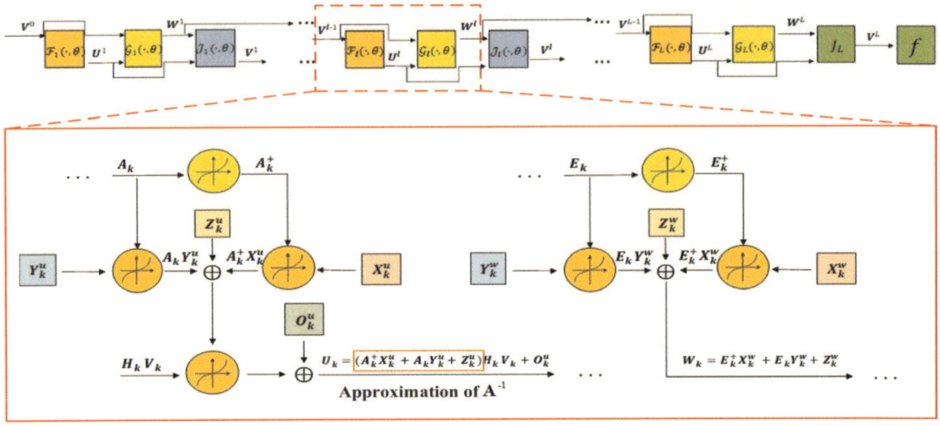


**Figure 3.**
*The architecture of IAIDNN which unfolds the WMMSE algorithm into a layer-wise network.*

**Figure 3**, where the weighted minimum mean-square error (WMMSE) iterative algorithm is unfolded into a layer-wise network. $\mathcal{F}$, $\mathcal{G}$, and $\mathcal{J}$ denote the layers of the network for updating different variables. In particular, inspired by the first-order Taylor expansion, the matrix inversion $\mathbf{A}^{-1}$ is approximated by $\mathbf{A}^{\dagger}\mathbf{X} + \mathbf{A}\mathbf{Y} + \mathbf{Z}$, where $\mathbf{X}$, $\mathbf{Y}$, and $\mathbf{Z}$ are introduced trainable parameters. $\mathbf{A}^{\dagger}$ represents the proposed non-linear operation where the diagonal elements of $\mathbf{A}$ are taken the reciprocal and non-diagonal elements are set as 0. The computational complexity of matrix inversion is $\mathcal{O}(n^3)$ while that of the proposed approximation is $\mathcal{O}(n^{2.37})$. In the backpropagation, the authors derived the generalized chain rule (GCR) in matrix form and the trainable parameters are updated based on it. Simulations have shown that the proposed IAIDNN achieves the performance of the WMMSE algorithm with much less iterations. Besides, the authors of ref. [13] developed a deep-unfolding framework for the passive and active beamforming joint design in a RIS-assisted MIMO system, which outperforms the conventional iterative algorithms.

### 1.3 MIMO detection

In MIMO systems, the detector plays an important role in the receiver. Traditional iterative detection algorithms are designed based on the assumption that the channel model is subject to a specific distribution, thus the performance is unsatisfactory in variable environments. To tackle the issue, DL-based detectors have been proposed [1, 15–20]. In refs. [1, 15, 16], the authors proposed a DNN-based joint channel estimation and signal detection algorithm for the receiver design where the detectors are designed to adapt to different wireless channels.

Several model-driven DL-based methods based on conventional iterative detectors have also been investigated in recent years [17–20]. The authors of ref. [17] unfolded the orthogonal approximate message passing (OAMP) detector into a layer-wise structure named OAMP-Net2. It only introduces a few learnable parameters to improve the stability and speed of convergence, and the parameters are optimized to adapt to different channel environments. Besides, a deep detector named LoRD-Net has been proposed for signal detection [18]. The LoRD-Net incorporates domain knowledge in its architecture design, thus requiring much fewer parameters than data-driven NNs. Furthermore, a joint channel estimation and signal detection model-driven NN has been proposed in ref. [20] to reduce the effect of channel estimation errors on detection.

### 1.4 Channel decoding

With the development of fifth generation (5G), user data and system capacity have rapidly increased and a higher transmission rate means lower decoding latency demand. However, traditional decoders require high-complexity computation and a large number of iterations. To address the issue, DL-based decoding algorithms have been developed [21–26]. In ref. [21], researchers explored that it is easier for DL decoders to learn the structured codes than random codes and verified that NNs can learn a form of decoding algorithm, rather than only a classifier. The article [22] focuses on the issue that the successive interference cancelation (SIC) decoding is imperfect in the nonorthogonal multiple access (NOMA) system and proposes a novel DL-based scheme for decoding in MIMO-NOMA systems. A non-linear precoder and SIC decoder have been constructed by deep feedforward neural networks (FNNs) which help received signals decode accurately in the SIC manner.

The prior parameters play an important role in conventional iterative decoders but are usually set by experience. Thus, DL is a proper method to find the optimal value for the prior parameters and thus model-driven based decoding methods are promising techniques [24–26]. The authors of ref. [24] utilized the DL method to find the proper weights to the passing messages in the Tanner graph and achieved comparable performance with belief propagation (BP) decoders with less iterations. Furthermore, considering many expensive multiplication operations in ref. [24] which make it difficult to implement, Lugosch and Gross [25] proposed a neural offset min-sum decoding algorithm with no multiplications and less parameter computation. The proposed approach speeds up the training process and is friendly for hardware implementation. In addition, a model-driven low-density parity-check (LDPC) decoding network has been developed in ref. [26]. The iterative decoding progress between checking nodes and variable nodes is unfolded into a propagation network, which combines the advantages of deep learning and conventional normalized min-sum LDPC decoding methods.

### 1.5 End-to-end learning

The aforementioned DL-based approaches are optimized locally for individual modules, where global optimality cannot be guaranteed. The modules in the transceiver are usually highly correlated with each other, and thus a joint design can achieve better performance than a separate design. To fulfill the global optimization, several DL-based end-to-end communication systems have been proposed [27–32], where all the trainable parameters are updated based on an end-to-end loss. A DNN-based based end-to-end wireless communication system has been proposed in ref. [27], which includes channel encoding, decoding, modulation, and equalization. A conditional generative adversarial net (GAN) has been designed to model the channel distribution and the proposed end-to-end approach is effective on Rayleigh fading channels. In ref. [28], the authors proposed a DNN-based end-to-end joint transceiver design algorithm for FDD mmWave MIMO systems, which consists of the modules of pilot training, channel feedback and reconstruction, and hybrid beamforming. To avoid CSI mismatch caused by the transmission delay and feedback overhead, a two-timescale scheme has been considered. Specifically, a superframe is introduced as the long-timescale, where the CSI statistics remain constant. Each superframe consists of several frames, each of which contains a number of time slots that defined the short timescale. During each time slot, the instantaneous CSI remains unchanged. Correspondingly, a two-timescale DNN is developed as shown in **Figure 4**, which consists of a long-term DNN and a short-term DNN. The long-term DNN consists of modules of pilot training, high-dimensional CSI estimation and feedback, and hybrid beamforming, while the short-term DNN composes of modules of pilot training, low-dimensional equivalent CSI estimation and feedback, and digital beamforming. At the end of each frame, the long-term DNN is employed to obtain the high-dimensional full CSI to update the long-term analog beamformers. The short-term digital beamformers are updated based on the low-dimensional equivalent CSI acquired by the short-term DNN. The trainable parameters of all the modules are optimized to minimize the bit-error-rate (BER), which is the system's global optimization objective.

Inspired by ref. [28], the authors of ref. [31] designed a model-driven based end-to-end framework for joint transceiver design in time division duplexing (TDD) systems, which consists of a channel estimation deep-unfolding NN (CEDUN) and a hybrid beamforming deep-unfolding NN (HBDUN). As shown in **Figure 5**, the CEDUN is comprised of a pilot training NN and a recursive least squares (RLS) algorithm-induced deep-unfolding NN, where a set of trainable parameters are introduced to increase the degrees of freedom. For hybrid beamforming, the stochastic successive convex approximation (SSCA) algorithm is unfolded into a layer-wise structure in the HBDUN, which consists of an analog NN and a digital NN. Specifically, the phase of analog beamformers is set as trainable parameters in the analog NN. In the digital NN, two non-linear operations are introduced to approximate the matrix inversion. In addition, the authors consider the mixed-timescale scheme, where long-term analog beamformers are optimized based on the CSI statistics during a superframe, and short-term digital beamformers are updated in each time slot based on equivalent CSI. According to the mixed-timescale scheme, a novel two-stage training method is investigated to jointly train the framework. **Figure 5(a)** shows the first stage of training, where the trainable parameters of HBDUN are optimized with the loss function of the negative system sum rate. The second training stage is shown in **Figure 5(b)**, where the parameters of analog NN are fixed, and low-dimensional equivalent CSI is obtained. The parameters of CEDUN and digital NN are optimized in

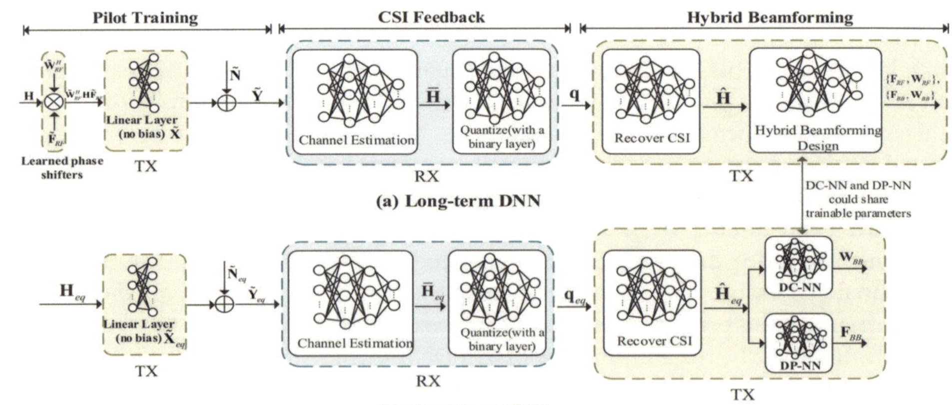


**Figure 4.**
*The architecture of the proposed DNN-based end-to-end framework.*

*DOI: http://dx.doi.org/10.5772/intechopen.112038*

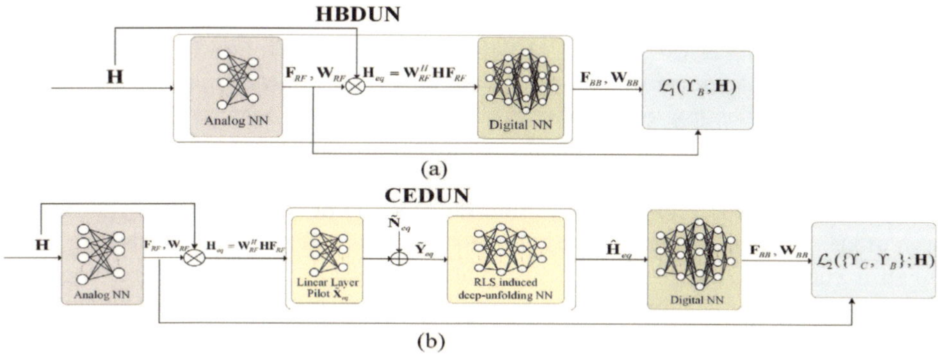


**Figure 5.**
*The architecture of the end-to-end deep-unfolding framework.*

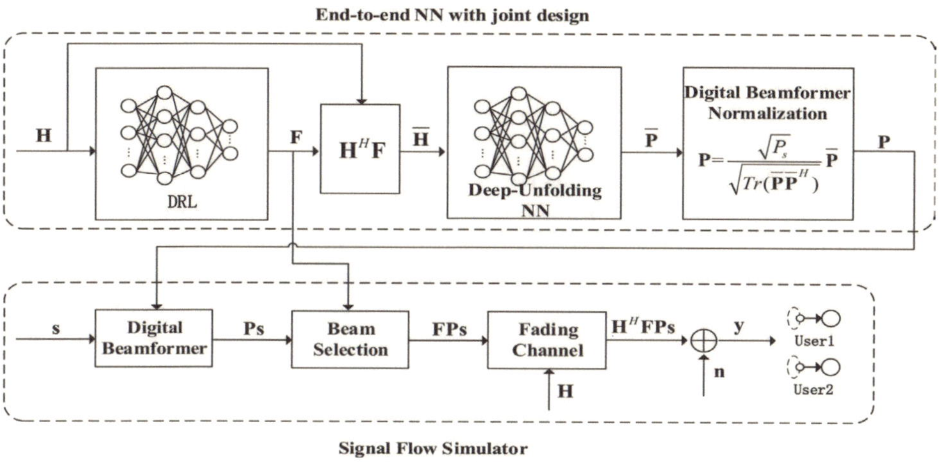


**Figure 6.**
*The architecture of the end-to-end DRL and deep-unfolding framework.*

this stage with the same loss function in the first stage. Simulation results have shown that the deep-unfolding NNs perform comparable with the traditional algorithms with reduced complexity and the joint design method achieves better performance than the separate design.

In addition, the authors of ref. [32] proposed an end-to-end DRL and deep-unfolding framework for joint beam selection and digital beamforming design, the architecture of which is shown in **Figure 6**. Specifically, the framework consists of a DRL-based NN for beam selection and a model-driven based NN for digital beamforming. A novel training method has been developed to jointly train the DRL-based NN and unfolding NN in an end-to-end way. This work indicates that the model-driven NNs can be trained with other DL methods such as DRL-based NN.

## 2. Deep learning for MIMO-based semantic communications

Future 6G wireless networks are expected to bridge the physical and cyber worlds, enabling human interactions with multiple intelligent devices through various data modalities like images and text [33]. This introduces a number of applications from autonomous driving to the Internet of Everything, which involves intelligent human-to-machine and machine-to-machine connections. These new fascinating applications have imposed challenging requirements on communication networks, including ultra-high reliability, ultra-low latency, and extremely high data rates [34]. However, supporting and enabling such applications will require coping with explosive growth in bandwidth and complexity, due to the transmission of these massive datasets and large models. Driven by the aforementioned requirements, there is a springing up of semantic communication research in both academia and industry. The growing trend of semantic communications aims at accurately recovering the statistical structure of the underlying information of the source signal and designing the communication transceiver in an end-to-end fashion, similar to joint source and channel coding (JSCC) by taking the source semantics into account. A data-aware communication transceiver with intelligence that is able to understand the relevance and meaning of data traffic is of paramount importance as it would significantly improve the transmission efficiency.

## 2.1 Formulation of semantic communications

In general, a semantic communication MIMO transceiver can be modeled as the framework shown in **Figure** 7, where an end-to-end communication system is developed to incorporate coding and modulation [35]. In particular, the encoding, decoding, and transmission procedures are parameterized by the DNNs, and the system is optimized in a back-propagation manner with the data-driven method. In particular, as shown in **Figure** 7, the transmitter maps the source, $\mathbf{s}$, into a symbol stream, $\mathbf{x}$, and then passes it through the physical channel with transmission impairments. The received symbol stream, $\mathbf{y}$, is decoded at the receiver to have an estimation of the source, $\hat{\mathbf{s}}$. Both the transmitter and the receiver are represented by DNNs. In particular, the DNNs at the transmitter consist of the semantic encoder and channel encoder, while the DNNs at the receiver consist of the semantic decoder and channel decoder. The semantic encoder learns to transform the transmitted data into an encoded feature vector while the semantic decoder learns to recover the transmitted data from the received signals. Moreover, the channel encoder and channel decoder aim at eliminating the signal distortion caused by the wireless channel.

We consider a MIMO system with $N_t$ transmit antennas and $N_r$ receive antennas. The encoded symbol stream can be represented by

$$\mathbf{x} = f_2(f_1(\mathbf{s};\boldsymbol{\theta}_1);\boldsymbol{\theta}_2), \tag{1}$$

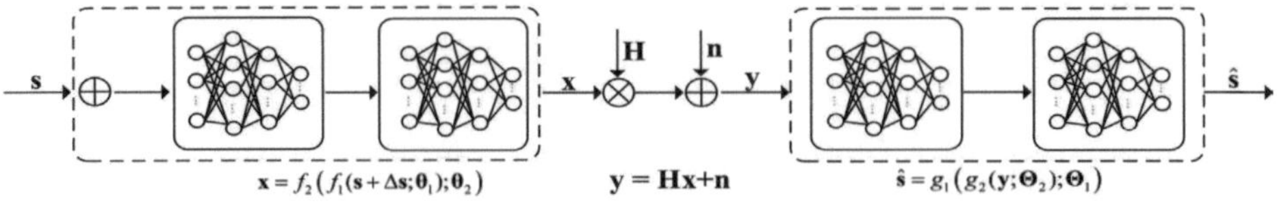


**Figure 7.**
*The framework of a typical semantic communication system.*

where $\mathbf{x} \in \mathbb{C}^{N_t \times 1}$, $\boldsymbol{\theta}_1$ and $\boldsymbol{\theta}_2$ denote the trainable parameters of the semantic encoder, $f_1(\cdot)$, and the channel encoder, $f_2(\cdot)$, respectively. Subsequently, the signal received at the receiver, $\mathbf{y} \in \mathbb{C}^{N_r \times 1}$, is given by

$$\mathbf{y} = \mathbf{H}\mathbf{x} + \mathbf{n}, \tag{2}$$

where $\mathbf{H} \in \mathbb{C}^{N_r \times N_t}$ denotes the channel matrix and $\mathbf{n} \sim \mathcal{CN}(\mathbf{0}, \sigma^2\mathbf{I})$ is the additive white Gaussian noise (AWGN). Correspondingly, the decoded signal is given as

$$\hat{\mathbf{s}} = g_1(g_2(\mathbf{y}; \boldsymbol{\Theta}_2); \boldsymbol{\Theta}_1), \tag{3}$$

where $\boldsymbol{\Theta}_1$ and $\boldsymbol{\Theta}_2$ denote the trainable parameters of the semantic decoder, $g_1(\cdot)$, and the channel decoder, $g_2(\cdot)$, respectively. For clarity, we denote $\boldsymbol{\theta}$ as the set of trainable parameters and $f_{\boldsymbol{\theta}}(\cdot)$ as the DNNs in the considered semantic communication systems. Thus, we have $\hat{\mathbf{s}} = f_{\boldsymbol{\theta}}(\mathbf{s})$.

### 2.2 Semantic importance-gudied design for MIMO transceivers

Regarding the physical layer transceiver, the modules of the transceiver are often optimized independently. In particular, the modulation, beamforming, and signal detection modules are designed to minimize the bit-error-rate (BER), and the channel feedback and estimation aim to optimize the mean-square-error (MSE). In the case of semantic communications, the MIMO transceiver can be designed by revising the modules in the traditional transceiver. Next, we will discuss some advancements in the MIMO transceiver design in semantic communications.

1. **Measure of Semantic Importance:** In semantic communications, the source data is mapped into different semantic features by the DNN models. Different semantic features are of different importance for completing target tasks, where semantic importance is defined as the correlation between the semantic features and the target task [36]. The method to measure the importance of semantic features can be variable in different semantic communication systems. Specifically, there are gradient-based approaches that draw on ideas related to the interpretability of DNNs [37], and the entropy-modeling approach that focuses on the regional complexity in terms of the source data content [38], etc. In semantic resource allocation, the model distinguishes the importance levels of different features and adaptively allocates the resource according to channel conditions and users' requirements. In particular, the semantic importance has shown great potential for the designs in DL-based multi-user MIMO systems, e.g., the design of precoding, the allocation of subcarriers, and various adaptive schemes.

2. **MIMO Precoding:** One of the main issues in multi-user MIMO systems is the mutual interference between the signals of different users. Limited by the number of receiving antennas, it is difficult for each user to eliminate the interference from other users alone. It is worth noting that the precoding algorithms such as singular value decomposition (SVD)-based precoding, convert the MIMO channel into a set of parallel subchannels with different SNRs. Intuitively, the performance of the MIMO systems can be significantly enhanced

by allocating subchannels with high SNRs to features with high importance levels, as these features would play a more important role in the target task. For instance, the semantic MIMO system designed in ref. [39] significantly outperforms traditional MIMO systems by jointly considering the CSI and entropy distribution of the semantic features, where the entropy can be regarded as a measure of semantic importance.

3. **Allocation of Subcarriers in OFDM Systems:** Orthogonal frequency-division multiplexing (OFDM) technique has been widely employed in MIMO systems to realize high-speed data transmission. In OFDM, each subcarrier is equivalent to a subchannel with a certain CSI. The CSI of each subcarrier is instrumental in power allocation to boost the communication rate. However, in the case of semantic communications, the CSI can also be exploited to determine the allocation of subcarriers to different semantic features according to their importance. A typical example is the dual-attention mechanism proposed in ref. [40], which employs both channel-wise attention and spatial attention, and jointly learns to transmit important features with better subcarriers, which achieves state-of-the-art performance among existing JSCC schemes.

4. **Adaptive Design Based on CSI:** Semantic importance can also be exploited to design adaptive schemes for MIMO systems based on CSI. These adaptive designs consider both CSI and semantic importance in an adaptable manner, which significantly improves the system performance and reduces the training overhead. For image transmission over MIMO channels, the authors in ref. [41] have employed the channel attention module proposed in ref. [42], to distinguish the importance of different features and adjust their weights according to different CSI scenarios. Based on feature importance and the complexity of image, the authors in ref. [43] have proposed an adaptive CSI feedback scheme for precoding, which improves the effectiveness by adjusting the feedback overhead.

## Author details

Yunlong Cai*, Qiyu Hu, Guangyi Zhang and Kai Kang
College of Information Science and Electronic Engineering, Zhejiang University, China

*Address all correspondence to: ylcai@zju.edu.cn

## References

[1] Ye H, Li GY, Juang B-H. Power of deep learning for channel estimation and signal detection in OFDM systems. IEEE Wireless Communications Letters. 2018; **7**(1):114-117

[2] Huang H, Yang J, Huang H, Song Y, Gui G. Deep learning for super-resolution channel estimation and DOA estimation based massive MIMO system. IEEE Transactions on Vehicular Technology. 2018;**67**(9):8549-8560

[3] Wen C-K, Shih W-T, Jin S. Deep learning for massive MIMO CSI feedback. IEEE Wireless Communications Letters. 2018;**7**(5): 748-751

[4] Wang T, Wen C-K, Jin S, Li GY. Deep learning-based CSI feedback approach for time-varying massive MIMO channels. IEEE Wireless Communications Letters. 2019;**8**(2):416-419

[5] He H, Wen C-K, Jin S, Li GY. Deep learning-based channel estimation for beamspace mmwave massive MIMO systems. IEEE Wireless Communications Letters. 2018;**7**(5):852-855

[6] Hu Q, Shi S, Cai Y, Yu G. DDPG-driven deep-unfolding with adaptive depth for channel estimation with sparse bayesian learning. IEEE Transactions on Signal Processing. 2022;**70**:4665-4680

[7] Ma X, Gao Z, Gao F, Di Renzo M. Model-driven deep learning based channel estimation and feedback for millimeter-wave massive hybrid MIMO systems. IEEE Journal on Selected Areas in Communications. 2021;**39**(8): 2388-2406

[8] Huang H, Song Y, Yang J, Gui G, Adachi F. Deep-learning-based millimeter-wave massive MIMO for hybrid precoding. IEEE Transactions on Vehicular Technology. 2019;**68**(3): 3027-3032

[9] Sohrabi F, Attiah KM, Yu W. Deep learning for distributed channel feedback and multiuser precoding in FDD massive MIMO. IEEE Transactions on Wireless Communications. 2021; **20**(7):4044-4057

[10] Huang C, Yang Z, Alexandropoulos GC, Xiong K, Wei L, Yuen C, et al. Multi-hop RIS-empowered terahertz communications: A DRL-based hybrid beamforming design. IEEE Journal on Selected Areas in Communications. 2021;**39**(6):1663-1677

[11] Wang Q, Feng K, Li X, Jin S. PrecoderNet: Hybrid beamforming for millimeter wave systems with deep reinforcement learning. IEEE Wireless Communications Letters. 2020;**9**(10): 1677-1681

[12] Hu Q, Cai Y, Shi Q, Xu K, Yu G, Ding Z. Iterative algorithm induced deep-unfolding neural networks: Precoding design for multiuser MIMO systems. IEEE Transactions on Wireless Communications. 2021;**20**(2): 1394-1410

[13] Liu Y, Hu Q, Cai Y, Yu G, Li GY. Deep-unfolding beamforming for intelligent reflecting surface assisted full-duplex systems. IEEE Transactions on Wireless Communications. 2022; **21**(7):4784-4800

[14] Shi S, Cai Y, Hu Q, Champagne B, Hanzo L. Deep-unfolding neural-network aided hybrid beamforming based on symbol-error probability minimization. IEEE Transactions on Vehicular Technology. 2023;**72**(1): 529-545

[15] Hua H, Wang X, Xu Y. Signal detection in uplink pilot-assisted multi-user MIMO systems with deep learning. In: Computing, Communications and IoT Applications. 2019. pp. 369–373

[16] Yi X, Zhong C. Deep learning for joint channel estimation and signal detection in OFDM systems. IEEE Communications Letters. 2020;**24**(12): 2780-2784

[17] He H, Wen C-K, Jin S, Li GY. Model-driven deep learning for MIMO detection. IEEE Transactions on Signal Processing. 2020;**68**:1702-1715

[18] Khobahi S, Shlezinger N, Soltanalian M, Eldar YC. LoRD-net: Unfolded deep detection network with low-resolution receivers. IEEE Transactions on Signal Processing. 2021; **69**:5651-5664

[19] Samuel N, Diskin T, Wiesel A. Learning to detect. IEEE Transactions on Signal Processing. 2019;**67**(10):2554-2564

[20] Zhang Y, Sun J, Xue J, Li GY, Xu Z. Deep expectation-maximization for joint MIMO channel estimation and signal detection. IEEE Transactions on Signal Processing. 2022;**70**:4483-4497

[21] Gruber T, Cammerer S, Hoydis J, Brink ST. On deep learning-based channel decoding. In: Annual Conference on Information Sciences and System. 2017. pp. 1–6

[22] Kang J-M, Kim I-M, Chun C-J. Deep learning-based MIMO-NOMA with imperfect SIC decoding. IEEE Systems Journal. 2020;**14**(3):3414-3417

[23] Cao C, Li D, Fair I. Deep learning-based decoding of constrained sequence codes. IEEE Journal on Selected Areas in Communications. 2019;**37**(11): 2532-2543

[24] Nachmani E, Be'ery Y, Burshtein D. Learning to decode linear codes using deep learning. In: Annual Allerton Conference on Communication, Control, and Computing. 2016. pp. 341–346

[25] Lugosch L. Gross WJ. Neural offset min-sum decoding. In: IEEE International Symposium on Information Theory. 2017. pp. 1361–1365

[26] Wang Q, Wang S, Fang H, Chen L, Chen L, Guo Y. A model driven deep learning method for normalized min-sum LDPC decoding. In: IEEE International Conference on Communications Workshops. 2020. pp. 1–6

[27] Ye H, Li GY, Juang B-HF, Sivanesan K. Channel Agnostic End-to-End Learning Based Communication Systems with Conditional GAN. AbuDhabi, UAE: Proc. IEEE Global Commun. Conf. Workshops; 2018. pp. 1-5

[28] Hu Q, Cai Y, Kang K, Yu G, Hoydis J, Eldar YC. Two-timescale end-to-end learning for channel acquisition and hybrid precoding. IEEE Journal on Selected Areas in Communications. 2022;**40**(1):163-181

[29] Dorner S, Cammerer S, Hoydis J, Brink ST. Deep learning based communication over the air. IEEE Journal of Selected Topics Signal Process. 2018;**12**(1):132-143

[30] Aoudia FA, Hoydis J. End-to-end learning of communications systems without a channel model. Asilomar Conference on Signals, Systems, and Computers. 2018:298-303

[31] Kang K, Hu Q, Cai Y, Yu G, Hoydis J, Eldar YC. Mixed-timescale deep-unfolding for joint channel estimation and hybrid beamforming.

IEEE Journal on Selected Areas in Communications. 2022;**40**(9): 2510-2528

[32] Hu Q, Liu Y, Cai Y, Yu G, Ding Z. Joint deep reinforcement learning and unfolding: Beam selection and precoding for mmwave multiuser MIMO with lens arrays. IEEE Journal on Selected Areas in Communications. 2021;**39**(8): 2289-2304

[33] Zhang G, Hu Q, Qin Z, Cai Y, Yu G. A unified multi-task semantic communication system with domain adaptation. In: IEEE Global Communications Conference. 2022. pp. 3971–3976

[34] Niu K, Dai J, Yao S, Wang S, Si Z, Qin X, et al. A paradigm shift toward semantic communications. IEEE Communications Magazine. 2022; **60**(11):113-119

[35] Hu Q, Zhang G, Qin Z, Cai Y, Yu G, Li GY. Robust semantic communications with masked VQ-VAE enabled codebook. IEEE Transactions Wireless Communications, early access. 2023

[36] Liu C, Guo C, Yang Y, Jiang N. Adaptable semantic compression and resource allocation for task-oriented communications. arXiv preprint arXiv: 2204.08910. 2022

[37] Selvaraju RR, Cogswell M, Das A, Vedantam R, Parikh D, Batra D. Grad-CAM: Visual explanations from deep networks via gradient-based localization. In: Proc. IEEE Int. Conf. Comput. Vis. (ICCV). Venice, Italy: IEEE; 2017

[38] Balle J, Chou PA, Minnen D, Singh S, Johnston N, Agustsson E, et al. Nonlinear transform coding. IEEE Journal of Selected Topics Signal Process. 2021; **15**(2):339-353

[39] Yao S, Wang S, Dai J, Niu K, Zhang P. Versatile semantic coded transmission over MIMO fading channels. arXiv preprint arXiv: 2210.16741. 2022

[40] Wu H, Shao Y, Mikolajczyk K, Gündüz D. Channel-adaptive wireless image transmission with OFDM. IEEE Wireless Communications Letters. 2022; **11**(11):2400-2404

[41] Bian C, Shao Y, Wu H, Gunduz D. Space-time design for deep joint source channel coding of images over MIMO channels. arXiv preprint arXiv: 2210.16985. 2022

[42] Xu J, Ai B, Chen W, Yang A, Sun P, Rodrigues M. Wireless image transmission using deep source channel coding with attention modules. IEEE Transactions on Circuits and Systems for Video Technology. 2022;**32**(4):2315-2328

[43] Zhang G, Hu Q, Cai Y, Yu G. Adaptive CSI feedback for deep learning-enabled image transmission. arXiv preprint arXiv:2302.13477. 2023

Section 2

# Antenna Techniques

Chapter 8

# Holographic Beamforming

*Ali Araghi and Mohsen Khalily*


## Abstract

This chapter presents the fundamentals of the holography technique to form the beam in electromagnetic (EM) structures. The application of holography in leaky-wave antennas, metasurface reflectors, and reconfigurable intelligent surfaces (RISs) is explained. Consequently, different methods to analyze and realize an EM hologram are presented. A comparison is made between forming the beam via holographic-based radiators, phased-array antennas, and MIMO systems. The thing which is common between these three is that all of them can contain a number of elements that are repeated in a fashion. However, the functionality of these elements in the three mentioned structures is totally different from each other. This concept is explained in detail in this chapter.


**Keywords:** holography technique, leaky-wave antenna, metasurface, periodic structures, MIMO systems, phased-array antenna

## 1. Introduction

Consider a case where an electromagnetic (EM) aperture with an arbitrary geometry is placed on the $xy$ plane as shown in **Figure 1(a)**. To have the ability to form the constructed beam, it is required to control both phase constant ($\beta$) and amplitude ($\alpha$) of the EM waves at different segments of the aperture. The beam tilt angle can typically be controlled by regulating $\beta$ whereas $\alpha$ distribution over the aperture controls side-lobe-level (SLL). Four segments are marked in **Figure 1(a)** as an illustrative example. In the case of a conventional phased-array antenna, these four segments represent four physical elements, i.e. antennas, where $\beta$ and $\alpha$ of each can be controlled in straightforward approaches by using phase shifters and attenuators respectively, or with a fully passive custom-built feeding network with proper delay lines and by applying the power-splitting technique. All these elements make the final EM aperture together with an engineered beam as $\beta$ and $\alpha$ are governed at different positions of the aperture.

Now consider the case where these four elements are located in such a way that they are not able to have a sensible impact on each other to build up a large EM aperture. Under this circumstance, each element acts as an individual aperture as shown in **Figure 1(b)** with its specific radiation properties. This configuration of elements can be applied in multiple input multiple output (MIMO) systems. With a dedicated port for each element, this configuration represents space diversity provided that the cross-correlation between the ports is kept low. It is also possible to use a single element and connect more than one port to it. In this case, each port belongs

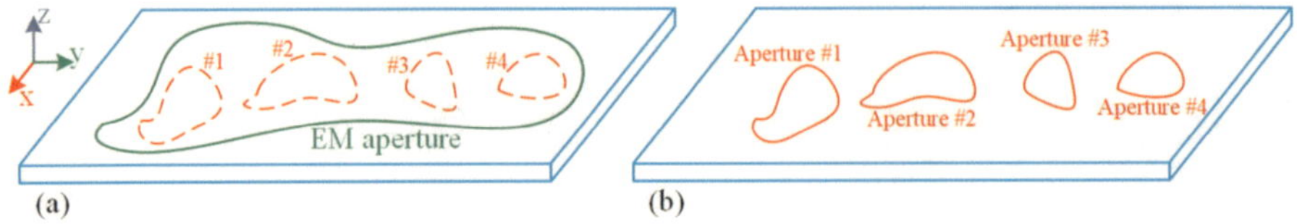


**Figure 1.**
*Aperture formation. (a) The case where four elements make a large aperture altogether and (b) the case where four elements create four separated apertures.*

to a specific radiation state or the so-called mode. Having orthogonal modes in this scenario will lead to a low cross-correlation between the ports, making the structure a good candidate for MIMO systems. This orthogonality can be obtained in radiation patterns (pattern diversity) or polarization (polarization diversity) or a combination of both.

Let us move back to the large EM-aperture of **Figure 1(a)**. The envision of such a large aperture is not limited to just phased-array antennas and can be obtained by several means including but not limited to leaky-wave structures, reflectarrays, and transmitarrays. Forming the beam in such structures is also fulfilled by regulating $\beta$ and $\alpha$ over the aperture but in approaches different from conventional phased arrays. One approach is to employ the holography technique [1] to govern $\beta$ on the structure and to correspondingly control the tilt angle of beam(s) which is known as "holographic beamforming".

This chapter presents the principles of the holography technique and then explores its capability to form the beam. To this end, some background information on leaky-wave structures and reflectarrays is required.

## 2. Holographic-based antennas

### 2.1 Holographic-based leaky-wave antennas

Let us start with the application of holography in leaky-wave antennas. A holographic leaky-wave antenna is a type of antenna that utilizes the principles of holography and leaky-wave propagation to construct the beam and achieve beam scanning capabilities [2]. A leaky-wave antenna (LWA) operates by "leaking" EM energy along its length, which leads to the formation of a propagating wave [3]. Unlike conventional resonating antennas that typically radiate energy perpendicular to the antenna's length, LWAs emit energy at an angle $\theta_m$ along their length. This tilt angle can be controlled by regulating the phase constant $\beta$ of the guided waves across the structure and formulated as [4]:

$$\theta_m \approx \sin^{-1}\left(\frac{\beta}{k_0}\right), \tag{1}$$

where $k_0$ being the free-space wavenumber. Considering (1), $\theta_m$ will have a real answer if and only if $|\beta| < k_0$. This means that LWAs support fast waves on the guide.

Holography, which originates from optics, is a technique to achieve a desired $\beta$ on the structure by governing the phase distribution on the structure. As $\beta$ is controlled in an LWA, the tilt angle of the constructed beam can be specified by (1). This technique is summarized below:

Having a dielectric slab on the $xy$ plane to design the aperture on, the first step is to define two field distributions on the structure known as reference wave $E_{\text{ref}}$ and object wave $E_{\text{obj}}$, generated by two hypothetical sources. To be more explicit, a source should be defined somewhere within the slab where it is aimed to place an actual surface-wave launcher (SWL). For a lossless structure, an ideal hypothetical source located at $(x = 0, y = 0)$ will generate a radially expanded field distribution on the slab for both TM and TE surface waves which can be formulated as below [5]:

$$E_{\text{ref}} = Ae^{-j\beta_r r}, \tag{2}$$

where $r = \sqrt{x^2 + y^2}$ and $A$ is the wave amplitude. This is schematically shown in **Figure 2(a)**. It is clear that the location and type of this source can be in a variety of forms. For example, it is possible to place a number of ports at one end of the slab to generate parallel phase lines as shown in **Figure 2(b)** with the formulated reference wave of $E_{\text{ref}} = Ae^{j\beta_y y}$. It is also possible to locate the source somewhere out of the slab as presented in **Figure 2(c)**. Under this circumstance, the final structure will not recognize as an LWA; this case is explained more in the next section.

The next step is to define an object wave $E_{\text{obj}}$ on the slab. To this end, another hypothetical source should be defined far from the slab toward the direction of the desired constructed beam. The slab is illuminated by this source and the corresponding induced waves should be calculated which represents $E_{\text{obj}}$. For example, for a beam desired to be formed toward $(\theta_m, \phi_m)$, the induced $E_{\text{obj}}$ is obtained by the mapping as below:

$$E_{\text{obj}} = Be^{jk_0\{\sin(\theta_m)\cos(\phi_m)x + \sin(\theta_m)\sin(\phi_m)y\}}, \tag{3}$$

where $B$ is the amplitude of the object waves.

The next step is to calculate the superposition of $E_s = E_{\text{ref}} \times E_{\text{obj}}$ as an interference pattern where $\angle E_s$ defines the desired EM hologram.

The aforementioned steps of calculating $E_{\text{ref}}$, $E_{\text{obj}}$, and $E_s$ make the "recording" process altogether which means to record the impact of the influencing parameters on the slab. For example, consider a case where an ideal source generates $E_{\text{ref}}$ as presented in **Figure 3(a)** on the slab at a specific frequency. With a defined object wave toward $(\theta_m = \pi/4, \phi_m = 2\pi/3)$, the obtained $E_{\text{obj}}$ on the slab is shown in

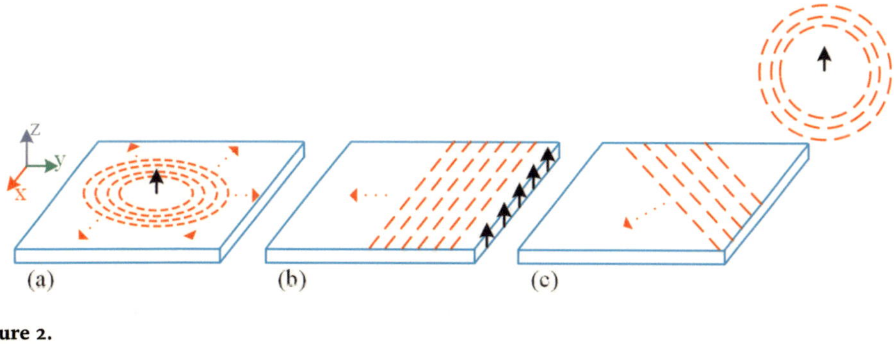


**Figure 2.**
*Different forms of reference wave on the slab. (a) Radial reference wave by an ideal single source at the center of the slab, (b) parallel reference wave formed by a number of sources at one edge of the slab and (c) induced reference wave from a source located outside the slab.*

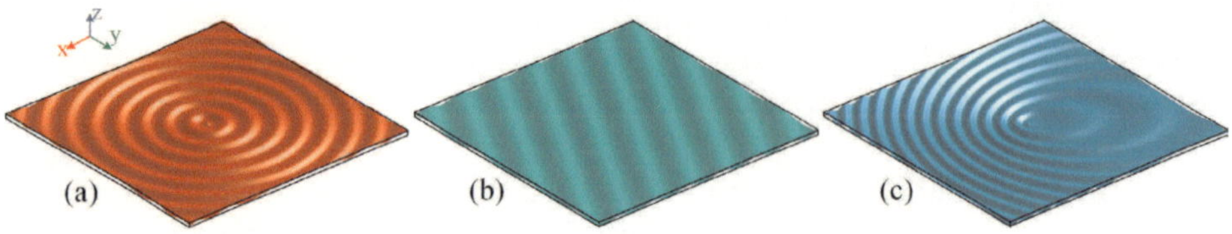


**Figure 3.**
*Recording process: (a) $E_{ref}$ with an ideal source at the center of the slab, (b) $E_{obj}$ in case of $(\theta_m = \pi/4, \phi_m = 2\pi/3)$, and (c) the obtained EM hologram.*

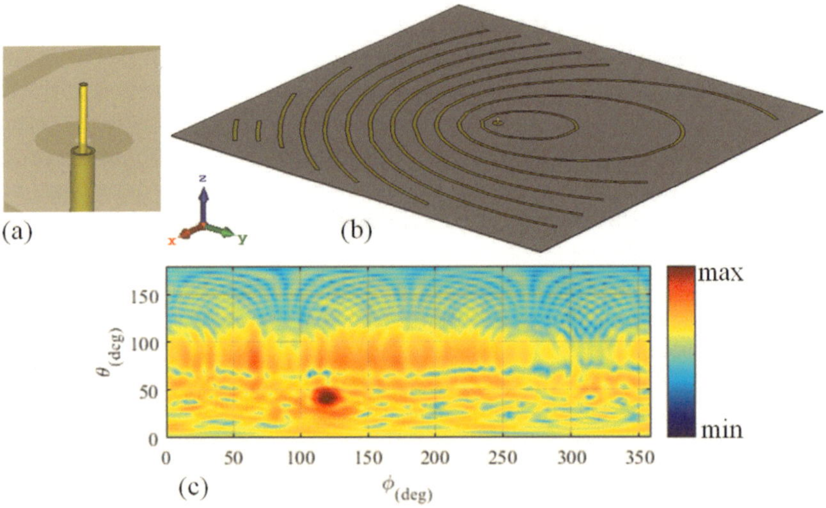


**Figure 4.**
*Reconstruction process: applying (a) surface-wave launcher, and (b) metal strips on the slab. (c) The constructed normalized radiation pattern.*

**Figure 3(b)** for the corresponding $k_0$. In this case, the pattern of the EM hologram is derived as presented in **Figure 3(c)**.

To embody a real-world structure from the calculated EM hologram, it is required to apply an SWL, exactly at the location where the hypothetical source has been placed in the recording process to be able to generate a field distribution as much similar to the derived $E_{ref}$ as possible. Then, a quasi-periodic pattern of scatterers, with the geometry and lattice inspired by $\angle E_s$ must be applied on the slab to locally sample the generated field distribution of the SWL. The scatterers can be printed metal-strips, sub-wavelength patches of arbitrary shape, dielectric cubes, or any other component that can scatter the launched surface waves on the slab. The process of applying the appropriate SWL and pattern of scatterers on the slab is called "reconstruction".

When the structure in hand is excited by its SWL, the induced surface waves will be leaked out to the open environment toward the predefined tilt angle of $(\theta_m, \phi_m)$ which makes a holographic-based LWA (HLWA).

As an example, an open-ended coaxial cable presented in **Figure 4(a)** can be applied on a grounded dielectric slab to generate TM surface wave distribution similar to **Figure 3(a)**. The surface-wave sampling can be performed by printing metal strips on the local maxima of the calculated EM hologram in **Figure 3(c)** which is presented in **Figure 4(b)**. When this structure is excited, the simulated normalized radiation pattern is obtained as presented in **Figure 4(c)**. This shows that the constructed beam is pointed well to the predefined angle of interest at the very first steps of the design which is $(\theta_m = \pi/4, \phi_m = 2\pi/3)$.

## 2.2 Holographic-based reflectors

As briefly pointed out in **Figure 2(c)**, the holography technique can be expanded to the case where the initial source is located outside the slab's body. In this case, the obtained structure will be a holographic-based reflector (HR) [6].

This time, let us sample the EM hologram by using a number of printed sub-wavelength squared-shape patches. These patches will form a quasi-periodic structure where their size is modulated based on the holography technique. In periodic structures, the smallest geometry that is repeated in a fashion is called a unit cell. In this case, the unit cell is a small portion of the dielectric slab with a single printed patch on one side and a full ground plane on the other side as shown in **Figure 5(a)**. The analysis of structure requires characterizing the surface impedance $Z_{\text{surf}} = E_t/H_t$ with $E_t$ and $H_t$ representing the tangential electric and magnetic fields respectively. The obtained structure is then an artificial impedance surface, commonly referred to as a metasurface.

The recording process in Section 2.1 is needed to be modified at the outset to reflect the location of the initial source, i.e. the feeder, on (2).

With an ideal feed located at $\left(x_f, y_f, z_f\right)$ and the slab on the $xy$ plane in a standard right-handed coordinate system, $E_{\text{ref}}$ is modified as

$$E_{\text{ref}} = Ae^{-jk_0 r}, \tag{4}$$

where $r = \sqrt{(x - x_f)^2 + \left(y - y_f\right)^2 + z_f^2}$.

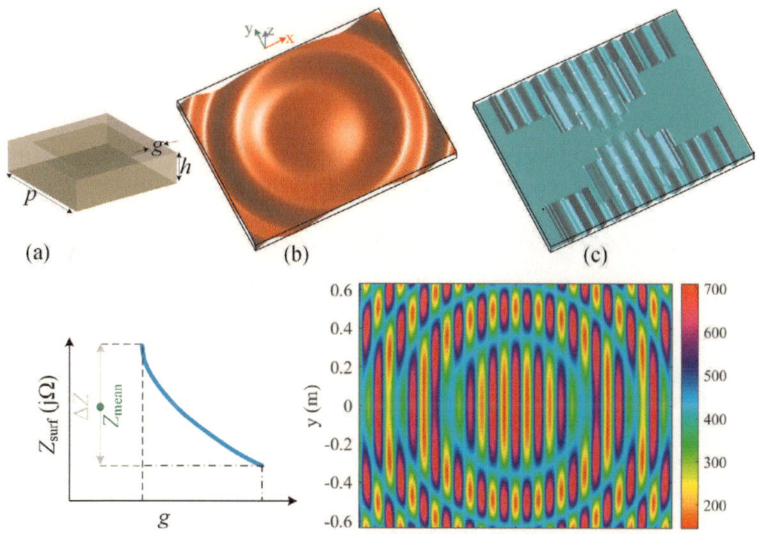


**Figure 5.**
*(a) The applied unit cell, a schema of (b) $E_{ref}$ and (c) $E_{obj}^{\kappa=1} + E_{obj}^{\kappa=2}$, (d) $Z_{surf}$ versus patch size variation and (e) the obtained $Z(x,y)$ on the surface [7].*

**Figure 5(b)** shows the obtained $E_{ref}$ when the feeder is placed at $\left(x_f, y_f, z_f\right) =$ (0,0,2.5m) for a 1.65 m × 1.25 m large dielectric sheet at $f = 3.5$ GHz [7].

In the holography technique, it is possible to define more than one main beam for the final constructed radiation pattern. Under this circumstance, a summation of the respective object waves will define the final distribution of $E_{obj}$ on the slab. Each object wave is derived by (3) toward the angle of interest. It is aimed in this structure to obtain two reflected beams to $(\theta_{\kappa=1}, \phi_{\kappa=1}) = (45°, 0)$ and $(\theta_{\kappa=2}, \phi_{\kappa=2}) = (-45°, 0)$. The calculated $E_{obj} = E_{obj}^{\kappa=1} + E_{obj}^{\kappa=2}$ is presented in **Figure 5(c)**.

In order to derive the EM hologram, it is now required to conduct a study on $Z_{\text{surf}}$ regarding the unit cell of the structure. This can be calculated by sweeping the phase delay ($\phi_D$) across the unit cell with periodicity of $p$ as follows [8]:

$$Z_{\text{surf}} = jZ_0\sqrt{\left(\frac{\phi_D}{k_0 p}\right)^2 - 1}, \tag{5}$$

where $Z_0$ is the free-space impedance. This can be fulfilled by using the eigenmode solver of a full-wave simulator for a specific size of the square patch. Then, the size of the patch must be varied and the calculation repeated to determine the span range of surface impedance $\Delta Z$ with the mean value $Z_{\text{mean}}$ over the range of patch size variation. This is schematically shown in **Figure 5(d)**.

Having all the above-mentioned information, it is possible to define the EM hologram pattern based on the impedance distribution as below [1]:

$$Z(x,y) = j\left(Z_{\text{mean}} + \frac{\Delta Z}{2m}\,\text{Re}\left[\left(\sum_{\kappa=1}^{m} E_{\text{obj}}^{\kappa}\right)E_{\text{ref}}^{*}\right]\right), \tag{6}$$

where $m$ is the number of beams which equals 2 in this case study.

It is shown that $Z_{\text{mean}} = 428.16\,j\Omega$ and $\Delta Z = 566.51\,j\Omega$ for the studied range of patch-size variation [7]. This results in an impedance distribution of **Figure 5(e)** as the EM hologram.

The reconstruction process is to use **Figure 5(d)** and **(e)** to modulate the size of patches on each unit cell and print them on the slab. This will lead to a structure shown in **Figure 6(a)**. When this metasurface reflector is illuminated by a feed horn

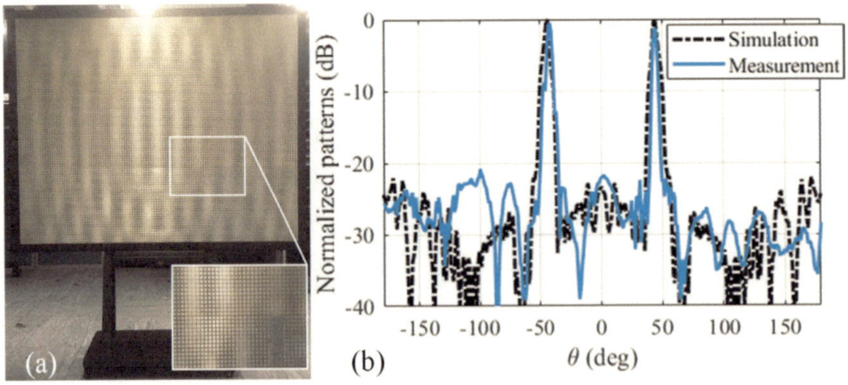


**Figure 6.**
*(a) The metasurface reflector and (b) simulated and measured normalized radiation patterns [7].*

located at the position defined during the recording process, the reflected beams from the surface are formed as presented in **Figure 6(b)** which is in line with the defined object waves.

## 3. Reconfigurable intelligent surface (RIS)

Another possible scenario is the case when the initial source is located outside the surface, but far from the structure. Assume an ideal initial source located at the angle of $(\theta_s = \pi/6, \phi_s = \pi/4)$ with respect to the surface normal far from the structure. Following the routine explains in Section 2.1 and 2.2, $E_{\text{ref}}$, $E_{\text{obj}}$, and the EM hologram patterns are calculated as shown in **Figure 7(a)**, **(b)**, and **(c)** respectively provided that the reflected beam is aimed to be pointed to $(\theta_m = \pi/3, \phi_m = \pi/6)$.

To translate this mechanism into a practical format, this is the case when the surface reflects the incoming waves from a far-located source to a desired direction. This is not a simple mirror reflection where there is no control over the angle of reflection; indeed, the reflection angle can be engineered in this case using the holography technique. This brings a new idea for the future generation of cellular networks.

Consider a case where there is a blind spot within the area under the coverage of a base station (BS). The conventional approach to providing coverage for this blind spot is to add a new BS in the network. However, this method can be expensive and sometimes very challenging regarding the environmental barriers in an area. The idea is to locate a surface in the line of sight (LoS) of the BS so that it can be illuminated by the BS. Then the receiving EM waves reflect back to that blind spot to recycle the waves and provide coverage without adding a new BS. It is possible to reconfigure the response of the surface by applying some components like Varactor or PIN diodes on each unit cell and recalculating the EM hologram for each state of reflection. Under this circumstance, the obtained structure is called a reconfigurable intelligent surface (RIS) [9]. This scenario is schematically shown in **Figure 8(a)**.

Note that holography is not the only technique to regulate the response of the RIS. The generalized Snell's law of reflection (GSR) can also be used in this regard [10]; a GSR-based RIS prototype [11] is shown in **Figure 8(b)**.

One of the biggest problems for RIS to be industrialized and practically applied in real-world networks is its very low aperture efficiency ($\eta_a$). This will be more challenging for the uplink scenario when the user attempt to connect BS via RIS. To have a more clear idea about this problem, consider a rectangular reflecting surface with a planar dimension of $X \times Y$, illuminated by a feed horn, distanced by $R$ as presented in **Figure 9(a)**. The feeder's radiation pattern can be expressed by $\cos^{q_E}(\theta)$ and $\cos^{q_H}(\theta)$

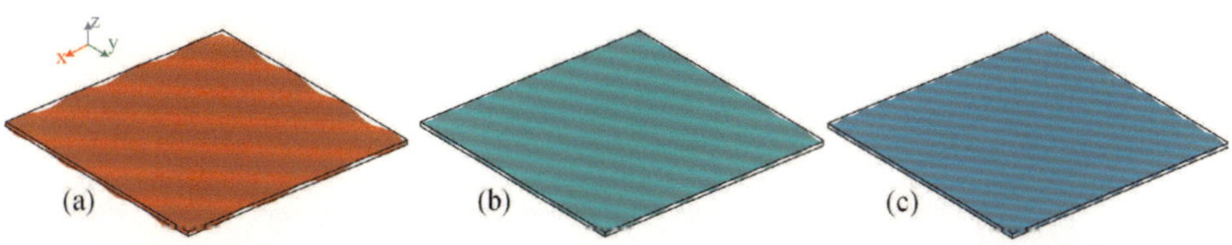


**Figure 7.**
*Recording process: (a) $E_{\text{ref}}$ with an ideal source at the angle of $(\theta_s = \pi/6, \phi_s = \pi/4)$ with respect to the surface normal located far from the structure, (b) $E_{\text{obj}}$ in case the reflected beam is aimed to be pointed to $(\theta_m = \pi/3, \phi_m = \pi/6)$, and (c) the obtained EM hologram.*

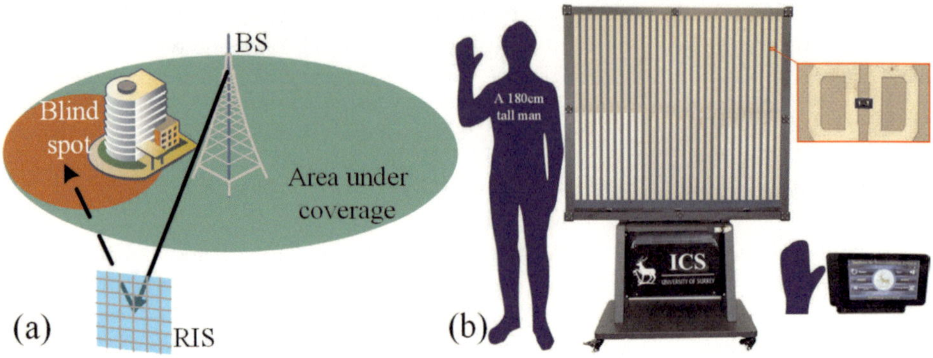


**Figure 8.**
*(a) The coverage provisioning via reconfigurable intelligent surface (RIS) for a blind spot and (b) a prototype example of RIS [11].*

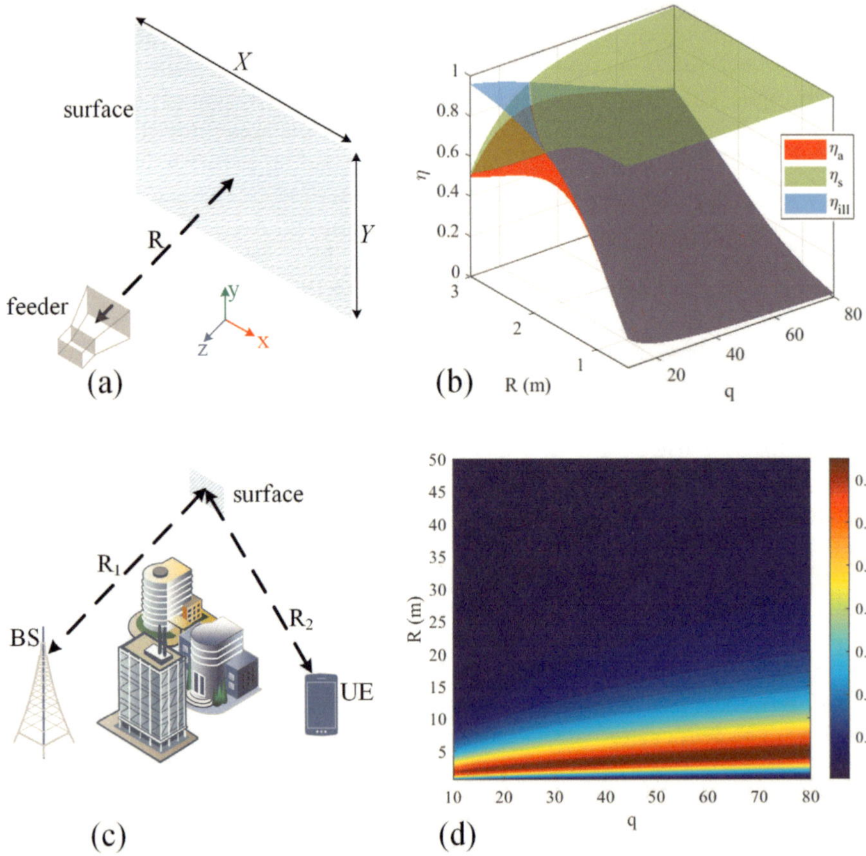


**Figure 9.**
*(a) The overall geometry of a reflective surface illuminated by a feeder, (b) the theoretical aperture efficiency when the feeder is not far from the aperture, (c) a schema of using reflective surfaces to provide the coverage for the user in the blind spot region of the BS, and (d) the theoretical aperture efficiency when the feeder is located relatively far from the aperture.*

at the E-plane and H-plane respectively. Product of the illumination ($\eta_{ill}$) and spillover ($\eta_s$) efficiencies is then defines $\eta_a$. In case of rectangular surfaces, $\eta_{ill}$ can be calculated by [12]:

$$\eta_{ill} = \frac{I^2}{sII}. \tag{7}$$

where $s = X \times Y$ and

$$I = \int_{x=-X/2}^{X/2} \int_{y=-Y/2}^{Y/2} \left\{ \frac{1}{\sqrt{R^2 + x^2 + y^2}} \left[ \left( \frac{R}{\sqrt{R^2 + x^2 + y^2}} \right)^{q_E + 2} \frac{y^2}{x^2 + y^2} + \left( \frac{R}{\sqrt{R^2 + x^2 + y^2}} \right)^{q_H + 1} \frac{x^2}{x^2 + y^2} \right] \right\} dydx, \tag{8}$$

with

$$II = \int_{x=-X/2}^{X/2} \int_{y=-Y/2}^{Y/2} \left[ \left( \frac{R}{\sqrt{R^2 + x^2 + y^2}} \right)^{2q_E} \frac{y^2}{x^2 + y^2} + \left( \frac{R}{\sqrt{R^2 + x^2 + y^2}} \right)^{2q_E} \frac{x^2}{x^2 + y^2} \right] \frac{R}{\left( \sqrt{R^2 + x^2 + y^2} \right)^3} dydx. \tag{9}$$

Under this circumstance, $\eta_s$ reads:

$$\eta_s = \frac{II}{III}, \tag{10}$$

with

$$III = \pi \left( \frac{1}{1 + 2q_E} + \frac{1}{1 + 2q_H} \right). \tag{11}$$

Now consider the rectangular aperture of **Figure 6**(a). Recall that the physical size of the aperture is $s = 1.65$ m $\times$ 1.25 m at 3.5 GHz. With a symetrical radiation pattern at the feeder, $q = q_E = q_H$ and $\eta_{ill}$ and $\eta_s$ are obtained by (7) and (10) respectively, followed by calculation of $\eta_a = \eta_{ill} \times \eta_s$. The result is shown in **Figure 9(b)** for different values of $q = 10 \sim 80$ and $R = 0.5 \sim 3$ m. This shows that it is possible to realize optimum values of $q$ and $R$ to use a specific feed horn and locate it at a specific distance from the surface to obtain the maximum possible $\eta_a$. This is a routine step of designing reflectarray antennas and metasurface reflectors. However, in the case of RIS where there is no control on $q$ and $R$, it is not possible to customize the structure to reach the optimum $\eta_a$.

**Figure 9(c)** shows a schema of applying the surface of **Figure 6(a)** for coverage provisioning purposes. Note that this surface has a very large size comparing to the operating wavelength which can potentially be a positive factor for $\eta_a$. We repeat the same calculation, but this time the distance range is expanded to $R = 0.5 \sim 50$ m. The result is shown in **Figure 9(d)**. As it is clear, a massive region of this plot shows a very low $\eta_a$ which can make the structure impractical for real-world applications. This will be more challenging when we pay attention to two factors, 1st: in cellular networks, the cell radius is much longer than 50 m, this means that the situation is even worse than **Figure 9(d)**; 2nd: for the uplink connection, the user equipment (UE) will have a very low gain (or low $q$) which will make the connection very challenging if not impossible (see **Figure 9(d)** for low values of $q$ and high $R$). Finally, it should be noted that these are all theoretical calculations; when it comes to practice, the obtained $\eta_a$ is expected to be relatively lower than the theory. This is also true even for reflectarray antennas where the feeder is optimized.

## 4. Comparison between holographic, MIMO, and phased-array beamforming

Consider a case when printed patches are used in three different structures, i.e. a holographic-based metasurface, a phased-array antenna, and a MIMO system. The only common thing between these three is the repeating geometry of patches across the structure. The detailed functionality of these patches is as below:

- Holographic-based metasurface: in this case, the printed patches are used to sample an induced/launched surface wave on the structure. The size of these patches is in the sub-wavelength scope and varies. This variation in size is known as modulation which is governed based on the holography technique[1]. The obtained structure will be an HLWA or an HR with respect to the location of the initial feeder. It is possible to form more than one beam by applying the superposition technique and tilting each of them to the direction of interest. **Figure 10(a)** schematically shows a multi-beam HR. With respect to **Figure 1**, this structure is a single aperture.

- Phased-array antenna: the size of patches is relatively bigger in the scopes of half-wavelength. Therefore, each patch is a resonating element. The element spacing is more than the case of HR metasurface, but should not be more than a wavelength or half a wavelength to stop forming unwanted grating lobes. Each element has typically a port where the corresponding amplitude and phase can be controlled to not only control the tilt angle but also to regulate the obtained SLL. A 4-element planar phased-array antenna with a tilted beam is schematically shown in **Figure 10(b)**. Regarding **Figure 1**, these elements form a single aperture altogether.

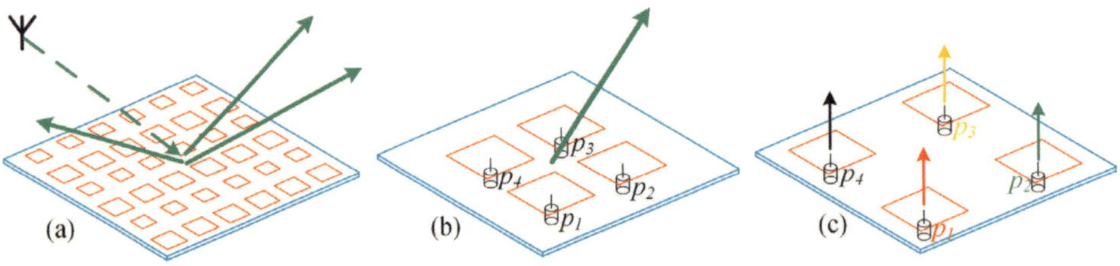


**Figure 10.**
*(a) A multibeam metasurface reflector, (b) a tilted-beam phased-array antenna and (c) a 4-element MIMO system.*

[1] Note that this is not just limited to the holography, other methods like GSR can also be used.

- MIMO system: the size of patches is again in the scopes of half-wavelength to make a resonating element. However, the spacing between elements is managed to decrease the coupling between elements. When there is a port for each element, this low mutual coupling can decrease the cross-correlation between the ports which can be a good candidate to be used in MIMO systems. As each element is functionally separated from the others, it is common to be called an embedded element. In this case, each embedded element has its own radiation characteristics. A schema of a 4-element MIMO system is presented in **Figure 10(c)**. This structure has four separate apertures considering **Figure 1**.

## 5. Conclusions

The holography technique and its application in forming the beam in electromagnetic structures are explained in this chapter. Leaky-wave antennas, metasurface reflectors, and reconfigurable intelligent surfaces (RISs) are assessed in this regard. The aperture efficiency of RIS is also studied which is one of the main barriers to industrializing this component in future cellular networks. A comparison is made between the functionality of the smallest building blocks in holographic-base metasurfaces, phased array antennas, and MIMO systems.

## Conflict of interest

The authors declare no conflict of interest.

## Author details

Ali Araghi[1]* and Mohsen Khalily[2]

1 University College London (UCL), London, United Kingdom

2 University of Surrey, Guildford, United Kingdom

*Address all correspondence to: a.araghi@ucl.ac.uk

## References

[1] Fong BH, Colburn JS, Ottusch JJ, Visher JL, Sievenpiper DF. Scalar and tensor holographic artificial impedance surfaces. IEEE Transactions on Antennas and Propagation. 2010;**58**(10): 3212-3221

[2] Araghi A, Khalily M, Xiao P, Tafazolli R. Holographic-based leaky-wave structures: Transformation of guided waves to leaky waves. IEEE Microwave Magazine. 2021;**22**(6):49-63

[3] Jackson DR, Caloz C, Itoh T. Leaky-wave antennas. Proceedings of the IEEE. 2012;**100**(7):2194-2206

[4] Xu F, Wu K. Understanding leaky-wave structures: A special form of guided-wave structure. IEEE Microwave Magazine. 2013;**14**(5):87-96

[5] Rusch C, Schäfer J, Gulan H, Pahl P, Zwick T. Holographic mmW-antennas with $TE_0$ and $TM_0$ surface wave launchers for frequency-scanning FMCW-radars. IEEE Transactions on Antennas and Propagation. 2015;**63**(4): 1603-1613

[6] Karimipour M, Komjani N. Holographic-inspired multibeam reflectarray with linear polarization. IEEE Transactions on Antennas and Propagation. 2018;**66**(6):2870-2882

[7] Araghi A, Khalily M, Xiao P, Wang F, Tafazolli R. Systematic design of a holographic-based metasurface reflector in the sub-6 GHz band. IEEE Antennas and Wireless Propagation Letters. 2022; **21**(10):1960-1964

[8] Patel AM, Grbic A. A printed leaky-wave antenna based on a sinusoidally-modulated reactance surface. IEEE Transactions on Antennas and Propagation. 2011;**59**(6):2087-2096

[9] Khalily M, Yurduseven O, Cui TJ, Hao Y, Eleftheriades GV. Engineered electromagnetic metasurfaces in wireless communications: Applications, research frontiers and future directions. IEEE Communications Magazine. 2022; **60**(10):88-94

[10] Díaz-Rubio A, Asadchy VS, Elsakka A, Tretyakov SA. From the generalized reflection law to the realization of perfect anomalous reflectors. Science Advances. 2017;**3**(8): e1602714

[11] Araghi A, Khalily M, Safaei M, Bagheri A, Singh V, Wang F, et al. Reconfigurable intelligent surface (RIS) in the sub-6 GHz band: Design, implementation, and real-world demonstration. IEEE Access. 2022;**10**: 2646-2655

[12] Zebrowski M. Illumination and spillover efficiency calculations for rectangular reflectarray antennas. High Frequency Design. 2012;**1**:28-38

Chapter 9

# Techniques for Compact Planar MIMO Antennas

*Yiying Wang*


## Abstract

MIMO Technology has promoted the developments of various antennas, then the planar antenna will be one of the main directions to satisfy the future compact requirement of the 5G+/6G communications. This chapter introduces different types of the planar antenna and summarizes the implicit compact techniques, where the related techniques like the diversity and the reconfigurable are not included owing to they are the inherent properties of the MIMO antennas. These antennas contain the patch antenna, slot antenna, dipole/monopole antenna, loop antenna, cavity antenna, Yagi-Uda antenna, fractal antenna, UWB antenna, PIFA etc., and their deformations to the specific purposes. On the contrary, the implicit compact techniques are not so explicit as the antenna configurations, but they are classified to be the close-spacing structure without decoupling, owing to the decoupling is not the necessary requirement of MIMO application, decoupling technique of spacing reduction, meandered line technique, multi-element method, co-radiator/co-location design, fractal antenna, and radiator-cutting antenna. Besides, the corresponding techniques for the compact design are also concluded, including the mode-cutting method, fractal technique, characteristic mode analysis, and the optimization algorithms.


**Keywords:** MIMO, 5G+/6G, planar antennas, compact techniques, integration

## 1. Introduction

Owing to the prominent advantages compared with the conventional single-input single-out (SISO) system, the MIMO technology has been extensively applied to many scenarios, in which the antenna with beamforming is one of the key features in order to realize the multiple path communications. Consequently, the multi-beam, the multi-polarization, or the related diversity or the reconfigurable techniques are the inherent properties of the MIMO antennas. To satisfy the requirement of MIMO communication, many antenna types have been employed, including the high-profile 3D antennas, such as the dielectric resonator antenna (DRA) [1–3], helix antenna [4], structure-loaded antenna [3, 5–7] or multi-element antenna [8, 9], and the other common 2D planar antennas. On the other hand, the 5G+/6G technology puts forwards the new compact, easy-fabricated and easy-integrated requirements for the antenna development resulting in the planar antennas will be one of the main directions in the future. Therefore, the focus of this chapter is on the introduction of planar antennas and especially the implicit techniques on how to design the compact structure.

 

Many planar antenna types have been proposed for the MIMO applications, but not all of them will be discussed in this chapter considering the related compact techniques. The planar antennas involved are patch antenna [10–22], slot antenna [23–30], dipole/monopole antenna [31–48], loop antenna [49–58], ultrawideband (UWB) antenna [59–71], Yagi-Uda antenna [72–77], cavity antenna [78–82], fractal antenna [83–90], and the planar inverted-F antenna (PIFA) [91–104]. These antennas do not appear in isolation, they often combine with other types for the specific purpose, such as, both the patch antenna and the dipole antenna were used to realize the linear and circular polarization design [11], the slot antenna [28] and the fractal antenna [83] also belong to the UWB antenna, and the radiator of UWB antennas [59–62] is monopole. However, we distinguish them according to their explicit features in this chapter as the above categories, and the relatively simple antenna structures are picked up from the similar works. Additionally, though these are planar structures, they can be used in the 3D situations [27, 43, 102] like in the mobile application. All selected types are the printed antennas, they will be good candidates for the future 5G+/6G applications from the view of easy fabrication and integration.

The compact design is always the research focus of MIMO antennas, many techniques have been employed to compress the volume of structure. However, we face a common problem that the antenna performance is affected because of coupling when they are close to each other. There are two general ways to solve this problem, one is that we need not care about the coupling but put them closer if the coupling is not too significant, which is because the MIMO antenna technique does not require the elements to work at the same time, the coupling will not affect the work status of MIMO system; the other is using the decoupling technique to realize the compact design, even so the antenna performance is also affected when the elements are close enough.

In addition to the above close-spacing compact techniques, changing the antenna shape is another conventional way to realize the compact design, such as, using meander line for the dipole/monopole or the slot antenna to save the spacing, and by the combinations with different antennas to change the shape, like the electric and magnetic dipoles, the patch and slot, the PIFA and slot etc. Besides, the fractal technique and the optimization algorithm with constraints are often implemented to change the antenna shape. And we can physically reduce the antenna size by performing the corresponding cutting based on the related modes.

In this chapter, we will focus on the introduction of the corresponding compact techniques of the planar antennas, including the close-spacing and the shape change methods. Therefore, the rest is organized as follows. Section 2 introduces the corresponding general compact methods implicit in different antenna types, including the close-spacing no-decoupling design, decoupling design, meander line method, multiple antenna structure, co-radiator/co-location design, fractal antenna, and the mode-cutting technique. The fundamentals for compact designs, including mode-cutting method, fractal technique, characteristic mode analysis (CMA), and optimization algorithm, are summarized in Section 3, which will be helpful to the future compact researches owing to the physical reduction of antenna size. Then, the conclusions are shown in Section 4.

## 2. Compact antenna techniques

Though different antenna types have been designed for the compact purpose depending on the present development trend, the compact techniques are similar accompanied by the types. We summarize the corresponding compact techniques in this section.

### 2.1 No-decoupling compact designs

The purpose of MIMO antenna is using the multiple path transmissions to realize the high-efficiency and high-capacity communication, which means the antenna may not work simultaneously and then the coupling is not a main concerned focus. In other words, we need not care about the mutual couplings among elements so seriously in some cases when we want to realize the compact design but put them closer. Moreover, the coupling can be reduced by properly arranging the positions of elements to form the orthogonal polarization etc.

In [101], even the cross line connected with four elements exists to improve the isolation, the minimum isolation is up to 9.7 dB. The similar phenomenon happens in [36, 80] where no-decoupling structures were used. The four-element 90 degrees rotated structure of [36] is shown in **Figure 1**, from which we know the coupling is significant and the authors gave that of about 12 dB. The shorting pins and the 90 degrees rotation also do not reduce the mutual coupling seriously at some frequencies in [80], which is about 13.3 dB, the corresponding configuration is shown in right subfigure.

When the spacings become larger, the coupling will be smaller [29, 35] where the orthogonal polarization techniques were employed as well. Using the shorting pins [78] or slot [81] to stop the current flowing to the neighbor element in the cavity antenna is an efficient way to reduce the coupling. And we can obtain the lower coupling by exciting the neighbor elements with the differential modes instead of additional decoupling structure [14, 46, 88].

### 2.2 Compact decoupling techniques

Generally, we should consider the mutual couplings in the MIMO antenna design which decreases the consequent undesirable problems of the related system. Except the above differential mode method, we often reduce the mutual coupling for the close-spacing elements from two aspects, one is from the source and the other in the transmission process. The decoupling techniques of patch antenna can illustrate these well [15–18]. When the spacing is large enough, the surface current on the ground plane affects the mutual coupling rarely so that a proper metamaterial absorber put between the patch is enough to stop the surface and the radiated waves in the decoupling process [17]. As the spacing becomes closer, the surface current on the

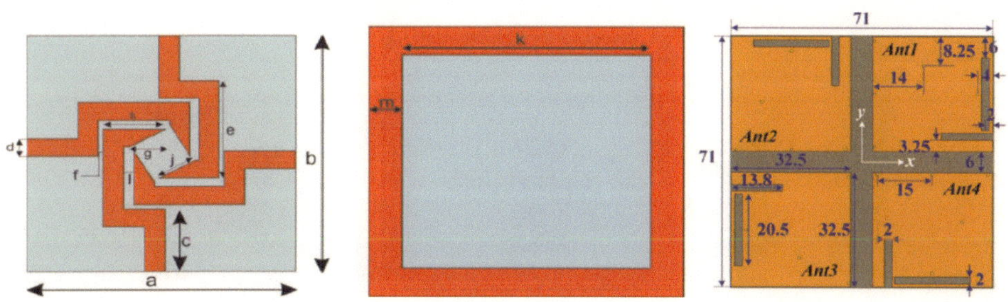


**Figure 1.**
*No-decoupling MIMO antennas: [36] (left) and [80] (right).*

ground plane diffuses to the neighbor element, and the near-field coupling to the other patch generates, so the defected ground structure (DGS) and/or resonator techniques between patches can reduce the couplings significantly [15, 16, 18]. **Figure 2** shows the simple decoupling structure in [15], in which the slot through the substrate and ground plane was curved. The surface current on the ground was cut off and the resonator was form between patches resulting in the reduction of mutual coupling. In order to reduce the mutual coupling of patches, literature [18] used another way, where the parasitic elements are put closer than the spacing between patches so as to induce the power to the parasitic metal rather than the neighbor patch. **Figure 3** shows the current distributions on the top layer of patch antenna before and after using parasitic technique. It is clear that the coupling to the neighbor element is suppressed.

The ideas were implemented into the PIFA antenna [92, 93, 96], loop antenna [49, 52, 53], slot antenna [69, 70], and UWB antenna [60, 62, 69, 70] and so on. **Figure 4** shows the corresponding antenna and the decoupling structures. Both the PIFA and the UWB suppress the coupling in the wave propagation process, and the other two cuts off the surface currents.

The neutral line technique is another normal method to reduce the coupling in the monopole [45] and UWB [60] antennas. It does not destroy the structure of ground plane but introduce the neutral line between elements. The decoupling structure of [60] is shown in **Figure 5**, where the circular disc of neutral line allows several

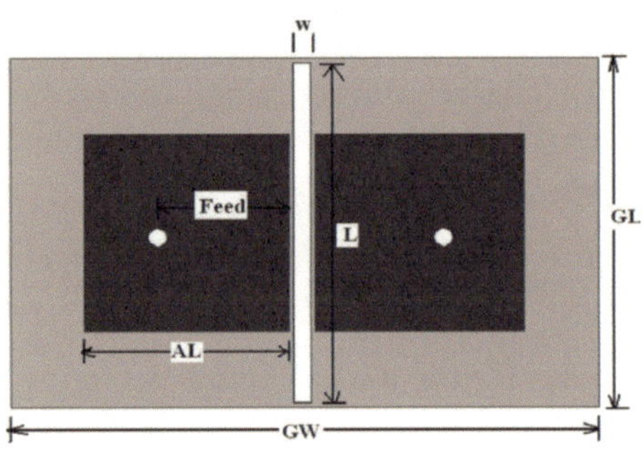


**Figure 2.**
*DGS and resonator techniques to reduce the coupling [15].*

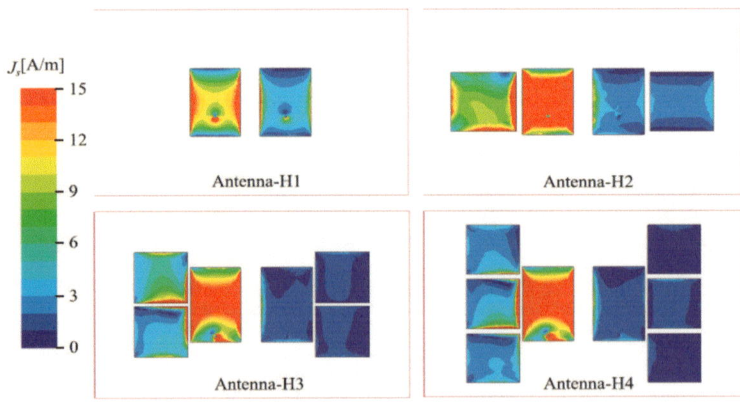


**Figure 3.**
*The comparisons of current distributions on the top layer [18].*

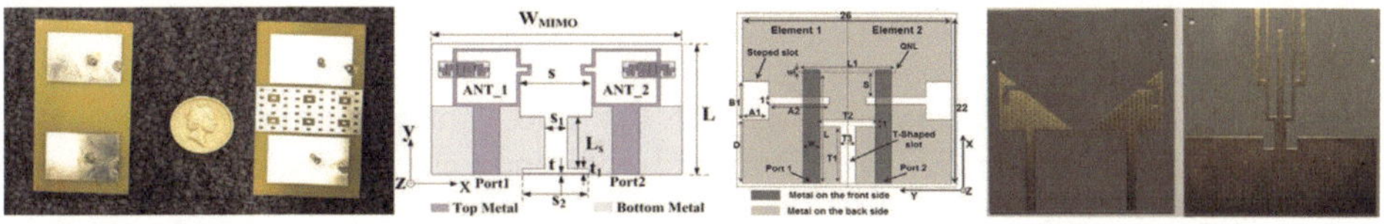


**Figure 4.**
*The antennas and decoupling structures (from left to right): PIFA [96], loop [52], slot [69], and UWB [62].*

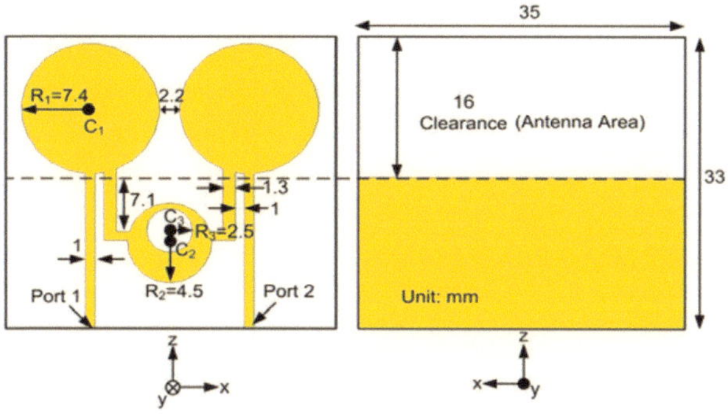


**Figure 5.**
*The antennas and the neutral line decoupling structures of [60].*

decoupling paths to cancel the coupling current on the ground so that the UWB decoupling is realized, and with the help of slot in the circular disc, the highest decoupling frequency can be tuned to 5 GHz.

If the cavity is cut properly, the antenna can realize the self-isolation without any additional structure. Such as in [81], the authors cut a slot symmetrically for the quarter cavity, the consequent 1/8 mode cavity antennas have good isolation. This self-isolated technique is also applied in the loop antenna, and the size is reduced by means of the introduction of two vertical stubs in the loop [53].

### 2.3 Meander line antennas

It is the conventional way to restrict an antenna to a fixed area by meandering the radiator. It is often adopted in the dipole/monopole, slot, and loop antennas, where the corresponding method is relatively simple, that is, we just adjust the meandered sections to resonate at the desired frequency as the straight one. We can find many meander line techniques employed in the MIMO antenna designs [25, 36, 38–40, 42, 47, 49, 56, 57, 67].

The dual-band is realized by two different length slots, the authors meandered the longer slot which makes the two slots have the similar length in the horizontal direction [25]. The arms of dipole/monopole are meandered as well for the compact design [36, 38–40, 42, 47, 67]. For the loop antenna, the authors put two loops on the different layers and the smaller one is embedded into the bigger one [49], while the loops meandered inward are implemented for the rectangular [57] and the Alford [56] loops, respectively.

### 2.4 Multiple antenna structures

Multiple antenna structure, or the hybrid structure with different antenna types, also can obtain the compact design for the MIMO antenna application. This combination not only saves the spacing, but also suits for the realization of multi-band or multi-polarization.

In [11], the patch antenna combines with the dipole antenna to realize the polarization diversity design, where the chamfered-edge square patch with an offset feed is used to obtain the circular polarization and the two dipoles are responsible for the linearly polarized radiation. Literature [32] also shows the combination of both patch antenna and the monopole antenna with the same ground, but the patch antenna is fed by the electromagnetic coupling.

Two pairs of slot antennas are etched in the patch to realize the dual-polarized radiation, and good isolation is obtained due to the proper feeding positions [19]. The authors put the IFA and the slot antenna together for different LTE bands in the mobile application [104]; and two slots with different lengths are put close for the dual-band radiation [25].

## 2.5 Co-radiator/co-location antennas

To some extent the co-radiator and the co-location antennas resemble the multiple antenna structure. For the co-radiator MIMO antenna, there exists one common radiator but excited by different ports, while the co-location antenna assembles different antennas in the fixed area. These two types are similar to some diversity antennas of one radiator but different feeds and the multi-band antenna with one radiator, respectively. In other words, the co-radiator/co-location antennas are not so unfamiliar things but they just have the common property for a kind of antennas.

The antenna of [105] shown in **Figure 6** explain the co-radiator antenna clear, where four ports excite the common radiator and the slot of ground plane is used to improve the isolation. The configuration of [106] resembles to this work, but use two ports to excite the co-radiator, the isolation is improved by means of the T-shape slot and the irregular stub extended from the ground plane.

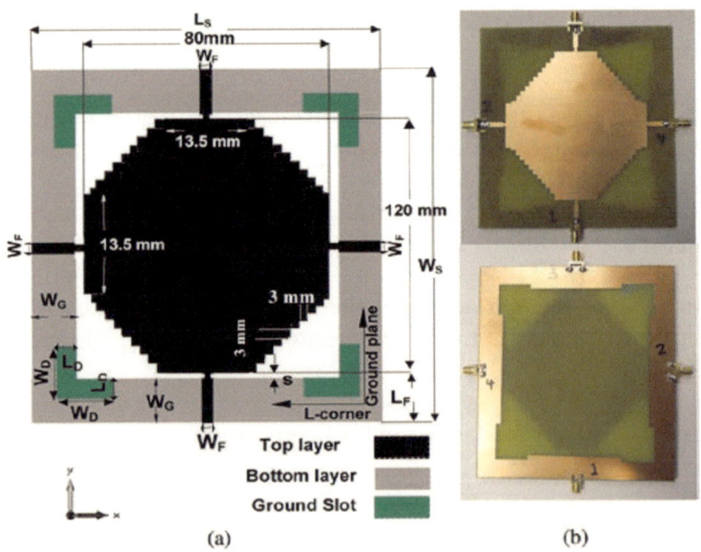


**Figure 6.**
*Co-radiator antenna of [105]: (a) Geometry and (b) prototype.*

The circular co-radiated patch is employed in [107], the configuration is shown in **Figure 7**. The authors analyzed the two close modes $TM_{02}$ and $TM_{11}$ with the monopole-like and patch-like radiations, respectively, and put several vias around the center to make sure the resonant frequencies are the same. Then, they used the center port to obtain the monopole-like radiation, the other two ports excite two orthogonal patch-like patterns.

The application of the co-radiator technique in the mobile terminal was investigated in [108], we repeat the configuration in **Figure 8**. The loop is the co-radiator of port 1 and port 2 as shown in **Figure 8c**. The two ports are put at the center of loop, and port 1 feeds the loop directly while port 2 feeds the loop by a microstrip line. Owing to the odd and even modes appear by the two ports, the high isolation is achieved in the design.

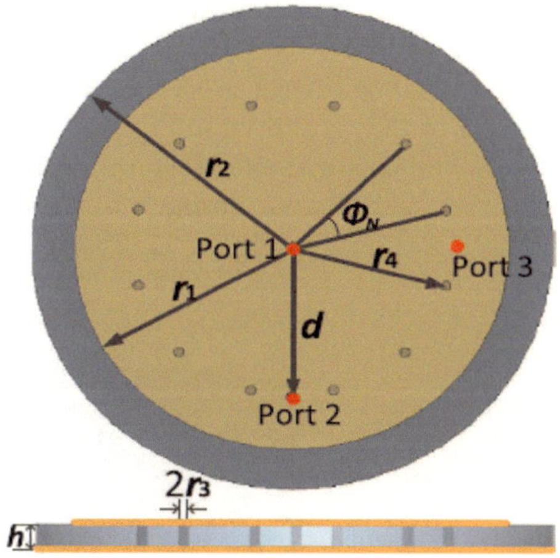


**Figure 7.**
*Circular co-radiator antenna of [107].*

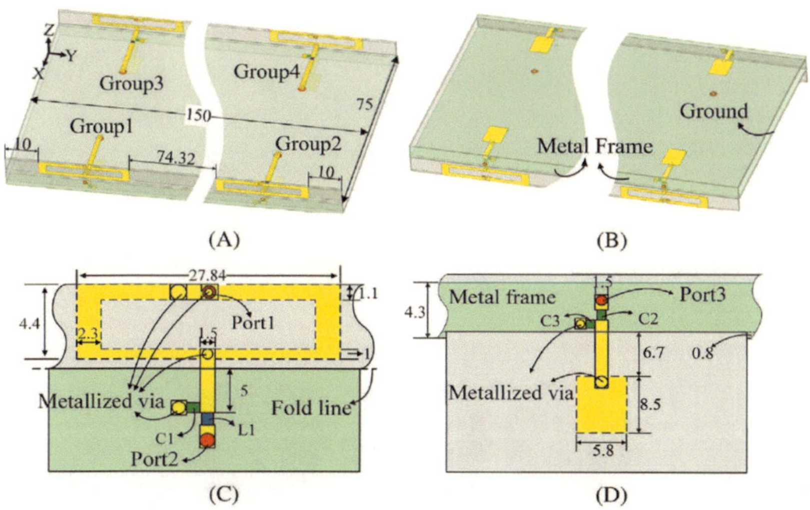


**Figure 8.**
*Co-radiator antenna applied in mobile terminal [108]: (A) front view, (B) back view, (C) details of the upper structure, and (D) details of the lower structure.*

The co-location antenna is the same as the co-radiator antenna in a way as in [107], where different radiation patterns are generated by the different ports at different positions but the radiator is not changed. However, we differentiate them in this section by introducing the co-location antenna with different radiators [30]. The corresponding antenna configuration and the current distributions at different frequencies are shown in **Figure 9**, where the two ports are placed on the top and left sides. When the lower frequency is excited at any port, the square-ring slot works, the edge branch radiates for the higher frequency.

### 2.6 Fractal antennas

The fractal technique is helpful to reduce the antenna size owing to the self-similar and space filling properties, thus it is used to design the MIMO antenna for the compact purpose.

In [83, 86, 88], the Koch fractal technique were adopted to design the MIMO antenna. The iteration process of [83] is shown in **Figure 10**. The initial shape is the octagon of **Figure 10a**, the fractal shape after first iteration is shown in **Figure 10b**, and the second iteration has satisfied the requirement of the UWB band. The corresponding MIMO antenna is shown in **Figure 11**. The C-shape slot is sliced on the

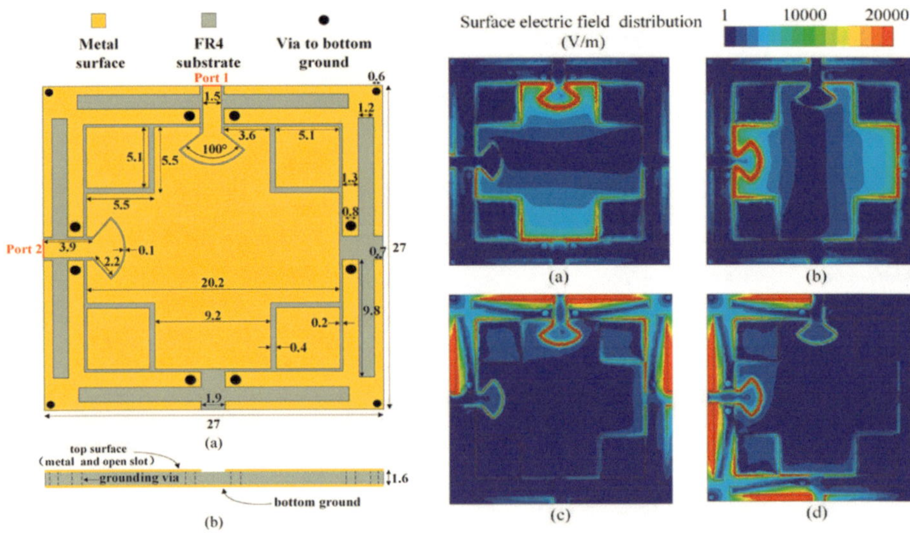


**Figure 9.**
*Antenna configuration (left: (a) Top view and (b) cross-sectional view) and current distributions at different frequencies (right: (a) Port 1 excited at 3.4 GHz, (b) Port 2 excited at 3.4 GHz, (c) Port 1 excited at 3.8 GHz, and (d) Port 2 excited at 3.8 GHz) of [30].*

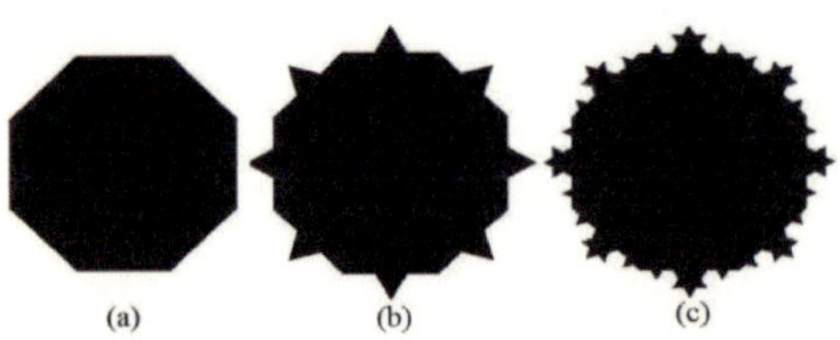


**Figure 10.**
*Iteration process of Koch fractal [83]: (a) initiator, (b) first iteration, and (c) second iteration.*

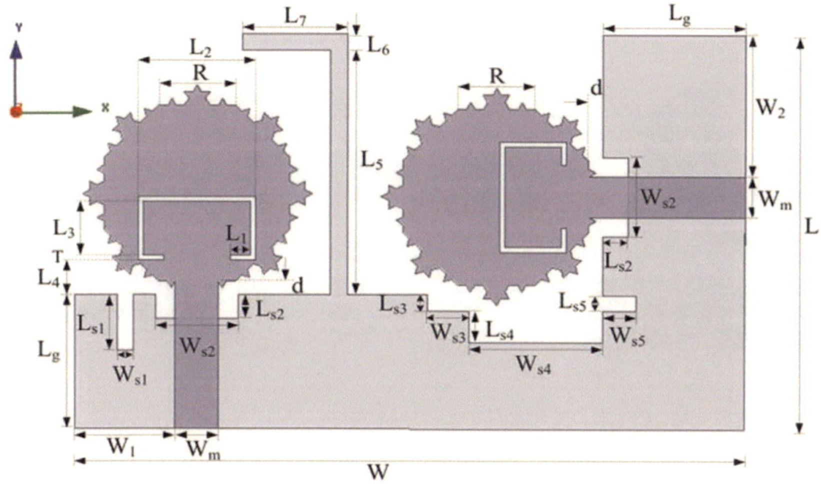


**Figure 11.**
*Fractal MIMO antenna of [83].*

fractal octagon to realize the rejection band, the orthogonal arrangement and a L-shape stub connected with the ground plane are used to increase the isolation.

The hybrid fractal technique is also used to design the UWB antenna [85], where both the Sierpinski and Koch fractal were applied, and a U-shape slot etched in the radiating element to notch the WLAN band. For the MIMO antenna design, the two antennas are put parallel, but the isolation is increased by both the stepped ground plane and the reflecting ground stub in between.

Using the similar idea as in [85], the authors in [87] combine the Koch and the Minkowski to get the dual-band MIMO antenna. The corresponding hybrid fractal idea and the MIMO antenna are shown in **Figure 12**. There is an initiator and a generator for the Minkowski fractal technique as shown in the left subfigure of **Figure 13**. This structure meets the dual-band requirements of 1.65–1.90 GHz and 2.68–6.25 GHz, and the high isolation is achieved with the help of the T-shape stub connected to the ground plane.

In contrast, the hybrid Quadric–Koch fractal antenna was designed in [89] for the multi-band requirement where the circular polarizations were obtained for the bands of 3.66–3.7 GHz and 5.93–6.13 GHz and the rest five bands are linear polarization. No additional structure was used in the MIMO antenna, where two elements are put symmetrically, and the isolation is better than 17 dB over the entire bands.

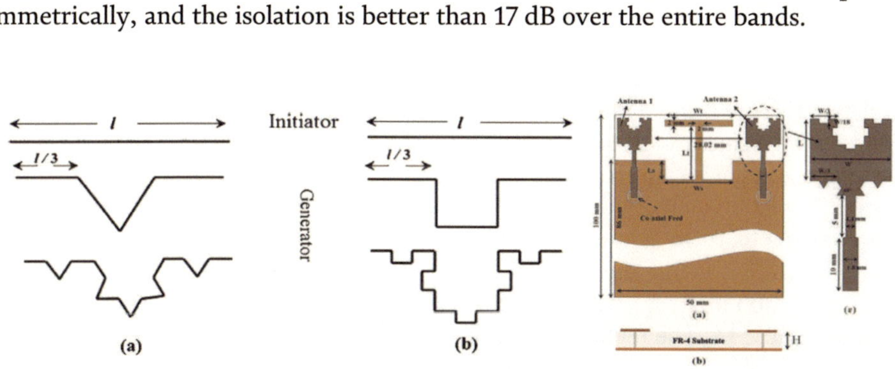


**Figure 12.**
*Hybrid fractal MIMO antenna of [85].*

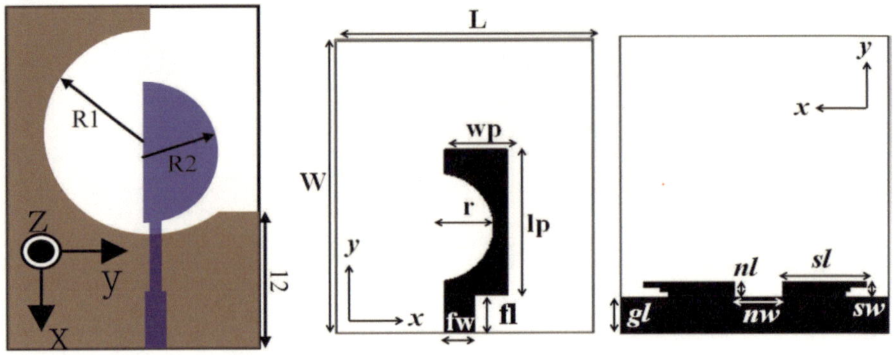


**Figure 13.**
*The elements of MIMO antennas in [109] (left) and [110] (right).*

The Hilbert fractal technique was employed in [90] to realize the dual-band property, the correlation coefficients lower than 0.1 when the two orthogonal-arranged antennas are put close.

## 2.7 Radiator-cutting antennas

Though the fractal technique reduces the antenna size significantly in physical, there is another way to reduce the antenna size physically, that is, by cutting the original antenna into a small piece. This method is suitable for the antenna with the symmetrical structure or the cavity antenna who has the symmetrical modes.

In the works of [109, 110], the radiator is cut into two pieces and leave one for the radiation. The corresponding element structures of the MIMO antennas are shown in **Figure 13**. It is clear that the antennas have no complete structures. It is the quasi-self-complementary monopole in [109] owing to the monopole has the semi-circle patch while the slot in the ground is not complete half-complementary structure. This cutting structure keeps the UWB property as that of the complete monopole. Through the symmetrical arrangement of two elements, the MIMO antenna has high isolation without any additional structure due to the asymmetry structure. While in [110], the rectangular monopole is not only cut into two pieces, but also a semi-circle slot is etched in the half-monopole to improve the optical transparency. However, the ground plane is modified by etching a slot and adding staircase stubs resulting in the improvement of the impedance bandwidth.

The non-integer order mode cutting technique is also adopted to reduce the antenna size physically. It is the same as the radiator-cutting method. That is because the complete structure has the integer mode, if we want to use its non-integer mode we have to cut the structure. In other words, the structure cutting means the mode cutting.

The non-integer mode structure of rectangular cavity in [80] has been shown in Section 2.1, where the length of the patch is half of the complete one so that the $TM_{1/2,0}$ mode etc. form. The same idea appears in [82], but one 1/8 mode is used for the circular cavity. While the semi-taper slot is etched on the top lay of the cavity in [111] in order to radiate the related power, where the four CPW-fed MIMO antenna is shown in **Figure 14**. The metallic vias are set to form five cavities, the outer four with the semi-taper slots are responsible for the radiation. The half mode of $TM_{110}$ will generate by the proper feeding, and this MIMO antenna has high isolation owing to the orthogonal arrangement and the existence of the metallic vias.

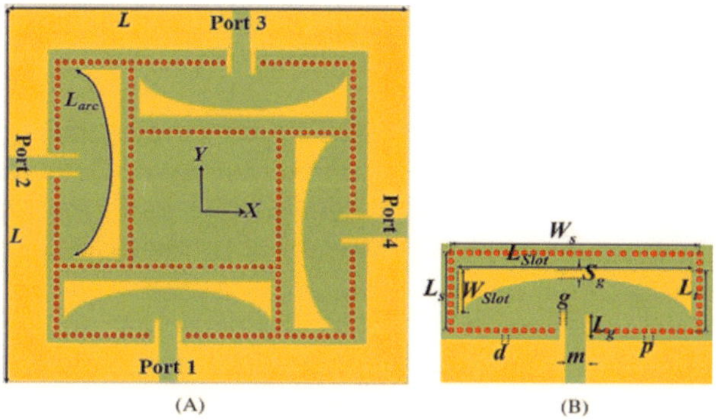


**Figure 14.**
*MIMO antenna of [111]: (A) geometry and (B) unit.*

## 3. Fundamentals of compact design

Section 2 has discussed the specific antennas and the corresponding compact techniques implicit in the design, but there are some methods leading to the compact design not only suit for one fixed structure, but also will be used into other types in the future. Thus, we discuss their applications in MIMO antennas and the related techniques in this section, including the detailed explanation of mode-cutting method, the fractal technique, the theory of characteristic mode, and the optimization algorithms.

### 3.1 Mode-antenna analyses

In Subsection 2.7, the mode-cutting methods for different antenna types has been shown, however, the cited references did not discuss the detailed modes so clear. Now we discuss the mode distributions according to the specific antenna types which have been studied in the MIMO antenna.

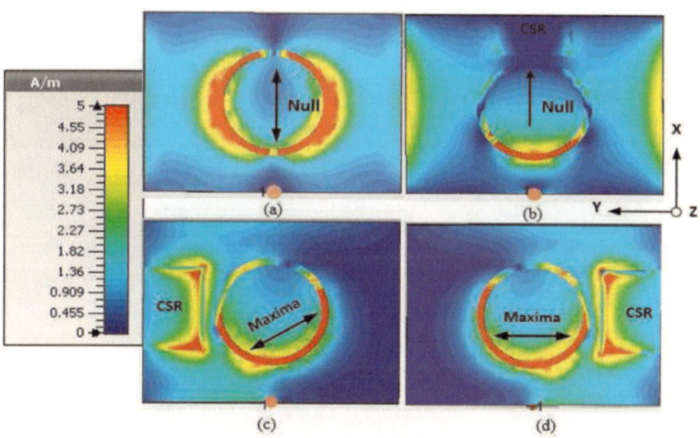


**Figure 15.**
*Current distributions of complete slot of [112]: (a) without CSR, (b) CSR at position 1, (c) CSR at position 2, and (d) CSR at position 3.*

If the mode-cutting method is implemented, the corresponding antenna has at least one integral modes, or we can say that the cutting method suits to the integral order mode of the related structures.

For the cutting of the semi-circle slot antenna, the authors in [112] discussed the current distribution of the complete slot as shown in **Figure 15**. Whether the complementary slot reflector exists or not, the current distributions are symmetrical. By optimizing the size and position of CSR, the cutting method can be implemented to get the half-loop slot, and the corresponding performances are affected rarely. Then, the two-element MIMO antenna is formed by symmetrical arrangement whose isolation less than 12 dB is achieved without any additional decoupling structure.

The literature [113] provides another way to design the cavity MIMO antenna which excites each sub-mode generated by the different shape cavities. The authors started the analysis for the closed circular cavity, then analyzed the characteristic modes for the open and sector cavities, and consequently obtained the methodology that the whole cavity can be divided into N sub-cavities. If a T-shape monopole is put at the proper position for each sub-cavity, the N-port MIMO antenna forms and high isolation is obtained. They took the 4-port circular open cavity as an example and fabricated an antenna prototype shown in **Figure 16**, whose measurements agree well with those of simulations. Thus, they continue discuss the related design for other shape cavity antennas.

### 3.2 Fractal techniques

The fractal technique has been employed in the MIMO antennas as in subsection 2.6 owing to it can reduce the size for the compact design. As we know the fractal technique needs several iterations, it is effective to reduce the size in finite iterations, but when the iteration reaches to a certain number, the size reduction will not so obvious until it does not work. That's because as the iteration increases, the length of fractal shape will become smaller and can not be comparable with the wavelength. Therefore, we have to consider the iteration number depending on the specific requirement. In this subsection, we introduce two common simple fractal techniques, the Koch and the Sierpiński fractals [114, 115].

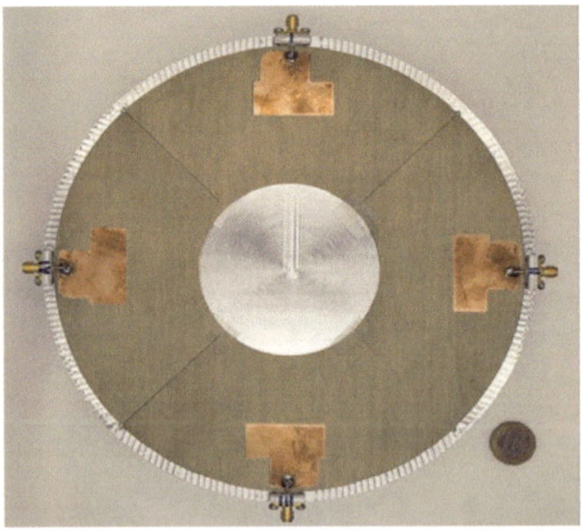

**Figure 16.**
*4-Port MIMO antenna of [113].*

1. Koch Fractal

   The **Figure 12** has presented the 2nd iteration process for the line segment, we do not repeat here. The detailed iteration method is, (1) the line segment is selected as the initiator, (2) divide the segment into three equal portions, then rotate the middle portion ±60 degrees to form another two new subsegments, and (3) repeat the process of last step. The length will increase 1/3 times for each iteration.

2. Sierpiński Fractal

   The Sierpiński iteration process of isosceles triangle is shown in **Figure 17** [114]. The coordinates for the next iteration can be obtained by

$$v_1(x,y) = \begin{pmatrix} \frac{1}{2} & 0 \\ 0 & \frac{1}{2} \end{pmatrix} \begin{pmatrix} x \\ y \end{pmatrix} + \begin{pmatrix} \frac{1}{2} \\ \frac{\sqrt{3}}{2} \end{pmatrix} \tag{1}$$

$$v_2(x,y) = \begin{pmatrix} \frac{1}{2} & 0 \\ 0 & \frac{1}{2} \end{pmatrix} \begin{pmatrix} x \\ y \end{pmatrix} + \begin{pmatrix} -\frac{1}{2} \\ \frac{\sqrt{3}}{2} \end{pmatrix} \tag{2}$$

$$v_3(x,y) = \begin{pmatrix} \frac{1}{2} & 0 \\ 0 & \frac{1}{2} \end{pmatrix} \begin{pmatrix} x \\ y \end{pmatrix} + \begin{pmatrix} 0 \\ 0 \end{pmatrix} \tag{3}$$

This iteration process described by formulas (1)–(3) can be summarized as, find the midpoint of each segment and connect them to form four same isosceles triangles, then remove the middle triangle, we will get the final iteration shape.

### 3.3 Characteristic mode analysis

The CMA permits the researcher to know the antenna performance at the initial stage without considering the specific excitations so that the user can have an insight into the radiation essence, which is extensively used in the application of mobile handset antenna where the large metal chassis exists [116]. And the MIMO antennas are also extensively investigated for the mobile terminal application, there are a lot of works using the CMA [47, 117, 118].

In [47], a meandered dipole working at 3.5 GHz was studied by using the CMA. The first 10 modes are compared and the symmetrical reverse modal currents are

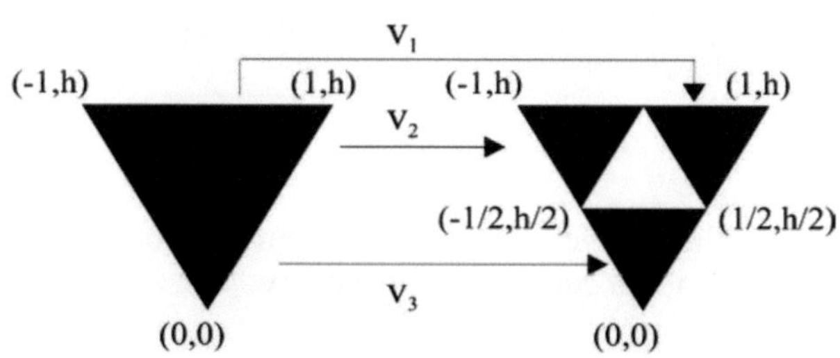


**Figure 17.**
*Iteration process of Sierpiński fractal [114]: left is initiator and right is the 1st iteration.*

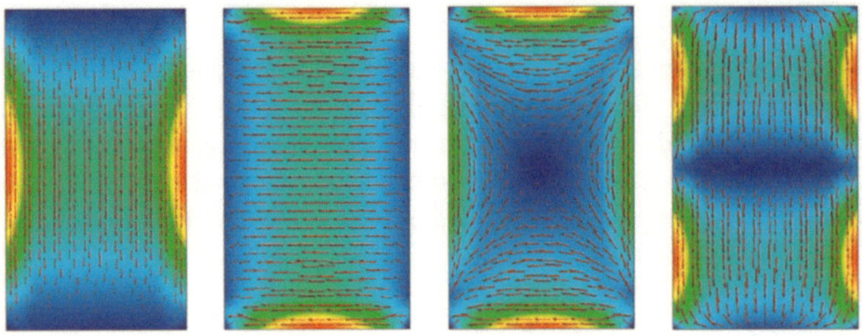

**Figure 18.**
*First four modes at 2.2 GHz of [118].*

excited to get the low SAR, and the isolation of 16 dB is achieved. On the contrary, the CMA is implemented on both the ground plane and the antenna in [117], but only on the chassis ground plane in [118]. The first four modes of [118] are shown in **Figure 18**. With the help of the current distributions, the antennas are put intensity-weak position of the corners for the MIMO antenna design.

Depending on the extensive application in MIMO antenna designs, we simply introduce the corresponding theory of the electric-field boundary conditions as follows [116, 117].

The related scattering field expression satisfy:

$$[L(\boldsymbol{J})]_{\tan} = \boldsymbol{E}^{i}_{\tan}(\boldsymbol{r}), \boldsymbol{r} \in S \tag{4}$$

where $L(\bullet)$ is the intetro-differential operator.

Owing to $L(\bullet)$ features with the impedance property, thus the formula (4) is rewritten as:

$$Z(\boldsymbol{J}) = [L(\boldsymbol{J})]_{\tan} \tag{5}$$

where $Z(\bullet)$ represent the tangential component of electric field related with $\boldsymbol{J}$.

The impedance matrix $\boldsymbol{Z}$ has the real and imaginary parts, we get the following related formulas:

$$\boldsymbol{Z} = \boldsymbol{R} + j\boldsymbol{X} \tag{6}$$

$$\boldsymbol{R} = (\boldsymbol{Z} + \boldsymbol{Z}^*)/2 \tag{7}$$

$$\boldsymbol{X} = (\boldsymbol{Z} - \boldsymbol{Z}^*)/2j \tag{8}$$

By means of Poynting's theorem, we get the straightforward relation for $\boldsymbol{R}$ and $\boldsymbol{X}$, which has clear physical meaning, as

$$\boldsymbol{X}\boldsymbol{J}_n = \lambda_n \boldsymbol{R}\boldsymbol{J}_n \tag{9}$$

where $\boldsymbol{J}_n$ and $\lambda_n$ are the real eigenvector and eigenvalue of $n$th mode, respectively.

The orthogonality of modal currents is defined by:

$$<\boldsymbol{J}_m, \boldsymbol{R}\bullet\boldsymbol{J}_n> \; = \; <\boldsymbol{J}^*_m, \boldsymbol{R}\bullet\boldsymbol{J}_n> \; = \delta_{mn} \tag{10}$$

$$<\boldsymbol{J}_m, \boldsymbol{X}\bullet\boldsymbol{J}_n> \; = \; <\boldsymbol{J}^*_m, \boldsymbol{X}\bullet\boldsymbol{J}_n> \; = \lambda_n\delta_{mn} \tag{11}$$

$$<\boldsymbol{J}_m, \boldsymbol{Z}\bullet\boldsymbol{J}_n> \; = \; <\boldsymbol{J}_m^*, \boldsymbol{Z}\bullet\boldsymbol{J}_n> \; = (1 + j\lambda_n)\delta_{mn} \tag{12}$$

where $\delta_{mn} = 1, m = n$ or $\delta_{mn} = 0, m \neq n$.

Depending on the theory of characteristic mode, we get the induced currents on the PEC body and the resultant fields:

$$\boldsymbol{J} = \sum_n \alpha_n \boldsymbol{J}_n \tag{13}$$

By using the Z impedance operator, formula (13) becomes:

$$\sum_n \alpha_n Z(\boldsymbol{J}_n) = \boldsymbol{E}^i_{\tan}(\boldsymbol{r}) \tag{14}$$

Taking the inner product by the current $\boldsymbol{J}_m$ for (14), we have:

$$\sum_n \alpha_n <Z(\boldsymbol{J}_n), \boldsymbol{J}_m> \; = \; <\boldsymbol{E}^i_{\tan}(\boldsymbol{r}), \boldsymbol{J}_m> \tag{15}$$

Applying the orthogonality of currents, we will get under the condition of m = n:

$$\alpha_n (1 + j\lambda_n) = \; <\boldsymbol{E}^i_{\tan}(\boldsymbol{r}), \boldsymbol{J}_m> \tag{16}$$

so we know:

$$\alpha_n = \frac{<\boldsymbol{E}^i_{\tan}(\boldsymbol{r}), \boldsymbol{J}_m>}{1 + j\lambda_n} \tag{17}$$

where $<\boldsymbol{E}^i_{\tan}(\boldsymbol{r}), \boldsymbol{J}_m>$ is called the modal excitation coefficient and the modal significance MS is defined as:

$$\text{MS} = \left| \frac{1}{1 + j\lambda_n} \right| \tag{18}$$

### 3.4 Optimization algorithms

The optimization algorithms are often used to reduce the antenna size, but for the MIMO antenna it becomes a multi-objective optimization problem due to we have to consider other parameters, like the effect of mutual coupling and the related position change etc. of antenna element. If we use the single optimization algorithm to optimize the MIMO antenna, it will take more time, so the hybrid algorithms are employed to process the complicated discrete and continuous mixed parameters [119–121].

In [119], both the antenna shape and the decoupling structure were considered, thus the hybrid algorithm of both the multiobjective evolutionary algorithm based on decomposition combined with differential evolution (MOEA/D-DE) and MOEA/D combined with genetic operator (MOEA/D-GO) were used, where the MOEA/D-DE is adopted to optimize the radiator while the MOEA/D-GO optimizes the isolated area shown in **Figure 19**. They divided the circle into 8 areas of the same size, each has the 45 degrees angle and divided into several small pieces. In the optimization process, "1" represents the existence of metal while "0" not. Literature [120] Integrates the particle

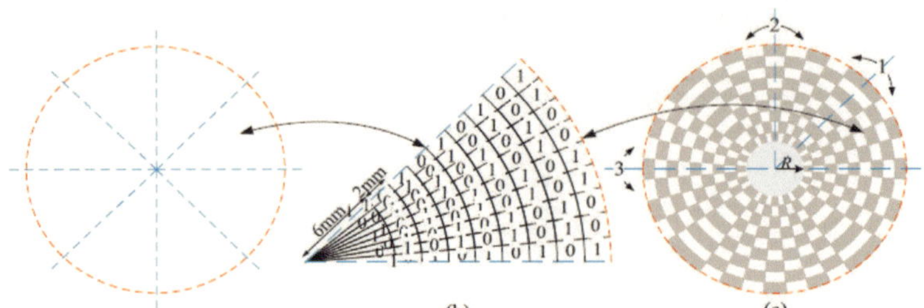


**Figure 19.**
*Fragment-type isolation of [119]: (a) split method, (b) discretization and assignment of "0" or "1", and (c) frqgment-type structure.*

swarm optimization (PSO) and binary PSO into multi-objective evolutionary algorithm based on decomposition (MOEA/D) to realize the optimization of antenna size and isolation, while the surrogate-based optimization was employed in [121].

Though different algorithms have been proposed to improve the optimization result, the corresponding optimization process is similar as in [119].

The optimized problem can be expressed as:

$$min\ F(X) = (f_1(X), f_2(X), \ldots, f_n(X))\ s.t. X \in \Omega \tag{19}$$

where $f_i(X)(i = 1, 2, \ldots, n)$ indicates the corresponding optimized objective, X is a decision variable, and $\Omega$ is the design space.

The authors of [119] presented three optimized objectives due to the four-port MIMO antenna so that they have to consider the mutual coupling of different ports. We can simplify the optimized objectives by two ports, they are:

$$f_1(X) = max\ (Q_1 - min\ |(S_{11})_{dB}|, 0)\omega \in [\omega_1, \omega_2] \tag{20}$$

$$f_2(X) = max\ (Q_2 - min\ |(S_{12})_{dB}|, 0)\ \omega \in [\omega_1, \omega_2] \tag{21}$$

where $[\omega_1, \omega_2]$ indicates the frequency band, $Q_1$ is the desired minimum of return loss, which is set to be 10 dB, and $Q_2$ desired minimum isolation.

With these optimized objectives and the constraints, the iteration process can be implemented by the hybrid utilization of EM simulator and the proposed algorithm.

## 4. Conclusions

Depending on the development trend of 5G+/6G, we focused on the summaries of the planar MIMO antennas and the related compact techniques in this chapter owing to they are easily fabricated and integrated into a system. These planar antennas contain several common antenna types, including the patch, dipole/monopole, slot etc. Even so, they still can be designed into the 3D structure and there are specific applications like in the mobile terminal. The compact techniques implicit in the designs are dug up and summarized into seven categories, including the no-decoupling and the decoupling designs, multiple antenna structure, meander line technique, co-radiator design, and the fractal and radiator-cutting antennas. Then, in Section 3, we discussed the related fundamentals for the compact designs though the antenna types are conventional, and showed the corresponding simple design methods, they are mode analyses, fractal techniques, characteristic analysis, and the optimization algorithms.

## Acknowledgements

Thanks so much to my master students Chenggui Zhou, Jiangkai Ma, and Yan Liang for their help of searching/classifying the papers and adjusting the format.

## Author details

Yiying Wang
Guilin University of Electronic Technology, Guilin, China

*Address all correspondence to: yiying@guet.edu.cn

## References

[1] Roslan SF, Kamarudin MR, Khalily M, Jamaluddin MH. An MIMO rectangular dielectric resonator antenna for 4G applications. IEEE Antennas and Wireless Propagation Letters. 2014;**13**: 321-324. DOI: 10.1016/s0014-5793(01) 03293-8

[2] Farahani M, Pourahmadazar J, Akbari M, Nedil M, Sebak AR, Denidni TA. Mutual coupling reduction in millimeter-wave MIMO antenna array using a metamaterial polarization-rotator wall. IEEE Antennas and Wireless Propagation Letters. 2017;**16**: 2324-2327. DOI: 10.1109/ LAWP.2017.2717404

[3] Li M, Cheung S. Isolation enhancement for MIMO dielectric resonator antennas using dielectric superstrate. IEEE Transactions on Antennas and Propagation. 2021;**69**(7): 4154-4159. DOI: 10.1109/ TAP.2020.3044683

[4] Chew M, Mavrakis S. Quadrifilar helix antenna for MIMO system. IEEE Antennas and Wireless Propagation Letters. 2004;**3**:197-199. DOI: 10.1109/ LAWP.2004.832642

[5] Hassan T, Khan MU, Attia H, Sharawi MS. An FSS based correlation reduction technique for MIMO antennas. IEEE Transactions on Antennas and Propagation. 2018;**66**(9): 4900-4905. DOI: 10.1109/ TAP.2018.2842256

[6] Boukarkar A, Lin XQ, Jiang Y, Nie LY, Mei P, Yu YQ. A miniaturized extremely close-spaced four-element dual-band MIMO antenna system with polarization and pattern diversity. IEEE Antennas and Wireless Propagation Letters. 2018;**17**(1):134-137. DOI: 10.1109/LAWP.2017.2777839

[7] Zhai G, Chen ZN, Qing X. Enhanced isolation of a closely spaced four-element MIMO antenna system using metamaterial mushroom. IEEE Transactions on Antennas and Propagation. 2015;**63**(8):3362-3370. DOI: 10.1109/TAP.2015.2434403

[8] Chiu C-Y, Yan J-B, Murch RD. Compact three-port orthogonally polarized MIMO Antennas. IEEE Antennas and Wireless Propagation Letters. 2007;**6**:619-622. DOI: 10.1109/ LAWP.2007.913272

[9] Pan Y, Cui Y, Li R. Investigation of a triple-band multibeam MIMO antenna for wireless access points. IEEE Transactions on Antennas and Propagation. 2016;**64**(4):1234-1241. DOI: 10.1109/TAP.2016.2526082

[10] Qin P-Y, Guo YJ, Weily AR, Liang C-H. A pattern reconfigurable U-slot antenna and its applications in MIMO systems. IEEE Transactions on Antennas and Propagation. 2012;**60**(2): 516-528. DOI: 10.1109/TAP.2011. 2173439

[11] Sharma Y, Sarkar D, Saurav K, Srivastava KV. Three-element MIMO antenna system with pattern and polarization diversity for WLAN applications. IEEE Antennas and Wireless Propagation Letters. 2017;**16**:1163-1166. DOI: 10.1109/LAWP.2016.2626394

[12] Narbudowicz A, Ammann MJ. Low-cost multimode patch antenna for dual MIMO and enhanced localization use. IEEE Transactions on Antennas and Propagation. 2018;**66**(1):405-408. DOI: 10.1109/TAP.2017.2767643

[13] Qian J-F, Chen F-C, Ding Y-H, Hu H-T, Chu Q-X. A wide stopband filtering patch antenna and its application in

MIMO system. IEEE Transactions on Antennas and Propagation. 2019;**67**(1): 654-658. DOI: 10.1109/TAP.2018.2874764

[14] Cheng B, Du Z. Dual polarization MIMO antenna for 5G mobile phone applications. IEEE Transactions on Antennas and Propagation. 2021;**69**(7): 4160-4165. DOI: 10.1109/TAP.2020.3044649

[15] OuYang J, Yang F, Wang ZM. Reducing mutual coupling of closely spaced microstrip MIMO antennas for WLAN application. IEEE Antennas and Wireless Propagation Letters. 2011;**10**: 310-313. DOI: 10.1109/LAWP.2011.2140310

[16] Ghannad AA, Khalily M, Xiao P, Tafazolli R, Kishk AA. Enhanced matching and Vialess decoupling of nearby patch antennas for MIMO system. IEEE Antennas and Wireless Propagation Letters. 2019;**18**(6): 1066-1070. DOI: 10.1109/LAWP.2019.2906308

[17] Garg P, Jain P. Isolation improvement of MIMO antenna using a novel flower shaped metamaterial absorber at 5.5 GHz WiMAX band. IEEE Transactions on Circuits and Systems II: Express Briefs. 2020;**67**(4):675-679. DOI: 10.1109/TCSII.2019.2925148

[18] Tran HH, Nguyen-Trong N. Performance enhancement of MIMO patch antenna using parasitic elements. IEEE Access. 2021;**9**: 30011-30016. DOI: 10.1109/ACCESS.2021.3058340

[19] Row J-S, Yeh S-H, Wong K-L. Compact dual-polarized microstrip antennas. Microwave and Optical Technology Letters. 2000;**27**(4):284-287. DOI: 10.1002/1098-2760(20001120)27:4 <284::AID-MOP21>3.0.CO;2-L

[20] Wong K-L, Chen J-Z, Li W-Y. Four-port wideband annular-ring patch antenna generating four decoupled waves for 5G multi-input–multi-output access points. IEEE Transactions on Antennas and Propagation. 2021;**69**(5): 2946-2951. DOI: 10.1109/TAP.2020.3025237

[21] Wong K-L, Jian M-F, Li W-Y. Low-profile wideband four-corner-fed square patch antenna for 5G MIMO mobile antenna application. IEEE Antennas and Wireless Propagation Letters. 2021;**20**(12):2554-2558. DOI: 10.1109/LAWP.2021.3119753

[22] Qian J-F, Chen F-C, Chu Q-X, Xue Q, Lancaster MJ. A novel electric and magnetic gap-coupled broadband patch antenna with improved selectivity and its application in MIMO system. IEEE Transactions on Antennas and Propagation. 2018;**66**(10):5625-5629. DOI: 10.1109/TAP.2018.2860129

[23] Karimian R, Oraizi H, Fakhte S, Farahani M. Novel F-shaped quad-band printed slot antenna for WLAN and WiMAX MIMO systems. IEEE Antennas and Wireless Propagation Letters. 2013;**12**:405-408. DOI: 10.1109/LAWP.2013.2252140

[24] Soltani S, Lotfi P, Murch RD. A dual-band multiport MIMO slot antenna for WLAN applications. IEEE Antennas and Wireless Propagation Letters. 2017;**16**: 529-532. DOI: 10.1109/LAWP.2016.2587732

[25] Nandi S, Mohan A. A compact dual-band MIMO slot antenna for WLAN applications. IEEE Antennas and Wireless Propagation Letters. 2017;**16**: 2457-2460. DOI: 10.1109/LAWP.2017.2723927

[26] Hu H-T, Chen F-C, Chu Q-X. A compact directional slot antenna and its

application in MIMO array. IEEE Transactions on Antennas and Propagation. 2016;**64**(12):5513-5517. DOI: 10.1109/TAP.2016.2621021

[27] Sun L, Li Y, Zhang Z. Wideband integrated quad-element MIMO antennas based on complementary antenna pairs for 5G smartphones. IEEE Transactions on Antennas and Propagation. 2021;**69**(8):4466-4474. DOI: 10.1109/TAP.2021.3060020

[28] Ren J, Hu W, Yin Y, Fan R. Compact printed MIMO antenna for UWB applications. IEEE Antennas and Wireless Propagation Letters. 2014;**13**: 1517-1520. DOI: 10.1109/LAWP.2014.2343454

[29] Srivastava G, Mohan A. Compact MIMO slot antenna for UWB applications. IEEE Antennas and Wireless Propagation Letters. 2016;**15**: 1057-1060. DOI: 10.1109/LAWP.2015.2491968

[30] Hu W, Chen Z, Qian L, Wen L, Luo Q, Xu R, et al. Wideband back-cover antenna design using dual characteristic modes with high isolation for 5G MIMO smartphone. IEEE Transactions on Antennas and Propagation. 2022;**70**(7):5254-5265. DOI: 10.1109/TAP.2022.3145456

[31] Anitha R, Vinesh PV, Prakash KC, Mohanan P, Vasudevan K. A compact quad element slotted ground wideband antenna for MIMO applications. IEEE Transactions on Antennas and Propagation. 2016;**64**(10):4550-4553. DOI: 10.1109/TAP.2016.2593932

[32] Oliveira JGD, D'Assunção Junior AG, Silva Neto VP, D'Assunção AG. New compact MIMO antenna for 5G, WiMAX and WLAN technologies with dual polarisation and element diversity. IET Microwaves, Antennas & Propagation. 2021;**15**(4):415-426. DOI: 10.1049/mia2.12057

[33] Sarkar D, Srivastava KV. A compact four-element MIMO/diversity antenna with enhanced bandwidth. IEEE Antennas and Wireless Propagation Letters. 2017;**16**:2469-2472. DOI: 10.1109/LAWP.2017.2724439

[34] Wang H, Liu L, Zhang Z, Li Y, Feng Z. A wideband compact WLAN/WiMAX MIMO antenna based on dipole with V-shaped ground branch. IEEE Transactions on Antennas and Propagation. 2015;**63**(5):2290-2295. DOI: 10.1109/TAP.2015.2405091

[35] Darvish M, Hassani HR. Quad band CPW-Fed monopole antenna for MIMO applications. In: Dans: 2012 6th European Conference on Antennas and Propagation (EUCAP). IEEE: Prague, Czech Republic; 2012. pp. 1-4

[36] Chouhan S, Panda DK, Gupta M, Singhal S. Meander line MIMO antenna for 5.8 GHz WLAN application. International Journal of RF and Microwave Computer-Aided Engineering. 2018;**28**(4):e21222. DOI: 10.1002/mmce.21222

[37] Im Y-T, Lee J-H, Bhatti RA, Park S-O. A spiral-dipole antenna for MIMO systems. IEEE Antennas and Wireless Propagation Letters. 2008;**7**:803-806. DOI: 10.1109/LAWP.2008.2001395

[38] Shoaib S, Shoaib I, Shoaib N, Chen X, Parini CG. MIMO antennas for mobile handsets. IEEE Antennas and Wireless Propagation Letters. 2015;**14**: 799-802. DOI: 10.1109/LAWP.2014.2385593

[39] Sharawi MS, Iqbal SS, Faouri YS. An 800 MHz 2×1 compact MIMO antenna system for LTE handsets. IEEE Transactions on Antennas and

Propagation. 2011;**59**(8):3128-3131. DOI: 10.1109/TAP.2011.2158958

[40] Mahmood F, Kazim J-R, Karlsson M, Gong S, Ying Z. Decoupling techniques of compact and broadband MIMO antennas for handheld devices. In: Dans: 2012 6th European Conference on Antennas and Propagation (EUCAP). IEEE: Prague, Czech Republic; 2012. pp. 1-5

[41] Khan MS, Shafique MF, Naqvi A, Capobianco A-D, Ijaz B, Braaten BD. A miniaturized dual-band MIMO antenna for WLAN applications. IEEE Antennas and Wireless Propagation Letters. 2015; **14**:958-961. DOI: 10.1109/LAWP.2014.2387701

[42] Thummaluru SR, Kumar R, Chaudhary RK. Isolation and frequency reconfigurable compact MIMO antenna for wireless local area network applications. IET Microwaves, Antennas & Propagation. 2019;**13**(4):519-525. DOI: 10.1049/iet-map.2018.5895

[43] Cui L, Guo J, Liu Y, Sim C-Y-D. An 8-element dual-band MIMO antenna with decoupling stub for 5G smartphone applications. IEEE Antennas and Wireless Propagation Letters. 2019; **18**(10):2095-2099. DOI: 10.1109/LAWP.2019.2937851

[44] Ren Z, Zhao A, Wu S. MIMO antenna with compact decoupled antenna pairs for 5G mobile terminals. IEEE Antennas and Wireless Propagation Letters. 2019;**18**(7): 1367-1371. DOI: 10.1109/LAWP.2019.2916738

[45] Serghiou D, Khalily M, Singh V, Araghi A, Tafazolli R. Sub-6 GHz Dual-Band 8 × 8 MIMO Antenna for 5G Smartphones. IEEE Antennas and Wireless Propagation Letters. 2020; **19**(9):1546-1550. DOI: 10.1109/LAWP.2020.3008962

[46] Xu Z, Deng C. High-isolated MIMO antenna design based on pattern diversity for 5G mobile terminals. IEEE Antennas and Wireless Propagation Letters. 2020;**19**(3):467-471. DOI: 10.1109/LAWP.2020.2966734

[47] Zhang HH, Yu GG, Liu XZ, Cheng GS, Xu YX, Liu Y, et al. Low-SAR MIMO antenna array design using characteristic modes for 5G mobile phones. IEEE Transactions on Antennas and Propagation. 2022;**70**(4):3052-3057. DOI: 10.1109/TAP.2021.3121174

[48] Ye Y, Zhao X, Wang J. Compact high-isolated MIMO antenna module with chip capacitive decoupler for 5G mobile terminals. IEEE Antennas and Wireless Propagation Letters. 2022;**21**(5): 928-932. DOI: 10.1109/LAWP.2022.3152236

[49] Zhou X, Quan XL, Li RL. A dual-broadband MIMO antenna system for GSM/UMTS/LTE and WLAN handsets. IEEE Antennas and Wireless Propagation Letters. 2012;**11**:551-554. DOI: 10.1109/LAWP.2012.2199459

[50] Wang Y-Y, Ban Y-L, Nie Z, Sim C-Y-D. Dual-loop antenna for 4G LTE MIMO smart glasses applications. IEEE Antennas and Wireless Propagation Letters. 2019;**18**(9):1818-1822. DOI: 10.1109/LAWP.2019.2930726

[51] Rhee C, Kim Y, Park T, Kwoun S-s, Mun B, Lee B, et al. Pattern-reconfigurable MIMO antenna for high isolation and low correlation. IEEE Antennas and Wireless Propagation Letters. 2014;**13**:1373-1376. DOI: 10.1109/LAWP.2014.2339012

[52] Nandi S, Mohan A. CRLH unit cell loaded triband compact MIMO antenna for WLAN/WiMAX applications. IEEE Antennas and Wireless Propagation

Letters. 2017;**16**:1816-1819. DOI: 10.1109/LAWP.2017.2681178

[53] Zhao A, Ren Z. Size reduction of self-isolated MIMO antenna system for 5G mobile phone applications. IEEE Antennas and Wireless Propagation Letters. 2019;**18**(1):152-156. DOI: 10.1109/LAWP.2018.2883428

[54] Kulkarni AN, Sharma SK. Frequency reconfigurable microstrip loop antenna covering LTE bands with MIMO implementation and wideband microstrip slot antenna all for portable wireless DTV media player. IEEE Transactions on Antennas and Propagation. 2013;**61**(2):964-968. DOI: 10.1109/TAP.2012.2223433

[55] Ahn C-H, Oh S-W, Chang K. A dual-frequency omnidirectional antenna for polarization diversity of MIMO and wireless communication applications. IEEE Antennas and Wireless Propagation Letters. 2009;**8**:966-969. DOI: 10.1109/LAWP.2009.2030135

[56] Hu PF, Leung KW, Pan YM, Zheng SY. Electrically small, planar, horizontally polarized dual-band omnidirectional antenna and its application in a MIMO system. IEEE Transactions on Antennas and Propagation. 2021;**69**(9):5345-5355. DOI: 10.1109/TAP.2021.3061096

[57] Fernandez SC, Sharma SK. Multiband printed meandered loop antennas with MIMO implementations for wireless routers. IEEE Antennas and Wireless Propagation Letters. 2013;**12**: 96-99. DOI: 10.1109/LAWP.2013.2243104

[58] Zhao X, Yeo SP, Ong LC. Planar UWB MIMO antenna with pattern diversity and isolation improvement for mobile platform based on the theory of characteristic modes. IEEE Transactions on Antennas and Propagation. 2018; **66**(1):420-425. DOI: 10.1109/TAP.2017.2768083

[59] Liu L, Cheung SW, Yuk TI. Compact MIMO antenna for portable devices in UWB applications. IEEE Transactions on Antennas and Propagation. 2013;**61**(8): 4257-4264. DOI: 10.1109/TAP.2013.2263277

[60] Zhang S, Pedersen GF. Mutual coupling reduction for UWB MIMO antennas with a wideband neutralization line. IEEE Antennas and Wireless Propagation Letters. 2016;**15**:166-169. DOI: 10.1109/LAWP.2015.2435992

[61] Wang Y, Zhu F, Gao S. Design of planar ultra-wideband antenna with polarization diversity and high isolation. In: Dans: 2016 IEEE International Conference on Ubiquitous Wireless Broadband (ICUWB). Nanjing, China: IEEE; 2016. pp. 1-3

[62] Zhang S, Ying Z, Xiong J, He S. Ultrawideband MIMO/diversity antennas with a tree-like structure to enhance wideband isolation. IEEE Antennas and Wireless Propagation Letters. 2009;**8**:1279-1282. DOI: 10.1109/LAWP.2009.2037027

[63] Saleem R, Bilal M, Bajwa KB, Shafique MF. Eight-element UWB-MIMO array with three distinct isolation mechanisms. Electronics Letters. 2015; **51**(4):311-313. DOI: 10.1049/el.2014.4199

[64] Gautam AK, Yadav S, Rambabu K. Design of ultra-compact UWB antenna with band-notched characteristics for MIMO applications. IET Microwaves, Antennas & Propagation. 2018;**12**(12): 1895-1900. DOI: 10.1049/iet-map.2018.0012

[65] Li J-F, Chu Q-X, Li Z-H, Xia X-X. Compact dual band-notched UWB

MIMO antenna with high isolation. IEEE Transactions on Antennas and Propagation. 2013;**61**(9):4759-4766. DOI: 10.1109/TAP.2013.2267653

[66] Lee J-M, Kim K-B, Ryu H-K, Woo J-M. A Compact ultrawideband MIMO antenna with WLAN band-rejected operation for mobile devices. IEEE Antennas and Wireless Propagation Letters. 2012;**11**:990-993. DOI: 10.1109/LAWP.2012.2214431

[67] Deng J-Y, Guo L-X, Liu X-L. An ultrawideband MIMO antenna with a high isolation. IEEE Antennas and Wireless Propagation Letters. 2016;**15**: 182-185. DOI: 10.1109/LAWP.2015.2437713

[68] Saxena S, Kanaujia BK, Dwari S, Kumar S, Tiwari R. A compact dual-polarized MIMO antenna with distinct diversity performance for UWB applications. IEEE Antennas and Wireless Propagation Letters. 2017;**16**: 3096-3099. DOI: 10.1109/LAWP.2017.2762426

[69] Luo C-M, Hong J-S, Zhong L-L. Isolation enhancement of a very compact UWB-MIMO slot antenna with two defected ground structures. IEEE Antennas and Wireless Propagation Letters. 2015;**14**:1766-1769. DOI: 10.1109/LAWP.2015.2423318

[70] Liu Y-Y, Tu Z-H. Compact differential band-notched stepped-slot UWB-MIMO antenna with common-mode suppression. IEEE Antennas and Wireless Propagation Letters. 2017;**16**: 593-596. DOI: 10.1109/LAWP.2016.2592179

[71] Roshna TK, Deepak U, Sajitha VR, Vasudevan K, Mohanan P. A compact UWB MIMO antenna with reflector to enhance isolation. IEEE Transactions on Antennas and Propagation. 2015;**63**(4): 1873-1877. DOI: 10.1109/TAP.2015.2398455

[72] Capobianco A-D, Pigozzo FM, Assalini A, Midrio M, Boscolo S, Sacchetto F. A compact MIMO array of planar end-fire antennas for WLAN applications. IEEE Transactions on Antennas and Propagation. 2011;**59**(9): 3462-3465. DOI: 10.1109/TAP.2011.2161557

[73] Hsu Y-W, Huang T-C, Lin H-S, Lin Y-C. Dual-polarized quasi Yagi–Uda antennas with endfire radiation for millimeter-wave MIMO terminals. IEEE Transactions on Antennas and Propagation. 2017;**65**(12):6282-6289. DOI: 10.1109/TAP.2017.2734238

[74] Jehangir SS, Sharawi MS. A miniaturized multi-wideband Quasi-Yagi MIMO antenna system. International Journal of RF and Microwave Computer-Aided Engineering. 2018;**28**(5):e21237. DOI: 10.1002/mmce.21237

[75] Jehangir SS, Sharawi MS. A wideband sectoral Quasi-Yagi MIMO antenna system with multibeam elements. IEEE Transactions on Antennas and Propagation. 2019;**67**(3): 1898-1903. DOI: 10.1109/TAP.2018.2889034

[76] Jehangir SS, Sharawi MS. A miniaturized UWB biplanar Yagi-like MIMO antenna system. IEEE Antennas and Wireless Propagation Letters. 2017; **16**:2320-2323. DOI: 10.1109/LAWP.2017.2716963

[77] Jehangir SS, Sharawi MS. A compact single-layer four-port orthogonally polarized Yagi-Like MIMO antenna system. IEEE Transactions on Antennas and Propagation. 2020;**68**(8):6372-6377. DOI: 10.1109/TAP.2020.2969810

[78] Sung Y. Closely spaced MIMO antenna based on substrate-integrated waveguide technology. Microwave and Optical Technology Letters. 2018;**60**(7): 1794-1798. DOI: 10.1002/mop.31249

[79] Niu B, Tan J. Compact four-element MIMO antenna using T-shaped and anti-symmetric U-shaped slotted SIW cavities. Electronics Letters. 2019;**55**(19): 1031-1032. DOI: 10.1049/el.2019.2142

[80] Chang L, Zhang G, Wang H. Triple-band microstrip patch antenna and its four-antenna module based on half-mode patch for 5G 4 × 4 MIMO operation. IEEE Transactions on Antennas and Propagation. 2022;**70**(1): 67-74. DOI: 10.1109/TAP.2021.3090572

[81] Niu B, Tan J. Compact self-isolated MIMO antenna system based on quarter-mode SIW cavity. Electronics Letters. 2019;**55**(10):574-576. DOI: 10.1049/el.2019.0606

[82] Nandi S, Mohan A. A compact eighth-mode circular SIW cavity-based MIMO antenna. IEEE Antennas and Wireless Propagation Letters. 2021; **20**(9):1834-1838. DOI: 10.1109/LAWP.2021.3098711

[83] Tripathi S, Mohan A, Yadav S. A compact octagonal fractal UWBMIMO antenna with WLAN band-rejection. Microwave and Optical Technology Letters. 2015;**57**(8):1919-1925. DOI: 10.1002/mop.29220

[84] Singhal S. Four element ultra-wideband fractal multiple-input multiple-output antenna. Microwave and Optical Technology Letters. 2019;**61**(12): 2811-2818. DOI: 10.1002/mop.31980

[85] Sampath R, Selvan KT. Compact hybrid Sierpinski Koch fractal UWB MIMO antenna with pattern diversity. International Journal of RF and Microwave Computer-Aided Engineering. 2019;**e22017**:1-13. DOI: 10.1002/mmce.22017

[86] Rajkumar S, Anto Amala A, Selvan KT. Isolation improvement of UWB MIMO antenna utilising molecule fractal structure. Electronics Letters. 2019;**55**(10):576-579. DOI: 10.1049/el.2019.0592

[87] Choukiker YK, Sharma SK, Behera SK. Hybrid fractal shape planar monopole antenna covering multiband wireless communications with MIMO implementation for handheld mobile devices. IEEE Transactions on Antennas and Propagation. 2014;**62**(3):1483-1488. DOI: 10.1109/TAP.2013.2295213

[88] Rajkumar S, Srinivasan N, Natesan A, Selvan KT. A penta-band hybrid fractal MIMO antenna for ISM applications. International Journal of RF and Microwave Computer-Aided Engineering. 2018;**28**(2):e21185. DOI: 10.1002/mmce.21185

[89] Rajkumar S, Vivek Sivaraman N, Murali S, Selvan KT. Heptaband swastik arm antenna for MIMO applications. IET Microwaves, Antennas & Propagation. 2017;**11**(9):1255-1261. DOI: 10.1049/iet-map.2016.1098

[90] Peristerianos A, Theopoulos A, AnastasiosG K, Kaifas T, Siakavara K. Dual-band fractal semi-printed element antenna arrays for MIMO applications. IEEE Antennas and Wireless Propagation Letters. 2016;**15**:730-733. DOI: 10.1109/LAWP.2015.2470681

[91] Chattha HT, Nasir M, Abbasi QH, Huang Y, AlJa'afreh SS. Compact low-profile dual-port single wideband planar inverted-F MIMO antenna. IEEE Antennas and Wireless Propagation Letters. 2013;**12**:1673-1675. DOI: 10.1109/LAWP.2013.2293765

[92] See CH, Hraga HI, Noras JM, Abd-Alhameed RA, McEwan NJ. Compact multiple input and multiple output/diversity antenna for portable and mobile ultra-wideband applications. IET Microwaves, Antennas & Propagation. 2013;**7**(6):444-451. DOI: 10.1049/iet-map.2012.0574

[93] Lim J-H, Jin Z-J, Song C-W, Yun T-Y. Simultaneous frequency and isolation reconfigurable MIMO PIFA using PIN diodes. IEEE Transactions on Antennas and Propagation. 2012;**60**(12): 5939-5946. DOI: 10.1109/TAP.2012.2211552

[94] Lee B, Harackiewicz FJ, Wi H. Closely mounted mobile handset MIMO antenna for LTE 13 band application. IEEE Antennas and Wireless Propagation Letters. 2014;**13**:411-414. DOI: 10.1109/LAWP.2014.2307310

[95] Bhatti RA, Choi J-H, Park S-O. Quad-band MIMO antenna array for portable wireless communications terminals. IEEE Antennas and Wireless Propagation Letters. 2009;**8**:129-132. DOI: 10.1109/LAWP.2008.2012274

[96] Radhi AH, Nilavalan R, Wang Y, Al-Raweshidy H, Eltokhy AA, Aziz NA. Mutual coupling reduction with a novel fractal electromagnetic bandgap structure. IET Microwaves, Antennas & Propagation. 2019;**13**(2):134-141. DOI: 10.1049/iet-map.2018.5273

[97] Chiu C-Y, Murch RD. Compact four-port antenna suitable for portable MIMO devices. IEEE Antennas and Wireless Propagation Letters. 2008;**7**:142-144. DOI: 10.1109/LAWP.2008.919341

[98] Yao Y, Wang X, Chen X, Yu J, Liu S. Novel diversity/MIMO PIFA antenna with broadband circular polarization for multimode satellite navigation. IEEE Antennas and Wireless Propagation Letters. 2012;**11**:65-68. DOI: 10.1109/LAWP.2012.2183335

[99] Mun B, Jung C, Park M-J, Lee B. A Compact frequency-reconfigurable multiband LTE MIMO antenna for laptop applications. IEEE Antennas and Wireless Propagation Letters. 2014;**13**: 1389-1392. DOI: 10.1109/LAWP.2014.2339802

[100] Abbosh AI, Al-Rizzo H, Abushamleh S, Bihnam A, Khaleel HR. Flexible CPW-IFA antenna array with reduced mutual coupling. In: Dans: 2014 IEEE Antennas and Propagation Society International Symposium (APSURSI). Memphis, TN, USA: IEEE; 2014. pp. 1716-1717

[101] Chang L, Wang H. Miniaturized wideband four-antenna module based on dual-mode PIFA for 5G 4 × 4 MIMO applications. IEEE Transactions on Antennas and Propagation. 2021;**69**(9): 5297-5304. DOI: 10.1109/TAP.2021.3069490

[102] Yuan X-T, Chen Z, Gu T, Yuan T. A wideband PIFA-pair-based MIMO antenna for 5G smartphones. IEEE Antennas and Wireless Propagation Letters. 2021;**20**(3):371-375. DOI: 10.1109/LAWP.2021.3050337

[103] Liu DQ, Zhang M, Luo HJ, Wen HL, Wang J. Dual-band platform-free PIFA for 5G MIMO application of mobile devices. IEEE Transactions on Antennas and Propagation. 2018;**66**(11): 6328-6333. DOI: 10.1109/TAP.2018.2863109

[104] Barani IRR, Wong K-L. Integrated inverted-F and open-slot antennas in the metal-framed smartphone for 2×2 LTE LB and 4×4 LTE M/HB MIMO operations. IEEE Transactions on Antennas and Propagation. 2018;**66**(10):

5004-5012. DOI: 10.1109/TAP.2018.2854191

[105] MoradiKordalivand A, Rahman TA, Khalily M. Common elements wideband MIMO antenna system for WiFi/LTE access-point applications. IEEE Antennas and Wireless Propagation Letters. 2014;**13**:1601-1604. DOI: 10.1109/LAWP.2014.2347897

[106] Mao C-X, Chu Q-X. Compact coradiator UWB-MIMO antenna with dual polarization. IEEE Transactions on Antennas and Propagation. 2014;**62**(9): 4474-4480. DOI: 10.1109/TAP.2014.2333066

[107] Piao D, Wang Y. Tripolarized MIMO antenna using a compact single-layer microstrip patch. IEEE Transactions on Antennas and Propagation. 2019;**67**(3):1937-1940. DOI: 10.1109/TAP.2018.2889147

[108] Huang H, Jiang W, Zhang T, Zhu Y, Pang B, Hu W. Shared radiator based high-isolated tri-port mobile terminal antenna group design. International Journal of RF and Microwave Computer-Aided Engineering. 2022; **32**(7):e23177. DOI: 10.1002/mmce.23177

[109] Liu X-L, Wang Z-D, Yin Y-Z, Ren J, Wu J-J. A compact ultrawideband MIMO antenna using QSCA for high isolation. IEEE Antennas and Wireless Propagation Letters. 2014;**13**: 1497-1500. DOI: 10.1109/LAWP.2014.2340395

[110] Potti D, Tusharika Y, Alsath MGN, Kirubaveni S, Kanagasabai M, Sankararajan R, et al. A novel optically transparent UWB antenna for automotive MIMO communications. IEEE Transactions on Antennas and Propagation. 2021;**69**(7):3821-3828. DOI: 10.1109/TAP.2020.3044383

[111] Kumar K, Dwari S. Compact four-element MIMO SIW cavity backed slot antenna with high front-to-back ratio. International Journal of RF and Microwave Computer-Aided Engineering. 2019;**29**(1):e21512. DOI: 10.1002/mmce.21512

[112] Jehangir SS, Sharawi MS. A single layer semi-ring slot Yagi-like MIMO antenna system with high front-to-back ratio. IEEE Transactions on Antennas and Propagation. 2017;**65**(2):937-942. DOI: 10.1109/TAP.2016.2633938

[113] Molins-Benlliure J, Cabedo-Fabres M, Antonino-Daviu E, Ferrando-Bataller M. sector unit-cell methodology for the design of Sub-6 GHz 5G MIMO antennas. IEEE Access. 2022;**10**: 100824-100836. DOI: 10.1109/ACCESS.2022.3207163

[114] Anguera A, Jayasinghe C, Chowdary P, et al. Fractal antennas: an historical perspective. Fractal and Fractional. 2020;**4**(1):3. DOI: 10.3390/fractalfract4010003

[115] Dwivedy B, Das TK. Introduction to fractal antennas and their role in MIMO applications. In: Kumar Y, Tripathi S, Raj B, editors. Multifunctional MIMO Antennas: Fundamentals and Application. 1re éd ed. Boca Raton: CRC Press; 2022. pp. 1-25. DOI: 10.1201/9781003290230

[116] Chen Y, Wang CF. Characteristic mode theory for PEC bodies. In: Characteristic Modes: Theory and Applications in Antenna Engineering. 1st ed. Hoboken, NJ: Wiley; 2015. pp. 37-97. DOI: 10.1002/9781119038900.ch2

[117] Kumar Kishor K, Hum SV. A pattern reconfigurable chassis-mode MIMO antenna. IEEE Transactions on Antennas and Propagation. 2014;**62**(6):

3290-3298. DOI: 10.1109/TAP.2014.2313634

[118] Deng C, Lv X. Wideband MIMO antenna with small ground clearance for mobile terminals. IET Microwaves, Antennas & Propagation. 2019;**13**(9): 1419-1426. DOI: 10.1049/iet-map.2018.5972

[119] Lu D, Wang L, Yang E, Wang G. Design of high-isolation wideband dual-polarized compact MIMO antennas with multiobjective optimization. IEEE Transactions on Antennas and Propagation. 2018;**66**(3):1522-1527. DOI: 10.1109/TAP.2017.2784446

[120] Li Q-Q, Chu Q-X, Chang Y-L. Design of compact high-isolation MIMO antenna with multiobjective mixed optimization algorithm. IEEE Antennas and Wireless Propagation Letters. 2020; **19**(8):1306-1310. DOI: 10.1109/LAWP.2020.2997874

[121] Koziel S, Bekasiewicz A, Cheng QS. Conceptual design and automated optimisation of a novel compact UWB MIMO slot antenna. IET Microwaves, Antennas & Propagation. 2017;**11**(8): 1162-1168. DOI: 10.1049/iet-map.2016.0703

Chapter 10

# Recent Advances in the mm-Wave Array for Mobile Phones

*Yan Wang and Xiaoxue Fan*


## Abstract

With the development of communication system to the mm-wave band, the antenna design in the mm-wave band for mobile phones encounters new requirements and challenges. The mm-wave characteristics of short wavelength, high free-space path loss, and easy-to-be-blocking usually require mm-wave antennas with high gain and beam-scanning capability. Also, considering the very limited space occupied by antennas in mobile phones and the massive production of consumer electronics, small size, low cost, multiband, multi-polarization, and wide beam steering becomes the main key point of mm-wave array performance. In addition, as a special situation of the mobile antenna, the analysis of effect of the human tissue on the antenna performance is also important. So, in this chapter, a comprehensive summary on the recent advances in the mm-wave array for mobile phones including single-band, dual-band, and reconfigurable design of broadside array, horizontal polarized, vertical polarized, and dual-polarized design of endfire array, co-design of mm-wave array with lower band antenna, and user influence are summarized.


**Keywords:** mm-wave array, mobile phones, broadside array, endfire array, reconfigurable array, shared-aperture, beam steering, user influence

## 1. Introduction

Mobile communication has been updated greatly from the first-generation (1G) to recent fifth-generation (5G) and future promising sixth-generation (6G) mobile communication systems to meet the requirement of high data rate, large capacity, low latency, *et al.* for consumers [1]. As one of the key techniques for 5G and 6G communication system, millimeter-wave (mm-wave) frequency band with large bandwidth is adopted [2, 3]. Comparison with the centimeter-wave or decimeter-wave, the frequency band of mm-wave is much higher and thus shorter wavelength. However, due to its high frequency with short wavelength, the free-space loss of mm-wave is higher than that of lower frequency bands, and the mm-wave beam is usually blocked [4]. To mitigate the path loss and beam blockage, a high-gain antenna with wide-angle beam scanning is usually adopted to catch the strongest signal and ensure effective radiation in a 5G mm-wave system [5].

To ensure the wireless connection using mm-wave frequency band, the mm-wave array antenna should be applied in mobile phones. Besides the requirements of frequency bands, maximum output power, additional spectrum emission mask, and spurious emission from the 3GPP [6], four additional key difficulties should be considered for mm-wave array in mobile phones:

- Array size: Because most of the spaces are reserved for the displays, cameras, battery, printed circuit board (PCB), motor, speaker, and so on [7] for better user experience, the mm-wave array in mobile phones should only occupy a very small space. In addition, most of the antenna space in mobile phones has been occupied by the antennas working at the lower frequency band. The space for mm-wave arrays in mobile phones is extremely limited.

- Beam coverage: Because the posture of mobile phones is usually arbitrary in realistic scenarios [8], the mm-wave array in mobile phones should cover as wide as possible beam coverage with the required performance to catch the strongest signal for an effective signal connection. In practical applications, 2 or 3 mm-wave arrays are usually deployed in mobile phones to achieve the desired beam coverage.

- Integration with the mobile phones: Because of the requirements of mobile phones for the full-display, curved-display, metal-bezel, glass back cover, metal back cover [7], the mm-wave array in mobile phones should fit the industry design (ID) of mobile phones. In practical applications, the geometry, layout, thickness, and deployment of the mm-wave array are determined by the ID of the mobile phone.

- User influence: Because mobile phones are usually held by the human hands [8], the mm-wave array in mobile phones might also be covered by the human band. In practical applications, the effect of the human hands or human fingers on the antenna impedance, bandwidth, gain, and beam coverage should be studied.

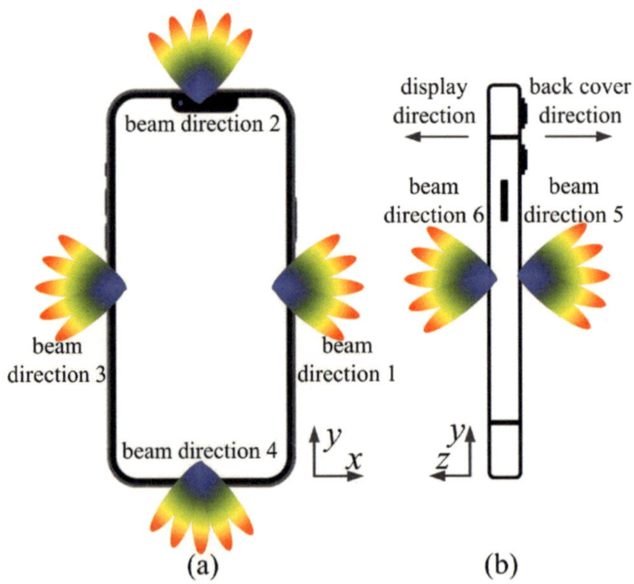


**Figure 1.**
*Conceptual diagram of the mm-wave array with desired spherical coverage for mobile phones. (a) Front view. (b) Side view [11].*

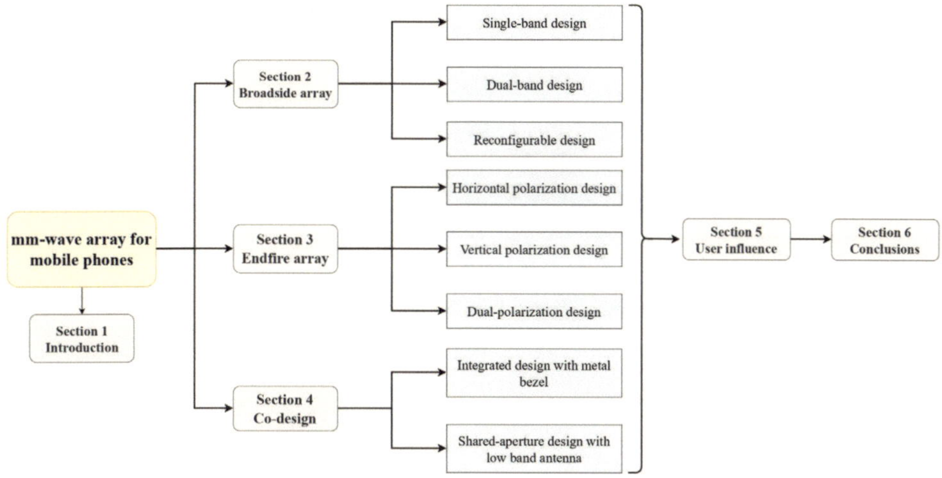


**Figure 2.**
*Block diagram of the mm-wave array for mobile phones.*

Since the first mm-wave array was designed for mobile phones in 2014 [9] and the first commercial mm-wave array module was adopted in Samsung Galaxy S20 in 2020 [10], significant progress has been achieved in addressing the above difficulties in recent years. The desired beam coverage as shown in **Figure 1** should be achieved with mm-wave arrays.

In this chapter, a comprehensive summary of the recent advances in the mm-wave array for mobile phone, such as broadside mm-wave array, endfire mm-wave array, co-design of the mm-wave array with metal-bezel and lower frequency band antenna, and user influence is conducted. **Figure 2** illustrates the block diagram of the mm-wave array for mobile phones. For the broadside mm-wave array, we focus on the single-band, dual-band, and reconfigurable designs. For the endfire mm-wave array, single-polarization and dual-polarization designs are summarized. For the co-design of the mm-wave array, integrating metal-bezel with mobile phone design and shared-aperture with lower band antenna design are summarized.

This chapter is organized as follows. In Section 2, the common antenna element types of broadside radiation are introduced, and the design challenges of mm-wave broadside arrays are analyzed. Then, the broadside arrays are divided into three parts: single-band design, dual-band design, and reconfigurable design, to be summarized respectively. In Section 3, the challenges of endfire mm-wave array design are first analyzed. Then, the typical horizontal polarized, vertical polarized, and dual-polarized mm-wave endfire arrays are summarized. In Section 4, the co-design of the mm-wave array in the mobile phone with a lower frequency band antenna is introduced, including an integrated design with metal-bezel and shared-aperture design, respectively. In Section 5, the user influence on the mm-wave array in the mobile phone is illustrated. Finally, conclusions are drawn in Section 6.

## 2. Broadside mm-wave array for mobile phone

Broadside array antenna is the array with the direction of maximum radiation, which is vertical to the array. As shown in **Figure 3**, the broadside array is achieved

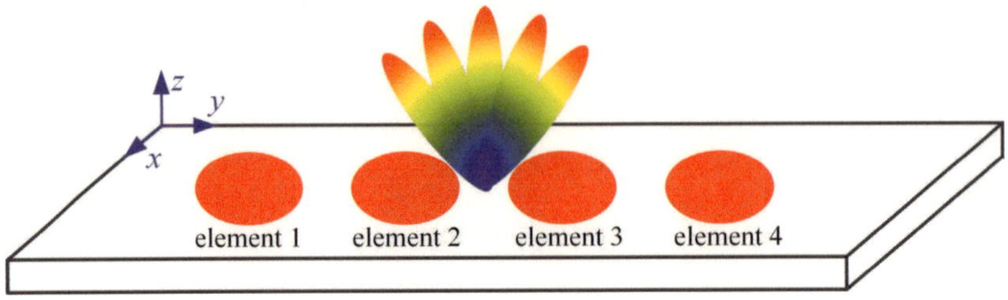


**Figure 3.**
*Conceptual diagram of the antenna array with broadside radiation pattern.*

**Figure 4.**
*Conceptual diagram of the broadside mm-wave array and desired beam directions. (a) Array placed horizontally with the mobile phone. (b) Array placed vertically with the mobile phone.*

with the array element in the $xy$-plane and maximum radiation direction along the $z$-axis. As shown in **Figure 4(a)**, when the array is placed horizontally with the mobile phone, the desired beam directions 5 and 6 in **Figure 1** can be achieved. In **Figure 4(b)**, when the array is placed vertically with the mobile phone, the desired beam directions 1, 2, 3, and 4 in **Figure 1** can be achieved. Usually, due to the thickness limitation of the mobile phone, beam directions 1, 2, 3, and 4 are mainly achieved by endfire arrays, which will be explained in detail in Section 3. The broadside arrays are mainly used to achieve beam directions 5 and 6.

For broadside array, the typical antenna elements are patch antenna [12], slot antenna [13], printed dipole antenna [14], dielectric resonant antenna [15], substrate integrated waveguide (SIW) cavity antenna [16], and so on. For the above antenna types, dual polarization can be achieved simply by quadrature feeding or by placing a pair of quadrature elements. The main challenge of broadside array design is how to achieve the array with the superior performance of small size, wide bandwidth, multiband, and wide beam coverage. In this section, the broadside arrays are classified into three parts: single-band, dual-band, and reconfigurable to summarize. At the same time, several solutions to the above challenges are also summarized.

### 2.1 Single-band broadside mm-wave array

The patch antenna is one of the most commonly used antenna elements in the broadside mm-wave wave array. The bandwidth of patch antennas is usually narrow, covering only a portion of the commercial mm-wave band. For example, as shown in **Figure 5(a)**, an optically invisible common patch antenna on display only has a bandwidth of 9% (27.1–29.7 GHz) [17]. As shown in **Figure 5(b)-(d)**, in order to improve the bandwidth of

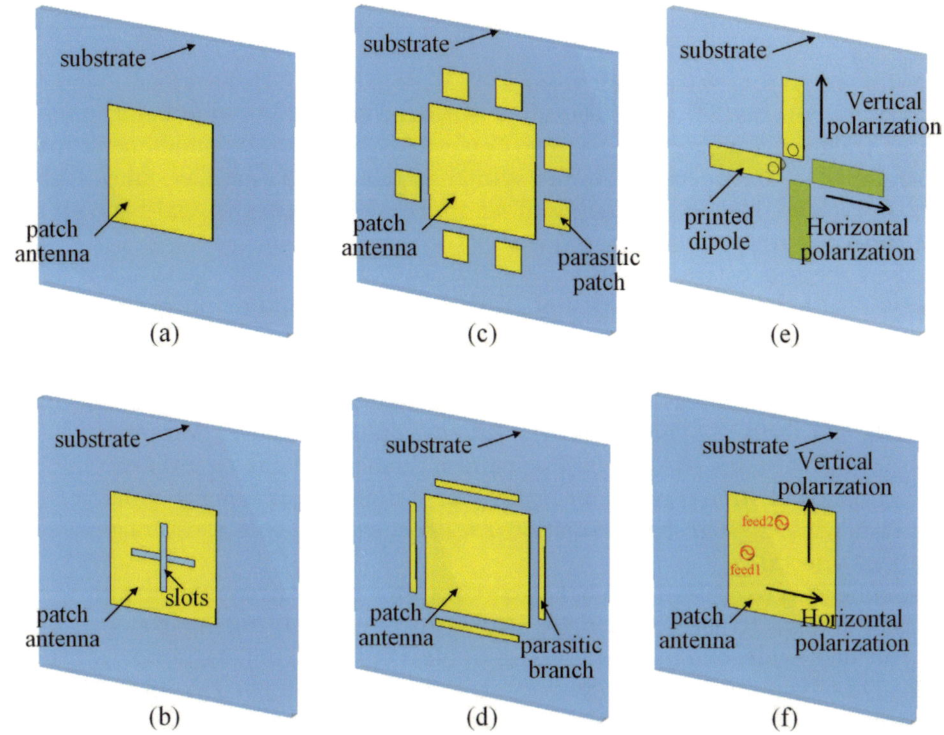


**Figure 5.**
*Typical design schematics of single-band broadside mm-wave array. (a) Common patch antenna element. (b) Slotting on the patches. (c) Using parasitic patches. (d) Using parasitic branches. (e) Printed dipole antenna, dual-polarized realized by orthogonal placement. (f) Dual-polarized. Realized by quadrature feeding.*

the patch antenna element, slotting on the patches [18], using parasitic patches [19], and using parasitic branches [20] are used to obtain more than 20% impedance bandwidth. Although the optimized patch antenna in [19] can cover the mm-wave band from 23 to 30.5 GHz (N257/258) band, it cannot cover the N259/N260 near 40GHz. In contrast, printed dipole antennas have a wider bandwidth. For example, the printed dipole antenna in [21] achieves a 50% (24–40 GHz) impedance bandwidth. In addition, as shown in **Figure 5(f)** and **(g)**, dual polarization can be easily achieved by placing a pair of antenna elements orthogonally [22] or by quadrature feeding [23].

The patch antenna is one of the most commonly used antenna elements in the broadside mm-wave wave array. The bandwidth of patch antennas is usually narrow, covering only a portion of the commercial mm-wave band. For example, similar to the common patch antenna element shown in **Figure 5(a)**, an optically invisible common patch antenna on display only has a bandwidth of 9% (27.1–29.7 GHz) [17]. As shown in **Figure 5(b)**-**(d)**, in order to improve the bandwidth of the patch antenna element, slotting on the patches [18], using parasitic patches [19], and using parasitic branches [20] are used to obtain more than 20% impedance bandwidth. Although the optimized patch antenna in [19] can cover the mm-wave band from 23 to 30.5 GHz (N257/258) band, it cannot cover the N259/N260 near 40GHz. In contrast, printed dipole antennas have a wider bandwidth. For example, the printed dipole antenna in [21] achieves a 50% (24–40 GHz) impedance bandwidth. In addition, as shown in **Figure 5(e)** and **(f)**, dual polarization can be easily achieved by placing a pair of antenna elements orthogonally [22] or by quadrature feeding [23].

## 2.2 Dual-band broadside mm-wave array

With several mm-wave bands around 28, 38, 45, and 60 GHz have been assigned for 5G development [24], mm-wave ultra-wideband antennas or multiband antennas are widely investigated to cover two or more frequency bands simultaneously to expand the available spectrum, improve antenna space utilization, save fabricated cost, and achieve high integration. And this part mainly focuses on dual-band broadside mm-wave array.

There are usually two ways to achieve dual-band antennas. One general way to achieve dual-band antennas is to combine two different structures operating at different frequency bands together [15, 25, 26]. For example, a hybrid antenna consisting of three resonators of strip, slot, and dielectric resonant antenna is proposed [15]. The strip and slot modes are used to cover the lower frequency band of 26.41–30.42 GHz, while the $TE_{111}$ and $TE_{131}$ modes of the DRA are employed to cover the upper-frequency band of 36.05–40.88 GHz. Two pairs of dipole antennas are proposed in [25]; the low-band radiation is generated by the pair of dipole arms along co-polarized direction, while the high-band radiation is realized by the dipole arms along cross-polarized direction.

Another way to achieve dual-band antennas is to adjust different modes of the same antenna structure to achieve dual resonance [14, 27–29]. For example, a compact dual-wideband magnetoelectric dipole is proposed in [14], the lower band of 24–29.3 GHz is achieved by 0.5λ mode, and the higher band of 35.5–43.5 GHz is achieved by 1λ mode. Also, the $TM_{10}$ mode and $TM_{20}$ mode of the gridded patches antenna are used to achieve dual-band coverage [27] .

## 2.3 Reconfigurable broadside mm-wave array

Considering the massive production of consumer electronics, the cost of each mobile antenna should be as low as possible. The reconfigurable design enables multiple operating modes of the mm-wave array through simple p-i-n diodes control and switching, which is one of the effective ways to save cost.

The reconfigurable design can be divided into two categories: direct control of the pattern reconfiguration, and control of the phased array excitation phase difference reconfiguration. As direct control of the pattern reconfiguration usually requires a large antenna design space and is not applicable for mobile phones [30], this part focuses on the reconfigurable design of the phase shifter. For the mm-wave arrays, the phase shifter is usually designed with the feeding network, and the excitation phase difference between elements is set by switching p-i-n diodes to achieve different

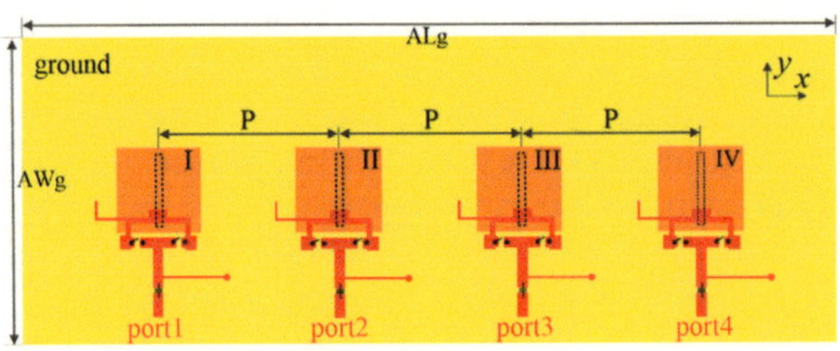


**Figure 6.**
*1-bit reconfigurable broadside mm-wave design [31].*

directions. As shown in **Figure 6**, in [31], a low-cost reconfigurable 1-bit patch antenna is designed with moderate performance. What is more, a series-fed beam-steerable 2-bit reconfigurable design is proposed in [32].

## 3. Endfire mm-wave array for mobile phone

Endfire array antenna is the array with the direction of maximum radiation, which lies along the line of the array. As shown in **Figure 7**, the endfire array is achieved with array element along the $y$-axis and maximum radiation direction in the $x$-axis. Compared with the broadside array, for the mobile phone with desired beam directions 1, 2, 3, and 4 in **Figure 1**, endfire array can be directly integrated into the PCB and save the thickness of the mobile phone. So, for mobile phones with a specific geometry of thin thickness, endfire array is preferable. However, due to the thin thickness of the mobile phone and thus the thin thickness of the endfire array, how to achieve the endfire array with the superior performance of thin thickness, small clearance, wide bandwidth, multiple polarization, and wide beam coverage is challenging. This section summarizes the typical design methods of horizontal-polarized endfire mm-wave array first. Then, the typical design methods of vertical-polarized endfire mm-wave array are summarized. Finally, the dual-polarized endfire mm-wave arrays are summarized.

### 3.1 Horizontal-polarized endfire mm-wave array

To better summarize the horizontal polarization endfire mm-wave array for mobile phones, **Figure 8** shows the conceptual diagram of some typical horizontal polarization endfire mm-wave arrays. As shown in **Figure 9(b)**, the simplest method to achieve the endfire radiation is to deploy the mm-wave array vertical to the system ground. As the radiation element is horizontal polarization, horizontal polarization endfire mm-wave array is achieved. A similar idea can be found in [33] where the broadband magnetoelectric dipole antenna is applied as the element where 109.5% relative bandwidth is achieved. However, the critical drawback of this method is that the width of the mm-wave array should be enough to achieve proper performance. Thus, the thickness of the mobile phone is affected by the mm-wave array. A dipole antenna closing to the system ground, as shown in **Figure 9(c)**, has endfire radiation with horizontal polarization. The system ground can redirect the radiation to achieve a high gain. The parallel double line [34] or the slot line with balun [35] can be applied to feed

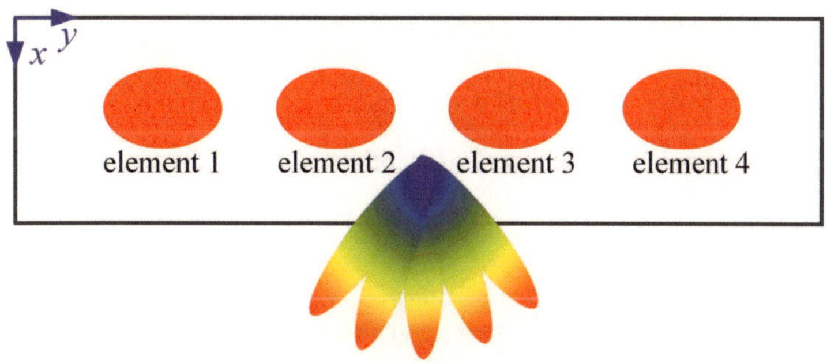


**Figure 7.**
*Conceptual diagram of the antenna array with endfire radiation pattern.*

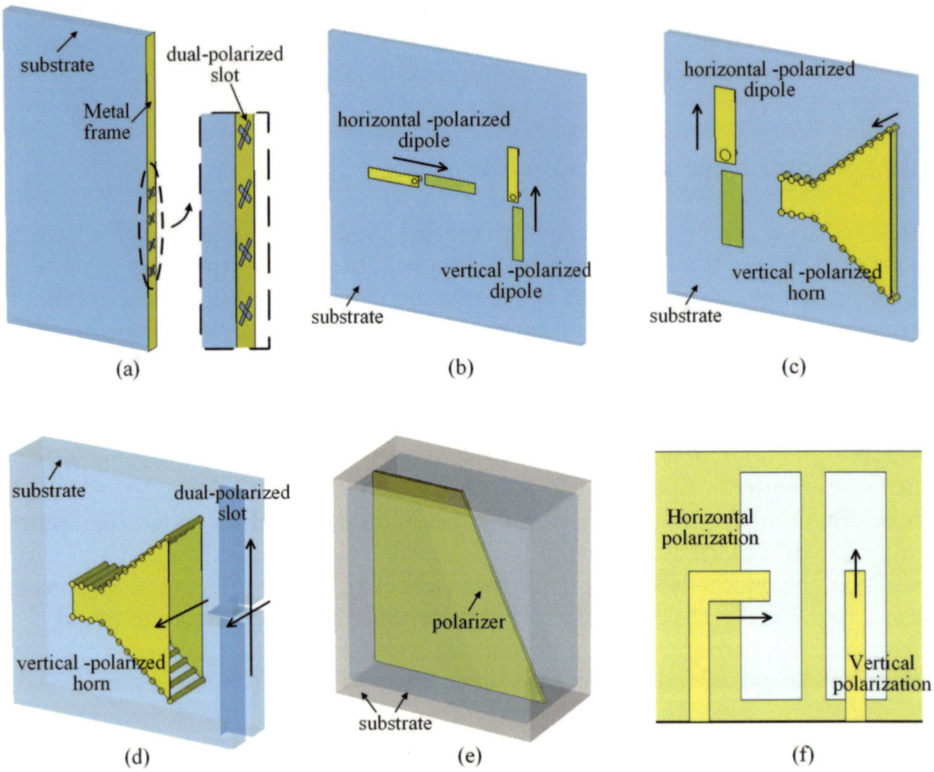


**Figure 8.**
*Typical design schematics of dual-polarized endfire mm-wave array. (a) Vertical deployment of dual-polarized element. (b) Horizontal-polarized dipole with vertical-polarized dipole. (c) Horizontal-polarized dipole with vertical-polarized horn. (d) Horizontal-polarized slot with vertical-polarized horn. (e) Two SIW horns with a polarizer. (f) Dual-polarized slot antennas.*

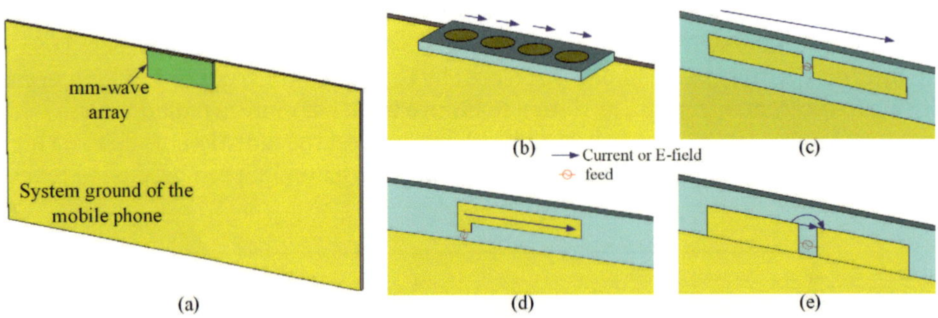


**Figure 9.**
*Conceptual diagram of the typical horizontal polarization endfire mm-wave array. (a) Mm-wave array on the mobile phone. (b) Vertical deployment for endfire radiation. (c) Dipole element for endfire radiation. (d) Monopole element for endfire radiation. (e) Open slot element for endfire radiation.*

the dipole. To achieve the wide bandwidth, the distance between the dipole antenna ground should be large. In [36], the arm length of the dipole antenna is tuned to different values to widen the bandwidth. Similar to the Yagi antenna, several directors are applied in [37–39] to further enhance the array gain and bandwidth. The monopole antenna of half-wavelength mode can also be placed near the system ground to achieve the horizontal polarization endfire radiation, as shown in **Figure 9(d)**. This structure has been studied in [40], working at 60 GHz with a bandwidth of 6 GHz. Also, the open-ended slot antenna can radiate the horizontal polarization endfire pattern, as shown in **Figure 9(e)**. This structure has been studied in [41] working at 28 GHz with a bandwidth of 5 GHz. To achieve good performance with wide bandwidth, the space of the monopole antenna and slot antenna in **Figure 8** should be large.

### 3.2 Vertical-polarized endfire mm-wave array

To better summary the vertical polarization endfire mm-wave array for mobile phone, **Figure 10** shows the conceptual diagram of some typical vertical polarization endfire mm-wave arrays. As shown in **Figure 10(b)**, the simplest method to achieve the endfire radiation is to deploy the mm-wave array vertically to the system ground. As the radiation element is vertical polarization, vertical polarization endfire mm-wave array is achieved. A similar idea can be found in [42, 43] where the slot, dielectric, and cavity resonators are applied simultaneously to achieve a wideband width of 47.1% [42] and 94.1 [43]. However, the critical drawback of this method is also that the width of the mm-wave array should be enough to achieve proper performance. Thus, the thickness of the mobile phone is affected by the mm-wave array. As shown in **Figure 10(c)**, the cavity slot element can be applied to radiate the vertical polarization pattern. Although the cavity slot element can achieve a low profile, the key technical difficulty is to achieve wide bandwidth. In [44], a substrate-integrated waveguide (SIW) endfire antenna array with zero clearance is designed. Three arbitrary pad-loading metallic vias are investigated to match the impedance within a relative bandwidth of 60%. Also, the slot on the SIW [45], taper slot [46], and the metasurface structure [47] can be applied to achieve a wide bandwidth of the SIW slot antenna. For the dipole element in **Figure 10(d)** and the monopole element in **Figure 10(e)**, vertical polarization with endfire radiation can be achieved. In [30], the monopole element with the parasitic element is applied to achieve the endfire radiation with steering beams. In [48], the compact vertically polarized endfire monopole-based Yagi antenna-in-package is proposed with a relative bandwidth of 16%. For the dipole or monopole element, the height should be large for a large bandwidth. By combing the cavity slot element and dipole/monopole element, the endfire magnetoelectric antenna with stable performance within a wide bandwidth can be achieved. For example, the SIW cavity slot antenna and dipole antenna in [49, 50]

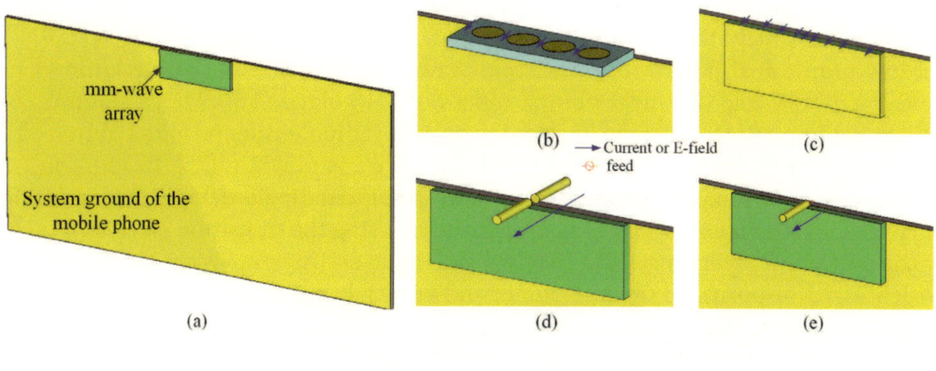


**Figure 10.**
*Conceptual diagram of the typical vertical-polarized endfire mm-wave array. (a) mm-wave array on the mobile phone. (b) Vertical deployment for endfire radiation. (c) Cavity slot element for endfire radiation. (d) Dipole element for endfire radiation. (e) Monopole element for endfire radiation.*

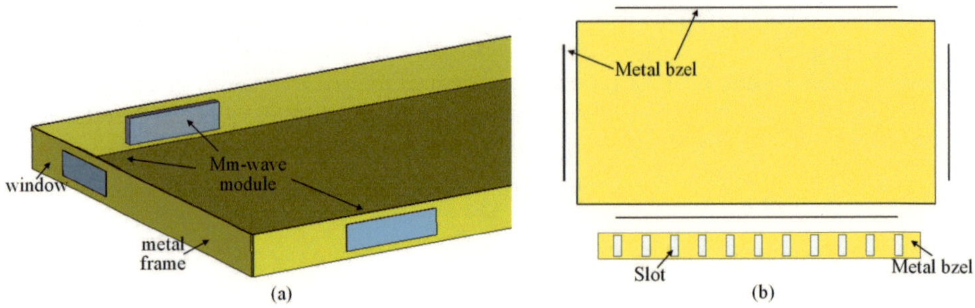


**Figure 11.**
*Typical design schematics of integrating the mm-wave array in the mobile phone. (a) Using a window to reduce the blockage from metal-bezel. (b) Using the metal-bezel to design the mm-wave array antenna.*

and the SIW cavity slot antenna and monopole antenna in [51] are combined to achieve the wideband endfire magnetoelectric antenna. Also, to achieve good performance with wide bandwidth, the profile of the dipole element or the monopole element in **Figure 11** should be large.

### 3.3 Dual-polarized endfire mm-wave array

With the horizontal-polarized endfire mm-wave array and vertical-polarized endfire mm-wave array, the dual-polarized endfire mm-wave array can be easily achieved. For example, if the mm-wave array vertical to the system ground in **Figure 9(b)** and **Figure 10(b)** can radiate dual-polarized patterns, dual-polarized endfire mm-wave array can be easily achieved. This idea can be found in [52], where a dual-polarized slot antenna is vertically deployed on a mobile phone, as shown in **Figure 8(a)**. Thus, a dual-polarized endfire mm-wave array with a − 10 dB impedance bandwidth from 23.2 to 29.7 GHz is achieved in [52]. This method also has the drawback of being high profile. Also, if the horizontal dipole element in **Figure 8(c)** and vertical dipole element in **Figure 10(d)** can be designed in the near space, dual-polarized mm-wave dipole array can also be achieved. In [40, 53], the dual-polarized endfire mm-wave dipole array is achieved by combing the vertical dipole and horizontal dipole as shown in **Figure 8(b)**. The dual-polarized endfire mm-wave array antenna should have a large profile to achieve the good performance. To reduce the profile, the horizontal-polarized dipole element in **Figure 9(c)** and the vertical-polarized horn element in **Figure 10(c)** can be combined to achieve the low-profile dual-polarized endfire mm-wave array. In [54], the low-profile dual-polarized endfire mm-wave array is realized by co-designing a horizontal-polarized yagi-uda antenna and a vertical-polarized SIW horn antenna, as shown in **Figure 8(c)**. Also, the horizontal-polarized dipole antenna with balun feeding can replace the yagi-uda antenna to achieve a wide bandwidth [55]. In addition, the clearance of the combination of horizontal-polarized dipole element and vertically polarized horn element can be further reduced by using the transition plates [56, 57] or the overlapped apertures [58]. The dual-polarized endfire mm-wave array antenna with horizontal-polarized dipole element and vertically polarized horn element usually has a large clearance to achieve a good performance. To reduce the clearance, the horizontal-polarized open slot element in **Figure 9(e)** and vertical-polarized horn element in **Figure 10(c)** can be applied. In [59], the dual-polarized endfire mm-wave array with a small clearance is realized by co-designing a horizontal-polarized open slot antenna and a vertical-polarized horn element, as shown in **Figure 8(d)**. The horizontal-polarized open slot antenna consists of two metal blocks with a slot, and the vertical-polarized horn element is a SIW horn antenna. In [60, 61], the horizontal-polarized open slot antenna is realized by using two SIW structures. In [41, 62], the horizontal-polarized open slot antenna and the vertical-polarized horn are integrated into a single SIW structure. Besides, two SIW horns with a polarizer can be applied to achieve the dual linearly polarized endfire antenna [63, 64] or 45° dual linearly polarized endfire antenna [65], as shown in **Figure 8(e)**. Also, in [66], the orthogonal slot can be excited simultaneously to achieve the dual-polarized mm-wave endfire chain-slot antenna, as shown in **Figure 8(f)**.

## 4. Co-design of the mm-wave array in the mobile phone with lower frequency band antenna

The broadside mm-wave arrays in Section 2 and the endfire mm-wave arrays in Section 3 have achieved superior performance which can cover the desired beam directions 1–6 in **Figure 1**. The commercial mm-wave array module has been adopted in Samsung Galaxy S20 in 2020 [10]. However, because the full-display, curved-display, metal-bezel, glass back cover, metal back cover requirement of the mobile phone, the mm-wave array in mobile phone should fit the ID of mobile phones. Also, because most of the spaces are reserved for the displays, cameras, battery, PCB, motor, speaker, and so on for better user experience, the antenna in mobile phones should only occupy a very small space. So, the co-design of the mm-wave array in the mobile phone with a lower frequency band antenna is also widely studied. In this section, we first analyze the design of integrating the mm-wave array with metal-bezel in mobile phones. Then, the shared-aperture design of the mm-wave array with the lower frequency band antenna is summarized.

### 4.1 Integrating the mm-wave array with metal-bezel in the mobile phone

As the mm-wave array is deployed in the mobile phone, the modules of the mobile phone might have a significant effect on the performance of the mm-wave array. In [67], the effect of different kinds of mobile phone housing on the performance of four canonical types of mm-wave antennas is studied. The effective beam-scanning efficiency is proposed to evaluate the coverage performance. In [68], the effect of the metal-bezel of the mobile phone on the radiation pattern of the mm-wave array is studied. The blockage of the metal-bezel to the horizontal-polarized antenna is more severe than for the vertical-polarized antenna. And the coupling metal strips [68, 69] can be applied to reduce the blockage from metal-bezel. Also, the slots on the metal-bezel [70, 71] can be also used to reduce the blockage from the metal-bezel. In [72], a rectangle window in the metal-bezel was used to install the mm-wave array so that the mm-wave array could radiate the power through the rectangle window directly as shown in **Figure 11(a)**. This practical solution has been adopted in some commercial 5G mobile terminals such as Apple iPhone 12 with a wee mm-wave window [73]. Apart from reducing the effect from the metal-bezel, in [74], the metal-bezel can be applied to design the mm-wave leaky-wave array as shown in **Figure 11(b)**. In addition, the mm-wave array antenna could be directly implemented via the slot on the metallic bezel of the mobile terminals [75, 76]. Thus, the blockage effect from the metal-bezel is solved.

## 4.2 Shared-aperture design of the mm-wave array with the lower frequency band antenna

To integrate the mm-wave array with the lower frequency band antenna, a low pass (or high pass) filter can be applied [77, 78]. As shown in **Figure 12(a)**, a 3.5 GHz lower band antenna is directly connected to a 28 GHz mm-wave antenna with a low-pass and high-stop (3.5 GHz pass and 28 GHz stop) filter [77]. Thus, the mm-wave antenna and the lower frequency band can be designed in a near space with a single feeding port. To reduce the occupied space of the mm-wave array and lower frequency band antenna, the mm-wave array can be integrated into the lower frequency band antenna [11, 79–83]. As shown in **Figure 12(b)**, the mm-wave slot array antenna is integrated into the clearance of the lower frequency band inverted-F antenna [79]. In addition, a notch on the lower frequency band can be applied to integrate the mm-wave array antenna [81, 82]. Also, the mm-wave array antenna can be deployed on the lower frequency band antenna [83]. To further reduce the occupied space of the mm-wave array and the lower frequency band antenna, the metal pattern of the lower frequency band antenna can be applied to design SIW structure for the mm-wave array antenna [84–87]. As shown in **Figure 12(c)** [84], the lower frequency band antenna is a simple patch antenna. To integrate the mm-wave array on the patch antenna, the patch of the lower frequency band antenna is designed as a SIW slot array. Thus, the mm-wave array antenna and the lower frequency band antenna is designed in the same aperture. In addition, the mm-wave SIW slot array can be integrated into the monopole antenna [85] or inverted-F antenna [86, 87] of the lower frequency band. Besides using the SIW structure, the higher order mode of the lower frequency band antenna can also be directly applied to design the mm-wave array [88–91]. Also in [88], the half-wavelength slot mode of the lower frequency band antenna has the higher order mode of the slot. And multiple feedings are applied to excite the higher-order mode of the slot, which is the connected slot array. Thus, a single slot is designed to work at the mm-wave frequency band and lower frequency band simultaneously. In [89], a single microstrip grid array is designed to cover the mm-wave frequency band and lower frequency band simultaneously. In [90], a surface is the integration of a metasurface at the lower frequency band and a partially reflective surface (PRS) at the higher frequency band. In [91], a single slot is designed to function as a decoupling slot at the lower frequency band and the taper slot antenna at the mm-wave frequency band.

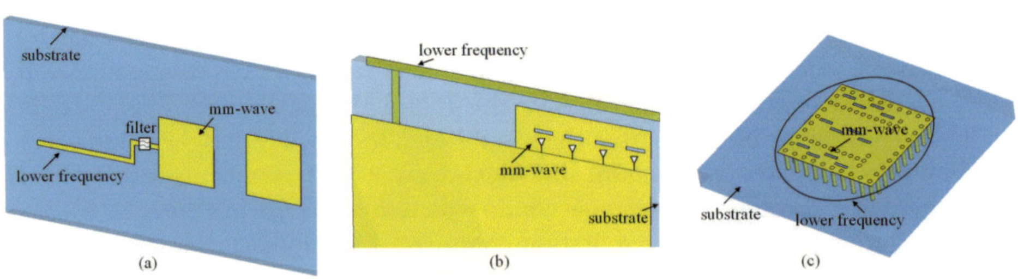


**Figure 12.**
*Typical design schematics of co-design of mm-wave array with lower frequency band antenna. (a) Using the filter to integrate the mm-wave and lower frequency band antennas. (b) Using the clearance of the lower frequency band antenna to deploy the mm-wave array. (c) Using the SIW structure to integrate the mm-wave and lower frequency band antennas.*

## 5. User influence

As a special situation for the mobile antenna, the effect of the human tissue on the antenna performance should be considered when designing the mobile antenna. Also, the electromagnetic field (EMF) exposure to the mobile antenna for the human tissue should also be studied for safety considerations. This is because the mobile antenna is usually close to the human head, body, or hands, as is used in practical situations. In this section, the mutual effect of the mm-wave array and the human tissue in the open literature is summarized.

The human body is electronically large at the mm-wave frequency band. To evaluate the mutual effect of the human tissue and the mm-wave array effectively and accurately, [92] developed the numerical and physical phantoms of a human body for evaluation of mobile antennas at 28 GHz. Thus, the ability to design antennas under practical operational conditions involving body effects is achieved. As for the EMF exposure to the mm-wave array, [93] compares the power density of a single antenna element, a four-element linear array, and an eight-element linear array at the near field region and the far field region. Zhao et al. [93] finds that, at the near field region, the power density is extremely high and it can be reduced as the number of array elements increases. At the far field region, the power density increases as the array elements increase.

As for the effect of the human on the mm-wave array, [76] shows that, for an eight-element mm-wave array located alongside the side edge of the mobile phone, the human hand results in a gain reduction of about 7.5 dB. However, for a four-element mm-wave array, if the human hand covers the mm-wave array, the loss from hand blockage on the antenna gains can be up to 20 to 25 dB [94]. To reduce the effect from the human hand, the mm-wave array should be deployed away from the human hand. Ojaroudiparchin et al. [95] illustrates that, if the mm-wave array is not close to the human hand, the gain loss from the human hand can be reduced to about 1.5 dB. Therefore, to achieve a good performance, multiple mm-wave arrays should be deployed in the mobile phone at different positions. In [96], it is concluded that, in the talk mode, the mm-wave array should be placed at the top of the mobile phone (close to the index finger). Also, it the talk mode, the user hand shadowing can be significantly reduced by placing the mm-wave array at the bottom of the chassis (close to the palm). In the data mode, the mm-wave array achieves less gain loss when deployed at the top of the mobile phone.

For the human body shadowing, [97] illustrates that the shadowing by the user's body might decrease the gain about 20–30 dB if the mm-wave array is close to the user. Zhao et al. [98] also observes a strong shadowing effect from the human body in the mm-wave band, which is around 20–25 dB at 15 GHz. Zhao et al. [99] finds that the equivalent isotropic radiated power (EIRP) values at cumulative distribution function (CDF) of 50% drop about 5–10 dB compared to the case that without the user body. To reduce the shadowing from the human body, the effect of the displacement of the mm-wave array on the shadowing from the human body is studied [100]. Syrytsin et al. [100] finds that the corner positions of the mobile phone achieve the best performance in terms of spatial coverage. Syrytsin et al. [101] compares the coverage efficiency and user shadowing from the mm-wave phased array and mm-wave switch diversity array and finds that the mm-wave phased array has superior performance. Also, in [102], it is found that, for the difference between the user and free-space cases, the circularly polarized array coverage efficiency is relatively less sensitive to user effects.

## 6. Conclusions

In this chapter, we summary recently mm-wave arrays for mobile phones. For broadside mm-wave array, the state-of-the-art single-band, dual-band, and reconfigurable designs are proposed in small size, low cost, and have moderate performance. For endfire array, the typical designs of horizontal-polarized, vertical-polarized, and dual-polarized arrays are analyzed, providing good reference solutions for endfire array design. For co-design of mm-wave array in the mobile phone, several ingenious solutions and practical solutions adopted in some commercial terminals are introduced that will contribute to the integrated design of mm-wave antennas with lower band antenna. In addition, the human body model evaluation, the effect of the human body on the mm-wave array, and the human body shadowing are also illustrated. Various designs are being used to solve the mm-wave array challenge in mobile phones, with promising applications.

## Acknowledgements

This work was supported in part by the National Natural Science Foundation of China under Grant 62101133, in part by the Natural Science Foundation of Shanghai under Grant 21ZR1406800, and in part by Shanghai Rising Star Program under Grant 22QC1400100.

## Author details

Yan Wang* and Xiaoxue Fan
Fudan University, Shanghai, China

*Address all correspondence to: yanwang_fd@fudan.edu.cn

## References

[1] 6G Flagship research program. Key drivers and research challenges for 6G ubiquitous wireless intelligence, White Paper. 2019

[2] Rappaport TS, Sun S, Mayzus R. Millimeter wave Mobile communications for 5G cellular: It will work! IEEE Access. 2013;**1**:335-349. DOI: 10.1109/access.2013.2260813

[3] Marcus MJ. 5G and "IMT for 2020 and beyond". IEEE Wireless Communications. 2015;**22**:2-3. DOI: 10.1109/mwc.2015.7224717

[4] Hong W, Jiang ZH, Yu C. The role of Millimeter-wave technologies in 5G/6G wireless communications. IEEE Journal of Microwaves. 2021;**1**:101-122. DOI: 10.1109/jmw.2020.3035541

[5] Roh W, Seol J-Y, Park J. Millimeter-wave beamforming as an enabling technology for 5G cellular communications: Theoretical feasibility and prototype results. IEEE Communications Magazine. 2014;**52**:106-113. DOI: 10.1109/mcom.2014.6736750

[6] Technical Specification Group Radio Access Network. New frequency range for NR (24.25-29.5 GHz) (Release 15), Document TR 38.815 v15.1.0, 3GPP Tech. Rep. 2021

[7] Huang H-C. Evolution of Millimeter-wave antenna solutions and designs to cellular phones. IEEE Access. 2020;**8**:187615-187622. DOI: 10.1109/access.2020.3027424

[8] General Aspects for User Equipment (UE) Radio Frequency (RF) for NR (Release 15), document 38.817-1, 3GPP Tech. Rep. 2019

[9] Hong W, Baek K, Lee Y. Design and analysis of a low-profile 28 GHz beam steering antenna solution for future 5G cellular applications. In: 2014 IEEE Mtt-S International Microwave Symposium; 01-06 July. Tampa, USA: IEEE. p. 2014

[10] Ifixit. Samsung Galaxy S20 Ultra Teardown [Online]. 2021. Available from: https://www.ifixit.com/Teardown/Samsung+Galaxy+S20+Ultra+Teardown/131607

[11] Wang Y, Xu F. Shared-aperture 4G LTE and 5G mm-wave antenna in Mobile phones with enhanced mm-wave radiation in the display direction. IEEE Transactions on Antennas and Propagation. 2023;**71**:4772-4782. DOI: 10.1109/tap.2023.3262971

[12] Wang J, Li Y, Wang J. A low-profile dual-mode slot-patch antenna for 5G Millimeter-wave applications. IEEE Antennas and Wireless Propagation Letters. 2022;**21**:625-629. DOI: 10.1109/lawp.2022.3140747

[13] Stanley M, Huang Y, Loh T. A high gain steerable millimeter-wave antenna array for 5G smartphone applications. In: 2017 11th European Conference on Antennas and Propagation (Eucap). 2017. pp. 1311-1314. DOI: 10.23919/EuCAP.2017.7928542

[14] Ni S, Li X, Qiao X. A compact dual-wideband Magnetoelectric dipole antenna for 5G Millimeter-wave applications. IEEE Transactions on Antennas and Propagation. 2022;**70**:9112-9119. DOI: 10.1109/tap.2022.3184550

[15] Cui L-X, Ding X-H, Yang W W. Communication compact dual-band hybrid dielectric resonator antenna for 5G Millimeter-wave applications. IEEE Transactions on Antennas and

Propagation. 2023;**71**:1005-1010. DOI: 10.1109/tap.2022.3211389

[16] Guo J, Hu Y, Hong W. A 45° polarized wideband and wide-coverage patch antenna Array for Millimeter-wave communication. IEEE Transactions on Antennas and Propagation. 2022;**70**:1919-1930. DOI: 10.1109/tap.2021.3118705

[17] Park J, Lee SY, Kim J. An optically invisible antenna-on-display concept for Millimeter-wave 5G cellular devices. IEEE Transactions on Antennas and Propagation. 2019;**67**:2942-2952. DOI: 10.1109/tap.2019.2900399

[18] Chen J, Berg M, Rasilainen K. Broadband cross-slotted patch antenna for 5G Millimeter-wave applications based on characteristic mode analysis. IEEE Transactions on Antennas and Propagation. 2022;**70**:11277-11292. DOI: 10.1109/tap.2022.3209217

[19] Sim C-Y-D, Lo J Jr, Chen Z-N. Design of a broadband millimeter-wave array antenna for 5G applications. IEEE Antennas and Wireless Propagation Letters. 2022;**22**:1030-1034. DOI: 10.1109/lawp.2022.3231358

[20] Zhang T, Li L, Xie M. Low-cost aperture-coupled 60-GHz-phased Array antenna package with compact matching network. IEEE Transactions on Antennas and Propagation. 2017;**65**:6355-6362. DOI: 10.1109/tap.2017.2722867

[21] Fan F-F, Chen Q-L, Xu Y-X. A wideband compact printed dipole antenna Array with SICL feeding network for 5G application. IEEE Antennas and Wireless Propagation Letters. 2023;**22**:283-287. DOI: 10.1109/lawp.2022.3209326

[22] Wang M, Chan CH. Dual-polarized, low-profile dipole-patch Array for wide bandwidth applications. IEEE Transactions on Antennas and Propagation. 2022;**70**:8030-8039. DOI: 10.1109/tap.2022.3164416

[23] Tong X, Jiang ZH, Yu C. Low-profile, broadband, dual-linearly polarized, and wide-angle Millimeter-wave antenna arrays for Ka-band 5G applications. IEEE Antennas and Wireless Propagation Letters. 2021;**20**:2038-2042. DOI: 10.1109/lawp.2021.3102375

[24] Use of Spectrum Bands Above 24 GHz for Mobile Radio Services,document GN Docket 14-177, Notice Proposed Rulemaking, FCC Record 89A1. 2016

[25] Yang SJ, Yao SF, Ma R-Y. Low-profile dual-wideband dual-polarized antenna for 5G Millimeter-wave communications. IEEE Antennas and Wireless Propagation Letters. 2022;**21**:2367-2371. DOI: 10.1109/lawp.2022.3193808

[26] Huang D, Xu G, Wu J. A microstrip dual-Split-ring antenna Array for 5G Millimeter-wave dual-band applications. IEEE Antennas and Wireless Propagation Letters. 2022;**21**:2025-2029. DOI: 10.1109/lawp.2022.3189209

[27] Sun W, Li Y, Chang L. Dual-band dual-polarized microstrip antenna Array using double-layer gridded patches for 5G Millimeter-wave applications. IEEE Transactions on Antennas and Propagation. 2021;**69**:6489-6499. DOI: 10.1109/tap.2021.3070185

[28] Lu R, Yu C, Zhu Y. Millimeter-wave dual-band dual-polarized SIW cavity-fed Filtenna for 5G applications. IEEE Transactions on Antennas and Propagation. 2022;**70**:10104-10112. DOI: 10.1109/tap.2022.3209265

[29] Hong T, Zhao Z, Jiang W. Dual-band SIW cavity-backed slot Array using TM020 and TM120 modes for 5G applications. IEEE Transactions on Antennas and Propagation.

2019;**67**:3490-3495. DOI: 10.1109/tap.2019.2900394

[30] Zhang S, Syrytsin I, Pedersen GF. Compact beam-steerable antenna Array with two passive parasitic elements for 5G Mobile terminals at 28 GHz. IEEE Transactions on Antennas and Propagation. 2018;**66**:5193-5203. DOI: 10.1109/tap.2018.2854167

[31] Wang Y, Xu F, Jin Y-Q. Low-cost reconfigurable 1 bit Millimeter-wave Array antenna for Mobile terminals. IEEE Transactions on Antennas and Propagation. 2022;**70**:4507-4517. DOI: 10.1109/tap.2022.3140508

[32] Deng C, Liu D, Yektakhah B. Series-fed beam-steerable Millimeter-wave antenna design with wide spatial coverage for 5G Mobile terminals. IEEE Transactions on Antennas and Propagation. 2020;**68**:3366-3376. DOI: 10.1109/tap.2019.2963583

[33] Yin J, Wu Q, Yu C. Broadband Endfire Magnetoelectric dipole antenna Array using SICL feeding network for 5G Millimeter-wave applications. IEEE Transactions on Antennas and Propagation. 2019;**67**:4895-4900. DOI: 10.1109/tap.2019.2916463

[34] Mao C, Khalily M, Xiao P. High-gain phased Array antenna with Endfire radiation for 26 GHz wide-beam-scanning applications. IEEE Transactions on Antennas and Propagation. 2021;**69**:3015-3020. DOI: 10.1109/tap.2020.3028181

[35] Zhang J, Zhang S, Ying Z. Radiation-pattern reconfigurable phased Array with p-i-n diodes controlled for 5G Mobile terminals. IEEE Transactions on Microwave Theory and Techniques. 2020;**68**:1103-1117. DOI: 10.1109/tmtt.2019.2949790

[36] Syrytsin I, Zhang S, Pedersen GF. Compact quad-mode planar phased Array with wideband for 5G Mobile terminals. IEEE Transactions on Antennas and Propagation. 2018;**66**:4648-4657. DOI: 10.1109/tap.2018.2842303

[37] Brar RS, Vaughan RG. mmWave Yagi-Uda element and Array on liquid crystal polymer for 5G. IEEE Open Journal of Antennas and Propagation. 2023;**4**:34-45. DOI: 10.1109/ojap.2022.3228541

[38] Hwang I-J, Ahn B, Chae S-C. Quasi-Yagi antenna Array with modified folded dipole driver for mmWave 5G cellular devices. IEEE Antennas and Wireless Propagation Letters. 2019;**18**:971-975. DOI: 10.1109/lawp.2019.2906775

[39] Di Paola C, Zhang S, Zhao K. Wideband beam-switchable 28 GHz quasi-Yagi Array for Mobile devices. IEEE Transactions on Antennas and Propagation. 2019;**67**:6870-6882. DOI: 10.1109/tap.2019.2925189

[40] Hong W, Baek K-H, Ko S. Millimeter-wave 5G antennas for smartphones: Overview and experimental demonstration. IEEE Transactions on Antennas and Propagation. 2017;**65**:6250-6261. DOI: 10.1109/tap.2017.2740963

[41] Sun L, Li Y, Zhang Z. Wideband dual-polarized Endfire antenna based on compact open-ended cavity for 5G mm-wave Mobile phones. IEEE Transactions on Antennas and Propagation. 2022;**70**:1632-1642. DOI: 10.1109/tap.2021.3113701

[42] Omar AA, Park J, Kwon W. A compact wideband vertically polarized end-fire Millimeter-wave antenna utilizing slot, dielectric, and cavity resonators. IEEE Transactions on Antennas and Propagation. 2021;**69**:5234-5243. DOI: 10.1109/tap.2021.3061111

[43] Park J, Seong H, Whang YN. Energy-efficient 5G phased arrays incorporating vertically polarized Endfire planar folded slot antenna for mmWave Mobile terminals. IEEE Transactions on Antennas and Propagation. 2020;**68**:230-241. DOI: 10.1109/tap.2019.2930100

[44] Zhang J, Akinsolu MO, Liu B. Design of Zero Clearance SIW Endfire antenna Array using machine learning-assisted optimization. IEEE Transactions on Antennas and Propagation. 2022;**70**:3858-3863. DOI: 10.1109/tap.2021.3137500

[45] Khajeim MF, Moradi G, Shirazi RS. Wideband vertically polarized antenna with Endfire radiation for 5G Mobile phone applications. IEEE Antennas and Wireless Propagation Letters. 2020;**19**:1948-1952. DOI: 10.1109/lawp.2020.3009097

[46] Yang B, Yu Z, Dong Y. Compact tapered slot antenna Array for 5G Millimeter-wave massive MIMO systems. IEEE Transactions on Antennas and Propagation. 2017;**65**:6721-6727. DOI: 10.1109/tap.2017.2700891

[47] Federico G, Hubrechsen A, Coenen SL. A wide-scanning Metasurface antenna Array for 5G Millimeter-wave communication devices. IEEE Access. 2022;**10**:102308-102315. DOI: 10.1109/access.2022.3208597

[48] Kim H, Jang TH, Park CS. 60-GHz compact vertically polarized end-fire monopole-based Yagi antenna-in-package for wideband Mobile communication. IEEE Access. 2022;**10**:111077-111086. DOI: 10.1109/access.2022.3216264

[49] Li Y, Luk K-M. A multibeam end-fire Magnetoelectric dipole antenna Array for Millimeter-wave applications. IEEE Transactions on Antennas and Propagation. 2016;**64**:2894-2904. DOI: 10.1109/tap.2016.2554601

[50] Li A, Luk K-M. Millimeter-wave end-fire magneto-electric dipole antenna and arrays with asymmetrical substrate integrated coaxial line feed. IEEE Open Journal of Antennas and Propagation. 2021;**2**:62-71. DOI: 10.1109/ojap.2020.3044437

[51] Wang J, Li Y, Wang J. A low-profile vertically polarized magneto-electric monopole antenna with a 60% bandwidth for Millimeter-wave applications. IEEE Transactions on Antennas and Propagation. 2021;**69**:3-13. DOI: 10.1109/tap.2020.3030907

[52] Kong X, Jin X, Wang X. Design of Switchable Frequency-Selective Rasorber with A-R-A-T or A-T-A-R operating modes. IEEE Antennas and Wireless Propagation Letters. 2023;**22**:69-73. DOI: 10.1109/lawp.2022.3202219

[53] Montoya Moreno R, Ala-Laurinaho J, Khripkov A. Dual-polarized mm-wave Endfire antenna for Mobile devices. IEEE Transactions on Antennas and Propagation. 2020;**68**:5924-5934. DOI: 10.1109/tap.2020.2989556

[54] Hsu Y-W, Huang T-C, Lin H-S. Dual-polarized quasi Yagi–Uda antennas with Endfire radiation for Millimeter-wave MIMO terminals. IEEE Transactions on Antennas and Propagation. 2017;**65**:6282-6289. DOI: 10.1109/tap.2017.2734238

[55] Lu R, Yu C, Zhu Y. Compact Millimeter-wave Endfire dual-polarized antenna Array for low-cost multibeam applications. IEEE Antennas and Wireless Propagation Letters. 2020;**19**:2526-2530. DOI: 10.1109/lawp.2020.3038790

[56] Faizi Khajeim M, Moradi G, Sarraf SR. Broadband dual-polarized

antenna Array with Endfire radiation for 5G Mobile phone applications. IEEE Antennas and Wireless Propagation Letters. 2021;**20**:2427-2431. DOI: 10.1109/lawp.2021.3113993

[57] Zhang J, Zhao K, Wang L. Dual-polarized phased Array with end-fire radiation for 5G handset applications. IEEE Transactions on Antennas and Propagation. 2020;**68**:3277-3282. DOI: 10.1109/tap.2019.2937584

[58] Li H, Li Y, Chang L. A wideband dual-polarized Endfire antenna Array with overlapped apertures and small clearance for 5G Millimeter-wave applications. IEEE Transactions on Antennas and Propagation. 2021;**69**:815-824. DOI: 10.1109/tap.2020.3016512

[59] Sun K, Wang B, Yang T. Dual-polarized Millimeter-wave Endfire Array based on substrate integrated mode-composite transmission line. IEEE Transactions on Antennas and Propagation. 2022;**70**:341-352. DOI: 10.1109/tap.2021.3098551

[60] Li A, Luk K-M, Li Y. A dual linearly polarized end-fire antenna Array for the 5G applications. IEEE Access. 2018;**6**:78276-78285. DOI: 10.1109/access.2018.2884946

[61] Zhu Y, Deng C. Wideband dual-polarized end-fire phased Array antenna with small ground clearance for 5G mmWave Mobile terminals. IEEE Transactions on Antennas and Propagation. 2023;**71**:5469-5474. DOI: 10.1109/tap.2023.3263007

[62] Zhang J, Zhao K, Wang L. Wideband low-profile dual-polarized phased Array with Endfire radiation patterns for 5G Mobile applications. IEEE Transactions on Vehicular Technology. 2021;**70**:8431-8440. DOI: 10.1109/tvt.2021.3095560

[63] Li A, Luk K-M. Millimeter-wave dual linearly polarized Endfire antenna fed by 180° hybrid coupler. IEEE Antennas and Wireless Propagation Letters. 2019;**18**:1390-1394. DOI: 10.1109/lawp.2019.2917660

[64] Zhu Y, Deng C. Millimeter-wave dual-polarized multibeam Endfire antenna Array with a small ground clearance. IEEE Transactions on Antennas and Propagation. 2022;**70**:756-761. DOI: 10.1109/tap.2021.3098545

[65] Xia X, Wu F, Yu C. Millimeter-wave ±45° dual linearly polarized end-fire phased Array antenna for 5G/B5G Mobile terminals. IEEE Transactions on Antennas and Propagation. 2022;**70**:10391-10404. DOI: 10.1109/tap.2022.3185496

[66] Montoya Moreno R, Kurvinen J, Ala-Laurinaho J. Dual-polarized mm-wave Endfire chain-slot antenna for Mobile devices. IEEE Transactions on Antennas and Propagation. 2021;**69**:25-34. DOI: 10.1109/tap.2020.3001434

[67] Xu B, Ying Z, Scialacqua L. Radiation performance analysis of 28 GHz antennas integrated in 5G Mobile terminal housing. IEEE Access. 2018;**6**:48088-48101. DOI: 10.1109/access.2018.2867719

[68] Rodriguez-Cano R, Zhang S, Zhao K. Reduction of Main beam-blockage in an integrated 5G Array with a metal-frame antenna. IEEE Transactions on Antennas and Propagation. 2019;**67**:3161-3170. DOI: 10.1109/tap.2019.2900407

[69] Samadi Taheri MM, Abdipour A, Zhang S. Integrated Millimeter-wave wideband end-fire 5G beam steerable Array and low-frequency 4G LTE antenna in Mobile terminals. IEEE Transactions on Vehicular Technology. 2019;**68**:4042-4046. DOI: 10.1109/tvt.2019.2899178

[70] Rodriguez-Cano R, Zhang S, Zhao K. Mm-wave beam-steerable Endfire Array embedded in a slotted metal-frame LTE antenna. IEEE Transactions on Antennas and Propagation. 2020;**68**:3685-3694. DOI: 10.1109/tap.2020.2963915

[71] Rodriguez-Cano R, Zhao K, Zhang S. Handset frame blockage reduction of 5G mm-wave phased arrays using hard surface inspired structure. IEEE Transactions on Vehicular Technology. 2020;**69**:8132-8139. DOI: 10.1109/tvt.2020.2996360

[72] Kurvinen J, Kahkonen H, Lehtovuori A. Co-Designed mm-Wave and LTE Handset Antennas. IEEE Transactions on Antennas and Propagation. 2019;**67**:1545-1553. DOI: 10.1109/tap.2018.2888823

[73] Ifixit. iPhone 12 and 12 Pro Teardown. [Online]. 2021. Available from: https://www.ifixit.com/Teardown/iPhone+12+and+12+Pro+Teardown/137669.

[74] Li H, Wu M, Cheng Y. Leaky-wave antennas as metal rims of Mobile handset for mm-wave communications. IEEE Transactions on Antennas and Propagation. 2021;**69**:4142-4147. DOI: 10.1109/tap.2020.3044369

[75] Malfajani RS, Ashraf FB, Sharawi MS. A 5G enabled shared-aperture, dual-band, in-rim antenna system for wireless handsets. IEEE Open Journal of Antennas and Propagation. 2022;**3**:1013-1024. DOI: 10.1109/ojap.2022.3201627

[76] Yu B, Yang K, Sim C-Y-D. A novel 28 GHz beam steering Array for 5G Mobile device with metallic casing application. IEEE Transactions on Antennas and Propagation. 2018;**66**:462-466. DOI: 10.1109/tap.2017.2772084

[77] Islam S, Zada M, Yoo H. Highly compact integrated Sub-6 GHz and Millimeter-wave band antenna Array for 5G smartphone communications. IEEE Transactions on Antennas and Propagation. 2022;**70**:11629-11638. DOI: 10.1109/tap.2022.3209310

[78] Xiang BJ, Zheng SY, Wong H. A flexible dual-band antenna with large frequency ratio and different radiation properties over the two bands. IEEE Transactions on Antennas and Propagation. 2018;**66**:657-667. DOI: 10.1109/tap.2017.2786321

[79] Ding X-H, Zhang Q-H, Yang W-W. A dual-band antenna for LTE/mmWave Mobile terminal applications. IEEE Transactions on Antennas and Propagation. 2023;**71**:2826-2831. DOI: 10.1109/tap.2023.3237165

[80] Dan Z, He Z, Lin H. A patch Rectenna with an integrated impedance matching network and a harmonic recycling filter. IEEE Antennas and Wireless Propagation Letters. 2022;**21**:2085-2089. DOI: 10.1109/lawp.2022.3190879

[81] Ding X-H, Yang W-W, Qin W. A broadside shared aperture antenna for (3.5, 26) GHz Mobile terminals with steerable beam in Millimeter-waveband. IEEE Transactions on Antennas and Propagation. 2022;**70**:1806-1815. DOI: 10.1109/tap.2021.3118817

[82] Ding X-H, Yang W-W, Tang H. A dual-band shared-aperture antenna for microwave and Millimeter-wave applications in 5G wireless communication. IEEE Transactions on Antennas and Propagation. 2022;**70**:12299-12304. DOI: 10.1109/tap.2022.3209220

[83] Liang Q, Aliakbari H, Lau BK. Co-designed Millimeter-wave and Sub-6 GHz antenna for 5G smartphones. IEEE Antennas and Wireless Propagation

Letters. 2022;**21**:1995-1999. DOI: 10.1109/lawp.2022.3187782

[84] Zhang JF, Cheng YJ, Ding YR. A dual-band shared-aperture antenna with large frequency ratio, high aperture reuse efficiency, and High Channel isolation. IEEE Transactions on Antennas and Propagation. 2019;**67**:853-860. DOI: 10.1109/tap.2018.2882697

[85] Su Y, Lin XQ, Fan Y. Dual-band Coaperture antenna based on a single-layer mode composite transmission line. IEEE Transactions on Antennas and Propagation. 2019;**67**:4825-4829. DOI: 10.1109/tap.2019.2913706

[86] Ding YR, Cheng YJ. A tri-band shared-aperture antenna for (2.4, 5.2) GHz Wi-fi application with MIMO function and 60 GHz Wi-gig application with beam-scanning function. IEEE Transactions on Antennas and Propagation. 2020;**68**:1973-1981. DOI: 10.1109/tap.2019.2948571

[87] Liu Y, Li Y, Ge L. A compact Hepta-band mode-composite antenna for sub (6, 28, and 38) GHz applications. IEEE Transactions on Antennas and Propagation. 2020;**68**:2593-2602. DOI: 10.1109/tap.2019.2955206

[88] Ikram M, Abbas EA, Nguyen-Trong N. Integrated frequency-reconfigurable slot antenna and connected slot antenna Array for 4G and 5G Mobile handsets. IEEE Transactions on Antennas and Propagation. 2019;**67**:7225-7233. DOI: 10.1109/tap.2019.2930119

[89] Xu G, Peng H-L, Shao Z. Dual-band differential shifted-feed microstrip grid Array antenna with two parasitic patches. IEEE Transactions on Antennas and Propagation. 2020;**68**:2434-2439. DOI: 10.1109/tap.2019.2943409

[90] Li T, Chen ZN. Shared-surface dual-band antenna for 5G applications. IEEE Transactions on Antennas and Propagation. 2020;**68**:1128-1133. DOI: 10.1109/tap.2019.2938584

[91] Ikram M, Nguyen-Trong N, Abbosh A. Multiband MIMO microwave and Millimeter antenna system employing dual-function tapered slot structure. IEEE Transactions on Antennas and Propagation. 2019;**67**:5705-5710. DOI: 10.1109/tap.2019.2922547

[92] Vaha-Savo L, Cziezerski C, Heino M. Empirical evaluation of a 28 GHz antenna Array on a 5G Mobile phone using a body phantom. IEEE Transactions on Antennas and Propagation. 2021;**69**:7476-7485. DOI: 10.1109/tap.2021.3076535

[93] Zhao K, Ying Z, He S. EMF exposure study concerning mmWave phased Array in Mobile devices for 5G communication. IEEE Antennas and Wireless Propagation Letters. 2016;**15**:1132-1135. DOI: 10.1109/lawp.2015.2496229

[94] Alammouri A, Mo J, Ng BL. Hand grip impact on 5G mmWave Mobile devices. IEEE Access. 2019;**7**:60532-60544. DOI: 10.1109/access.2019.2914685

[95] Ojaroudiparchin N, Shen M, Zhang S. A switchable 3-D-coverage-phased Array antenna package for 5G Mobile terminals. IEEE Antennas and Wireless Propagation Letters. 2016;**15**:1747-1750. DOI: 10.1109/lawp.2016.2532607

[96] Zhang S, Chen X, Syrytsin I. A planar switchable 3-D-coverage phased Array antenna and its user effects for 28-GHz Mobile terminal applications. IEEE Transactions on Antennas and Propagation. 2017;**65**:6413-6421. DOI: 10.1109/tap.2017.2681463

[97] Ballesteros C, Vaha-Savo L, Haneda K. Assessment of mmWave handset arrays in the presence of the user body. IEEE Antennas and Wireless Propagation Letters. 2021;**20**:1736-1740. DOI: 10.1109/lawp.2021.3095352

[98] Zhao K, Helander J, Sjoberg D. User body effect on phased Array in user equipment for the 5G mmWave communication system. IEEE Antennas and Wireless Propagation Letters. 2017;**16**:864-867. DOI: 10.1109/lawp.2016.2611674

[99] Zhao K, Zhang S, Ho Z. Spherical coverage characterization of 5G Millimeter wave user equipment with 3GPP specifications. IEEE Access. 2019;**7**:4442-4452. DOI: 10.1109/access.2018.2888981

[100] Syrytsin I, Zhang S, Pedersen GF. User-shadowing suppression for 5G mm-wave Mobile terminal antennas. IEEE Transactions on Antennas and Propagation. 2019;**67**:4162-4172. DOI: 10.1109/tap.2019.2905685

[101] Syrytsin I, Zhang S, Pedersen GF. User impact on phased and switch diversity arrays in 5G Mobile terminals. IEEE Access. 2018;**6**:1616-1623. DOI: 10.1109/access.2017.2779792

[102] Syrytsin I, Zhang S, Pedersen GF. User effects on the circular polarization of 5G Mobile terminal antennas. IEEE Transactions on Antennas and Propagation. 2018;**66**:4906-4911. DOI: 10.1109/tap.2018.2851383

Chapter 11

# Interferometric Phase Transmitarray for Millimeter-Wave MIMO System

*Yu Luo and Xiaoxuan Guo*


## Abstract

A millimeter-wave (mmW) interferometric phase transmitarray for the multiple-input multiple-output (MIMO) system is proposed, and its phase distribution is the interference superposition of electromagnetic waves radiated by two patch antennas at different locations. Its characteristic is that when multiple EM waves illuminate the center of the array, the transmitted waves are formed into high-directivity beams. In addition, when the plane wave illuminates the interference phase transmitarray vertically, the transmissive plane wave will be scattered and focused to two different positions. A novel MIMO system can be implemented based on the above two characteristics. Compared with the conventional lens MIMO, the advantage of the MIMO system integrated by the interferometric phase transmitarray is that multiple antennas can share one transmitarray, which is beneficial to the miniaturization of the MIMO transceiver. More critically, all channels can efficiently transmit information and increase channel capacity.


**Keywords:** interferometric phase, transmitarray, MIMO, miniaturization, channel capacity

## 1. Introduction

Recently, mmW technology has been the top priority of the fifth-generation (5G) wireless network [1]. Multiple-input multiple-output (MIMO) has been recognized as the most effective application of 5G technology since it can increase the data transmission rate by expanding channel capacity [2]. Generally, elements in MIMO antennas are with wider beamwidth to ensure that each receiving antenna can receive the signals from each transmitting antenna. However, for mmW antennas, high-gain and narrower beamwidth antennas are employed to overcome channel fading. With high gain elements, conventional lens-based $2 \times 2$ MIMO systems are investigated [3–6] as shown in **Figure 1**. This kind of MIMO system requires multiple lenses, which is not conducive to the miniaturization of the MIMO transceiver. Moreover, the signals of Channel 1 and Channel 2 are robust and easy to be received, but the signals of Channel 3 and Channel 4 are weak and unable to transmit information effectively due to the narrow beamwidth of the elements. Besides, MIMO systems in the mmW band have

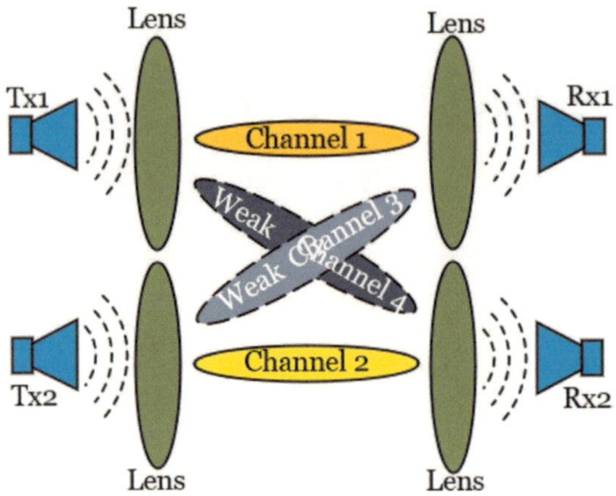


**Figure 1.**
*The 2 × 2 MIMO system integrated by the conventional lens.*

high requirements for antenna and circuit performance, which leads to high costs and complexity [7–9].

Current metamaterials have the characteristics of flexible control of electromagnetic (EM) waves and are widely used in mmW systems, such as radar, satellite communications, and imaging [10]. Metamaterials and their derivatives also have the advantages of thin thickness, small size, and lightweight, which can effectively solve problems such as cost and loss [11–13]. More importantly, metamaterials are widely used in transmitting and reflective arrays to realize beamforming and beam steering [14–21], which lays the foundation for realizing the miniaturization of the MIMO transceiver.

This paper proposes an interferometric phase transmitarray for the MIMO system. Its characteristic is that when multiple EM waves radiate toward the array, all the EM waves are beamforming into beams with high directivity. When the plane wave illuminates the interference phase transmitarray vertically, the transmissive plane wave will be scattered and focused to two different positions. Based on the above two characteristics, a novel MIMO system can be implemented, as shown in **Figure 2**. Compared with the MIMO system in **Figure 1**, the advantage of this novel MIMO system is that two antennas can share one transmitarray, which is beneficial to the miniaturization of the MIMO transceiver. All four channels can efficiently transmit information and increase channel capacity.

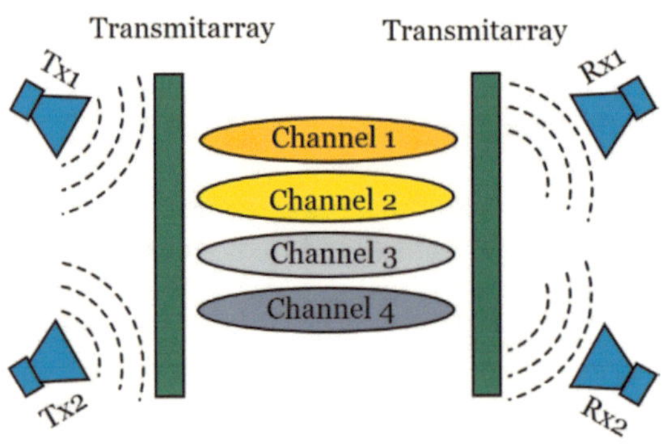


**Figure 2.**
*The MIMO system integrated by the proposed lens.*

## 2. Theoretical study

The theoretical study focuses on the phase distribution and incident field of the interferometric phase transmitarray.

### 2.1 The phase distribution

The ideal model of interferometric phase transmitarray is shown in **Figure 3**. Assuming the interferometric phase transmitarray is an N × N square array. When two EM waves are radiated by two feeds toward the array, the two transmitted EM waves are formed into a high-directivity beam.

For quantitative analysis, it is assumed that $O_{mn}$ ($x_m$, $y_n$, 0) is the position of the unit cell in row m and column n. The coordinate of the Feed 1 ($x_1$, $y_1$, $z_1$). Here, for the convenience of calculation, the ideal point source model is selected for Feed 1. The phase distribution caused by the propagation path of the EM Wave 1 radiated by Feed 1 to the array is calculated by Eq. (1),

$$\phi_1(x_m, y_n) = k_0\sqrt{(x_m - x_1)^2 - (y_n - y_1)^2 + z_1^2} \quad (1)$$

where $k_0 = 2\pi/\lambda_0$ is the wavenumber. To make the phase distribution of EM waves consistent passing through the transmitarray, the phase of the array itself should be -$\phi_1(x_m, y_n)$. Similarly, the phase distribution caused by the propagation path of the EM Wave 2 radiated by Feed 2 to the array is calculated by Eq. (2),

$$\phi_2(x_m, y_n) = k_0\sqrt{(x_m - x_2)^2 - (y_n - y_2)^2 + z_2^2} \quad (2)$$

When Feed 1 and Feed 2 are excited together, the phase distribution of the array aperture is interferometric superimposed by Eq. (3),

$$\Delta\phi(x_m, y_n) = \arg(A_1(x_m, y_n)\exp(j\phi_1(x_m, y_n)) + A_2(x_m, y_n)\exp(j\phi_2(x_m, y_n))) \quad (3)$$

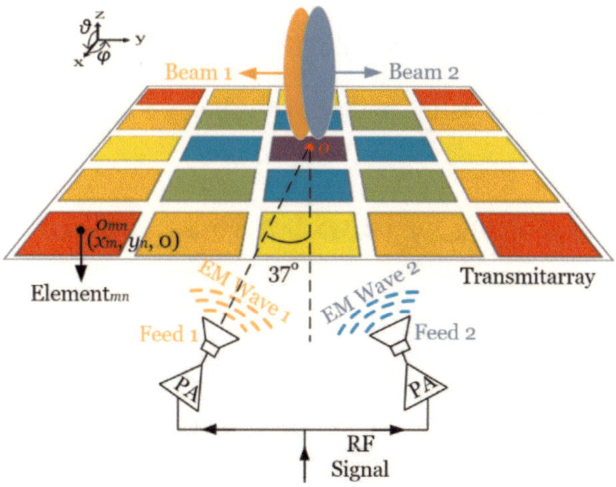


**Figure 3.**
*The theoretical model of interferometric phase transmitarray. Reprinted with permission from Ref. [22]; copyright 2022 IEEE.*

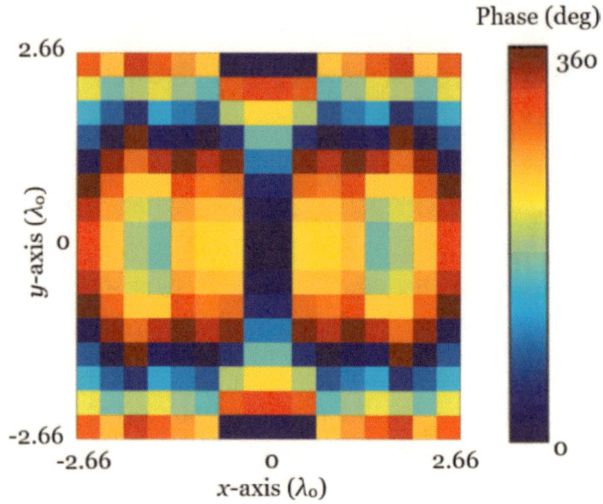


**Figure 4.**
*The phase distribution of the interferometric phase transmitarray. Reprinted with permission from Ref. [22]; copyright 2022 IEEE.*

where $A_1(x_m, y_n)$ and $A_2(x_m, y_n)$ are the amplitudes of E-field. To make the phase distribution of EM waves consistent through the transmitarray, the phase of the array itself should be $-\Delta\phi(x_m, y_n)$.

Establishing a MATLAB model. The two ideal point sources are set at ($-1.5\,\lambda_0$, 0, $-2\,\lambda_0$) and ($1.5\,\lambda_0$, 0, $-2\,\lambda_0$), and the phase distribution of the transmitarray is shown in **Figure 4**.

## 2.2 The incident field

Assume $f_{m,\,n}\,(\theta, \varphi)$ is the radiation pattern of the element in row $m$ and column $n$, the radiation pattern of the transmitarray can be expressed as Eq. (4),

$$F(\theta, \varphi) = f_{m,n}(\theta, \varphi) S_a(\theta, \varphi) \tag{4}$$

where $\theta$ and $\varphi$ are the elevation and azimuth angles. $S_a(\theta, \varphi)$ is expressed as Eq. (5),

$$S_a(\theta, \varphi) = \sum_{m=1}^{N} \sum_{n=1}^{N} \exp\left(-i\left(\phi(x_m, y_n) + kD \sin\theta \times ((m - 1/2)\cos\varphi + (n - 1/2)\sin\varphi)\right)\right) \tag{5}$$

where $l$ represents the length of the element. The phase $\phi(x_m, y_n)$ contains the phase of the unit itself $\phi_p(x_m, y_n)$ and the phase difference caused by the propagation distance $\phi_q(x_m, y_n)$. Furthermore, the directivity $Dir(\theta, \varphi)$ can be expressed as Eq. (6),

$$Dir(\theta, \varphi) = 4\pi |F(\theta, \varphi)|^2 \Big/ \int_0^{2\pi} \int_0^{\pi/2} |F(\theta, \varphi)|^2 \sin\theta d\theta d\varphi \tag{6}$$

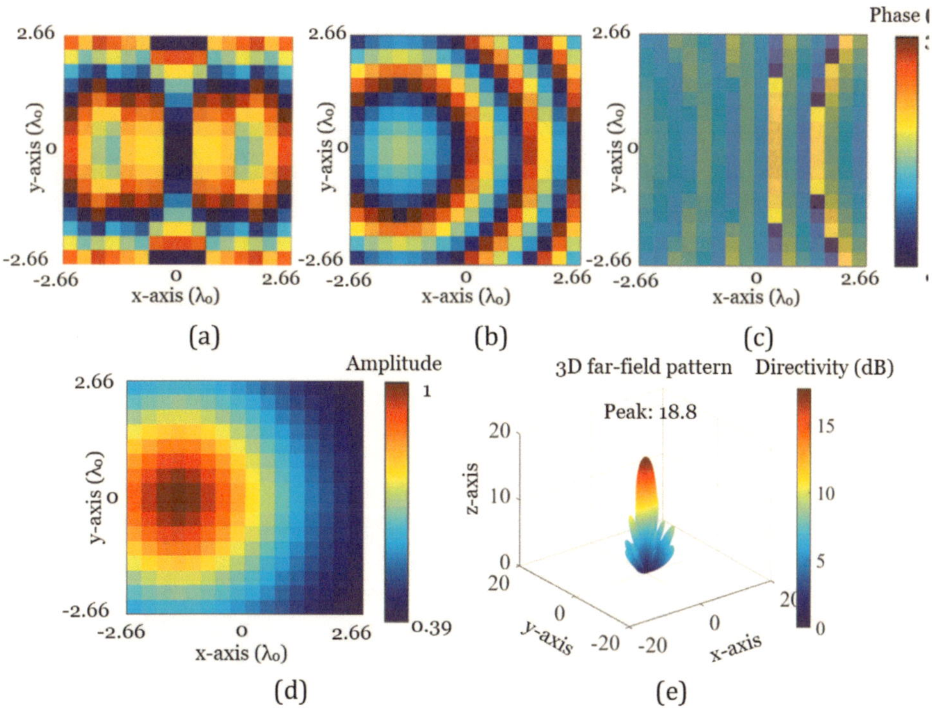


**Figure 5.**
*The incident field analysis of Feed 1: (a) the phase distribution of the transmitarray itself, (b) the phase distribution caused by the propagation distance, (c) the superimposed phase distribution, (d) the amplitude distribution, and (e) the 3-D pattern. Reprinted with permission from Ref. [22]; copyright 2022 IEEE.*

According to the above derivation, **Figure 5** shows the incident field of Feed 1. **Figure 5a** shows the phase distribution of the array itself, **Figure 5b** shows the phase distribution caused by the propagation distance, **Figure 5c** shows the superposition of **Figure 5a** and **5b**, **Figure 5d** shows the amplitude distribution, **Figure 5e** shows the 3-D pattern. Compared to **Figure 5b**, the phase distribution of **Figure 5c** is improved. Therefore, the beam shown in **Figure 5e** can achieve high directivity, and the peak is 18.8 dB.

**Figure 6** shows the incident field analysis of Feed 2. **Figure 6a** shows the phase distribution of the array itself, **Figure 6b** shows the phase distribution caused by the propagation distance, **Figure 6c** shows the superposition of **Figure 6a** and **6b**, **Figure 6d** shows the amplitude distribution, **Figure 6e** shows the 3-D pattern. Compared to **Figure 6b**, the phase distribution of **Figure 6c** is improved. Therefore, the beam shown in **Figure 6e** can achieve high directivity, and the peak is 18.8 dB.

## 3. Realization of the interferometric phase transmitarray

The unit of the interferometric phase transmitarray is shown in **Figure 7**. The metal layer contains a square ring and patch as shown in **Figure 7a**. The overall structure is five dielectric and six metal layers arranged alternately, as shown in **Figure 7b**.

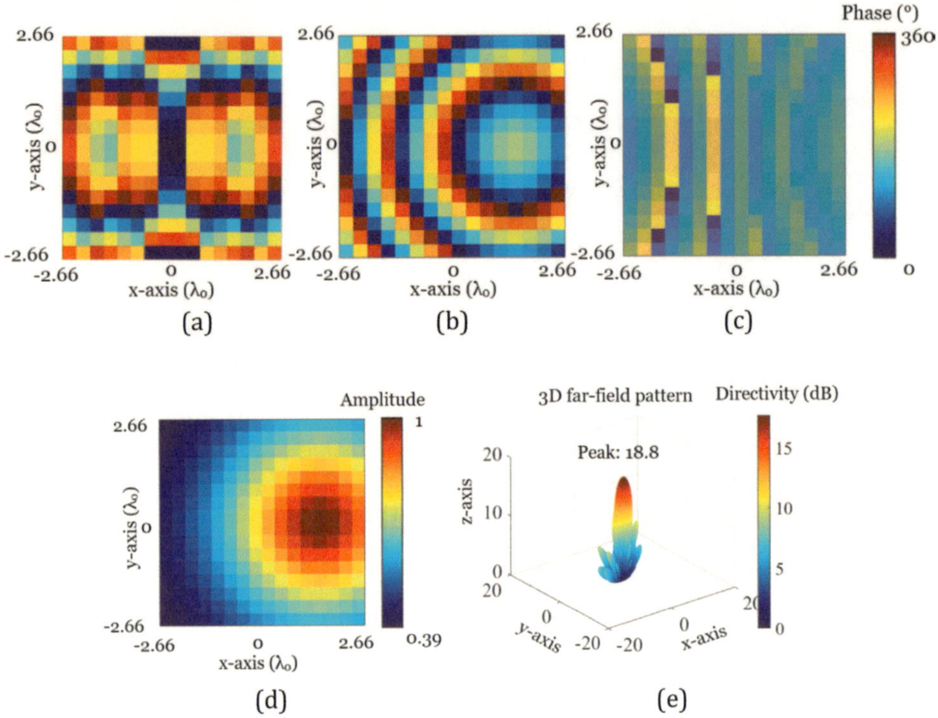


**Figure 6.**
*The incident field analysis of Feed 2: (a) the phase distribution of the transmitarray itself, (b) the phase distribution caused by the propagation distance, (c) the superimposed phase distribution, (d) the amplitude distribution, and (e) the 3-D pattern. Reprinted with permission from Ref. [22]; copyright 2022 IEEE.*

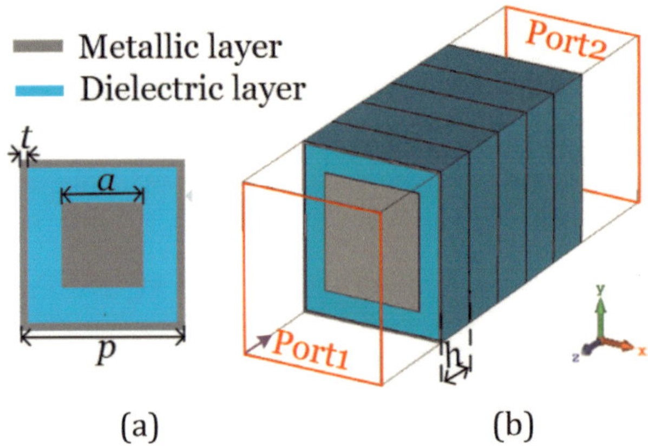


**Figure 7.**
*The view of the unit. (a) Top view. (b) Overall view. Reprinted with permission from Ref. [22]; copyright 2022 IEEE.*

The substrate adopts F4B, and its dielectric constant is 2.65. The thickness of the substrate is $h$ = 1.5 mm. $p$ = 4 mm is the unit period, and $t$ equals 0.05 mm. Size $a$ represents the side length of the square metal patch, which is related to the transmission characteristics of the unit. Eight units with different $a$ are for comparison. **Figure 8a** shows that the phase change of the unit contains 300°. It can be seen from

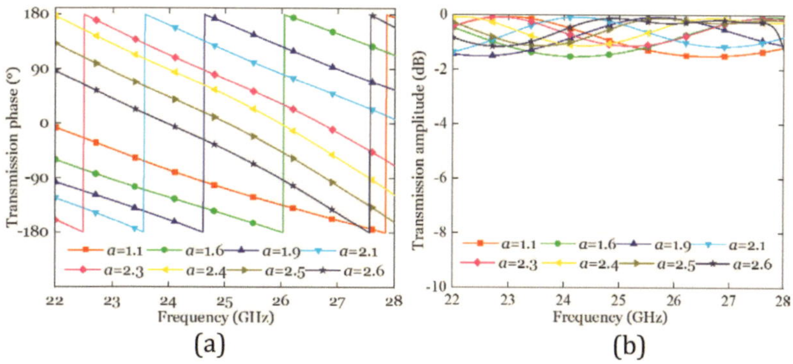


**Figure 8.**
*Transmission characteristics of the unit: (a) phases, (b) amplitudes. Reprinted with permission from Ref. [22]; copyright 2022 IEEE.*

**Figure 8b** that the transmission amplitude of the unit is higher than −2 dB, which indicates the EM waves can pass through the unit efficiently.

Utilize s a patch antenna as the feed. The substrate adopts RuiLong, and its dielectric constant is 2.2. **Figure 9a** shows the physical structure of the patch antenna. The parameters are $pa$ = 4.5 mm, $pb$ = 3.4 mm, and $pl$ = 14 mm. To connect the adapter, two air holes with a radius of 1 mm are drilled on the substrate. **Figure 9b** shows the radiation pattern of the patch antenna at 25 GHz, and the realized gain is 7.3 dBi.

Establishing a CST Microwave Studio model. **Figure 10a** shows the phase distribution of the interferometric phase transmitarray, and **Figure 10b** shows the simulation model. **Figure 10c** and **d** illustrate the 3-D radiation patterns when the two feeds are excited respectively, and the realized gain of the proposed transmitarray antenna is 18.4 dBi at 25 GHz. Moreover, the radiation patterns shown in **Figure 10c** and **d** are symmetric about the yoz-plane.

The above results show that when multiple EM waves radiate toward the array, all the transmitted waves are formed into high-directivity beams. This process is characteristic of the transmitter of the MIMO system integrated by interferometric phase transmitarray. Next, analyze the characteristics of the receiver. When the plane wave illustrates the interferometric phase transmitarray vertically, the transmitted plane

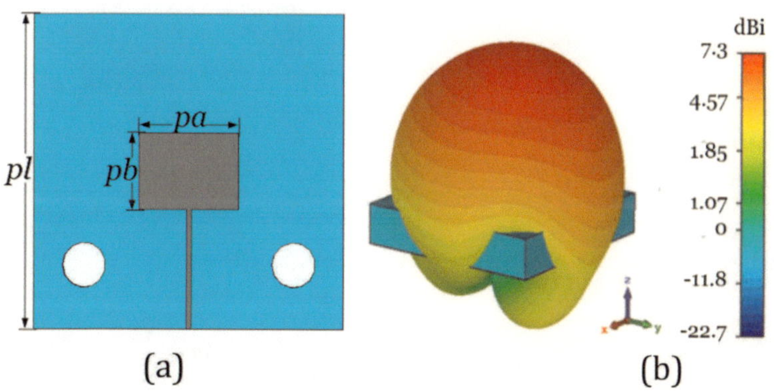


**Figure 9.**
*(a) The physical structure of the patch antenna. (b) The simulated radiation pattern. Reprinted with permission from Ref. [22]; copyright 2022 IEEE.*

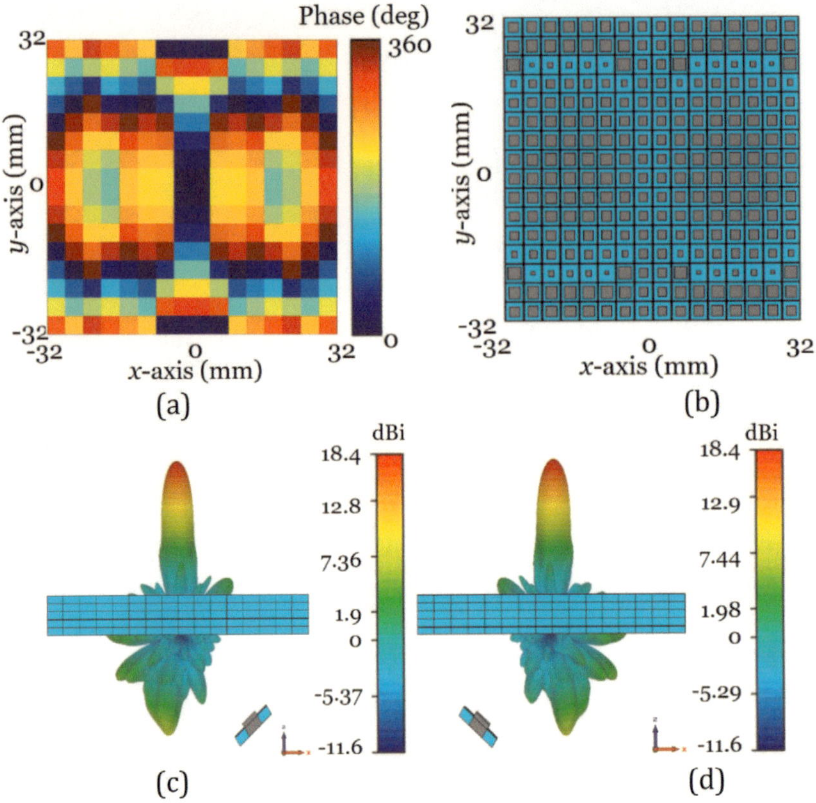


**Figure 10.**
*(a) The Phase distribution of the interferometric phase transmitarray, (b) the CST Microwave Studio model, (c) when the right feed is excited, the 3-D pattern at 25 GHz, (d) when the left feed is excited, the 3-D pattern at 25 GHz. Reprinted with permission from Ref. [22]; copyright 2022 IEEE.*

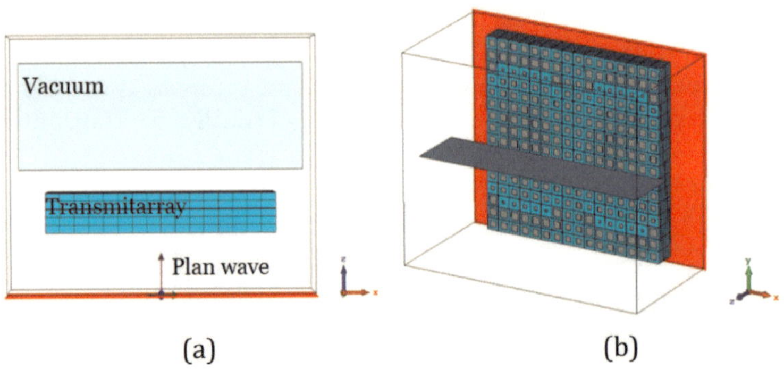


**Figure 11.**
*Simulation setup in CST time-domain solver. (a) xoz plane view. (b) 3D view.*

wave will be scattered and focused to two places. The proposed transmitarray is discretized and simulated using CST Studio Suite with the setup illustrated in **Figure 11**. Set the plane wave radiated along the z-axis to illuminate the transmitarray. A xoz plane square vacuum without thickness, as shown in **Figure 11a,** is set above the transmitarray to observe the focuses.

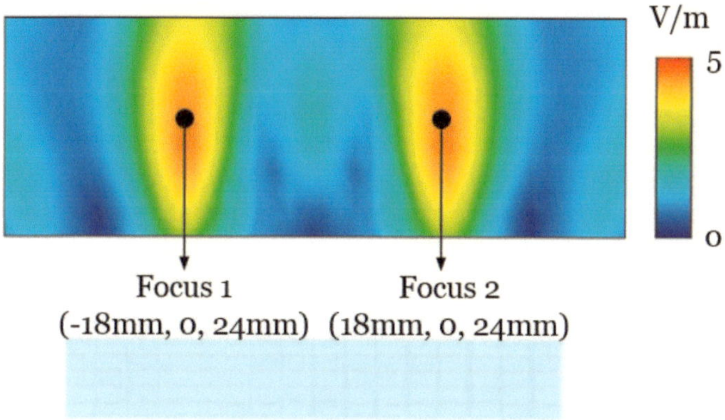


**Figure 12.**
*The E-field distribution of the interferometric phase transmitarray under plane wave illumination.*

Set up the e-field monitor and the simulated result is shown in **Figure 12**. It can be seen that when the plane wave illustrates the interferometric phase transmitarray vertically, the e-field will form two focuses, and its central coordinates are (−18 mm, 0, 24 mm) and (18 mm, 0, 24 mm).

In summary, when multiple EM waves radiate toward the transmitarray, all the EM waves are beamforming into high-directivity beams. When the plane wave illustrates the interferometric phase transmitarray vertically, the e-field will form two focuses, and its relative positions are consistent with the relative positions of the two feeds. These characteristics successfully enable the interferometric phase transmitarray to achieve the MIMO system, as shown in **Figure 2**.

## 4. Fabrication and experiment

To verify, a transmitarray and two patch antennas were fabricated. **Figure 13** shows that the prototype is verified in the anechoic chamber.

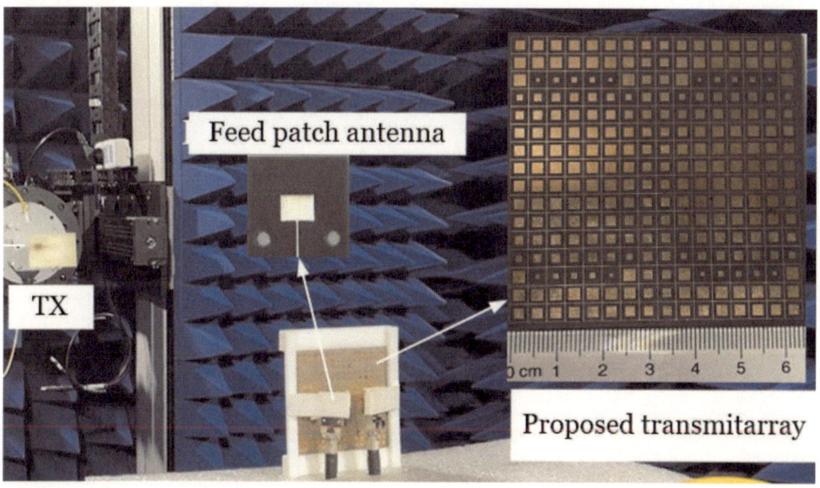


**Figure 13.**
*The prototype is verified in the anechoic chamber. Reprinted with permission from Ref. [22]; copyright 2022 IEEE.*

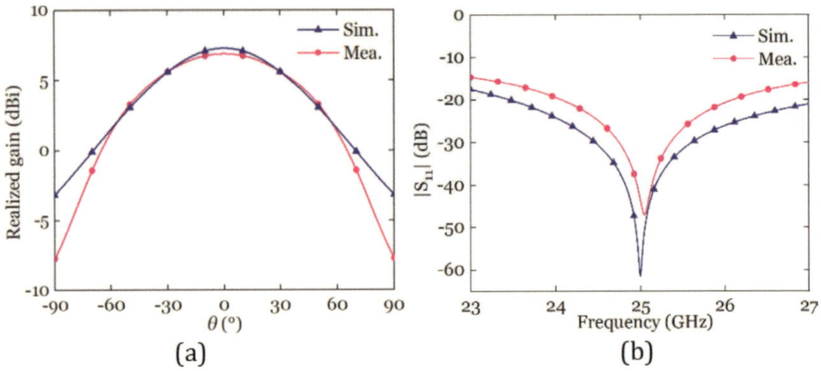


**Figure 14.**
*(a) The simulated and measured patterns of the patch antenna at 25 GHz. (b) The simulated and measured $|S_{11}|$ of the interferometric phase transmitarray excited by the patch antenna. Reprinted with permission from Ref. [22]; copyright 2022 IEEE.*

**Figure 14a** shows the patterns at 25 GHz of the patch antenna, and the simulated and measured curves are similar. **Figure 14b** shows the simulated and measured $|S_{11}|$ of the interferometric phase transmitarray excited by the patch antenna.

**Figure 15** shows the simulated and measured patterns at 25 GHz of the interferometric phase transmitarray excited by the patch antenna. The simulated realized gain is 18.4 dBi, which is 0.9 dBi higher than the measured realized gain. In addition, the simulated and measured sidelobe level is around -15 dB, within a reasonable range. Minor differences between curves are mainly caused by the measurement environment.

## 5. MIMO behavior

To evaluate the performance of the proposed interferometric phase MIMO system, a same-size unifocal transmitarray antenna with a focus on (18 mm, 0, −30 mm) is introduced for simple comparison. **Figure 16** shows the simulated pattern.

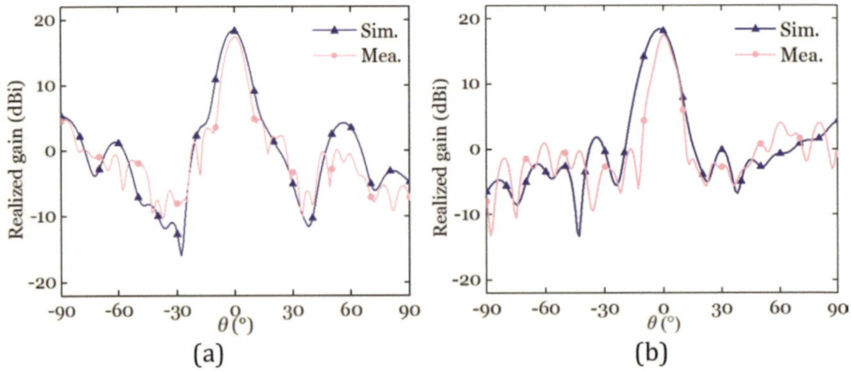


**Figure 15.**
*The simulated and measured patterns at 25 GHz of the interferometric phase transmitarray excited by the patch antenna. (a) xoz-plane. (b) yoz-plane. Reprinted with permission from Ref. [22]; copyright 2022 IEEE.*

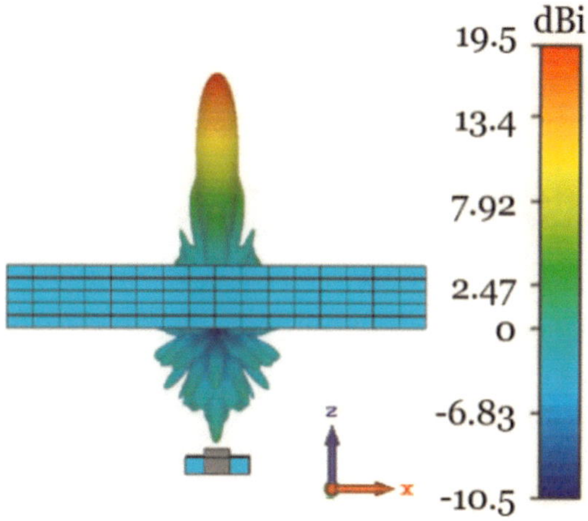


**Figure 16.**
*The simulated pattern of the unifocal transmitarray antenna at 25 GHz.*

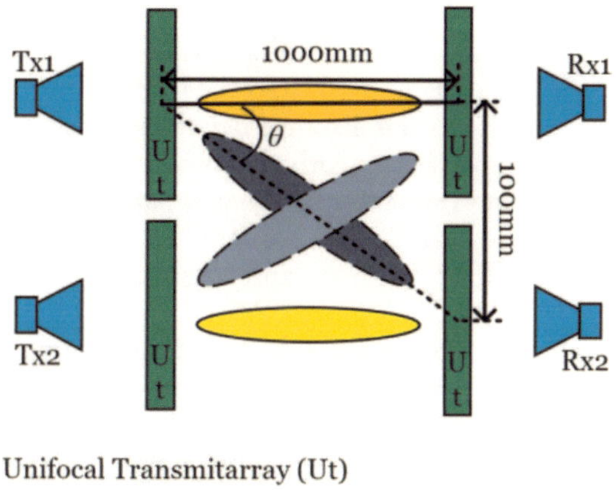


**Figure 17.**
*The simulated 3D pattern of the unifocal transmitarray antenna at 25 GHz.*

**Figure 17** shows the 2 × 2 MIMO system integrated by four unifocal transmitarray. It is assumed that the distance between two unifocal transmitarrays on one side is 100 mm and the transmission distance is 1000 mm, then the angle of the cross weak channels is $\theta$ = 6°. **Figure 18** shows the simulated xoz-plane pattern of the unifocal transmitarray antenna at 25 GHz. It can be seen from **Figure 18** that when $\theta$ = 6°, the gain is 15.2 dBi.

The channel capacity can be calculated by Eq. (7) as follows

$$C = B \log 2(1 + S/N) \tag{7}$$

where $B$ is bandwidth, and $S/N$ represents signal noise ratio. In MIMO links, the higher the receiving power, the greater the $S/N$, and the larger the channel capacity. The receiving power can be expressed as Eq. (8).

$$P_r = P_t \frac{G_t G_r {\lambda_0}^2}{16\pi^2 d^2} \tag{8}$$

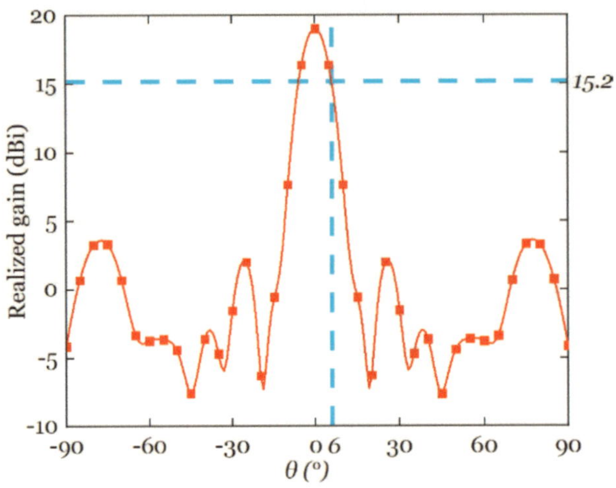


**Figure 18.**
*The simulated xoz-plane pattern of the unifocal transmitarray antenna at 25 GHz.*

where $P_r$ and $P_t$ represent receiving power and transmitting power, $G_r$ and $G_t$ represent the gain of the receiving antenna and the transmitting antenna, $\lambda_0$ is the wavelength in free space, and $d$ is the distance between transmitter and receiver.

Therefore, the channel capacity is positively correlated with the gain of the receiving antenna and transmitting antenna. To compare the two MIMO systems in **Figures 1** and **2**, it is assumed that all environments are the same, except for antenna gain differences. The gains of the receiving and transmitting antennas in the 2 × 2 MIMO system integrated by interferometric phase transmitarray are all 18.4 dBi. Normalize the channel capacity of each link to 1, and the channel capacity of the 2 × 2 MIMO system is 4. By comparison, there are two strong channels and two weak channels in the 2 × 2 MIMO system in **Figure 17**. The gains of the receiving and transmitting antennas are 19.5 dBi in strong channels and 15.2 dBi in weak channels. In the 2 × 2 MIMO system in **Figure 17**, the 18.4 dBi of the antenna gain is still used to normalize the channel capacity. After normalization, the channel capacity of the strong channel is 1.056, and the weak channel is 0.823 in **Figure 17**. Therefore, the channel capacity of the 2 × 2 MIMO system in **Figure 17** is 3.758, which is lower than the 2 × 2 MIMO system integrated by interferometric phase transmitarray. In addition, the distance between transmitter and receiver $d$ will also affect the channel capacity of the MIMO system in **Figure 17**. Since when $d$ decreases, the angle of the cross weak channels $\theta$ increases, resulting in a decrease in antenna gain in weak channels. When d increases, the result is the opposite.

Therefore, through simple comparison, it can be found that the channel capacity of the 2 × 2 MIMO system integrated by interferometric phase transmitarray is higher than the 2 × 2 MIMO system integrated by unifocal transmitarray.

## 6. Conclusions

The proposed interferometric phase transmitarray can adjust two EM waves to the boresight of the transmitarray. When the plane wave illustrates the proposed transmitarray, the transmitted plane wave will be scattered and focused on two positions. The MIMO system integrated by the interferometric phase transmitarray breaks the limitations of weak channels in conventional lens MIMO. In addition, the proposed method of the MIMO system can be extended to more channels.

## Author details

Yu Luo* and Xiaoxuan Guo
Tianjin Key Laboratory of Imaging and Sensing Microelectronic Technology, School of Microelectronics, Tianjin University, Tianjin, China

*Address all correspondence to: yluo@tju.edu.cn

## References

[1] Almasi MA, Mehrpouyan H, Vakilian V, Behdad N, Jafarkhani H. A new reconfigurable antenna MIMO architecture for mmwave communication. In: Proceedings of the IEEE International Conference Communication (ICC). Kansas City, MO, USA. 2018. pp. 1-7

[2] Loyka S. The capacity of Gaussian MIMO channels under total and per-antenna power constraints. IEEE Transactions on Communications. 2017; **65**(3):1035-1043

[3] Li X et al. 120 Gb/s wireless terahertz-wave signal delivery by 375 GHz-500 GHz multi-carrier in a 2 × 2 MIMO system. Journal of Lightwave Technology. 2019;**37**(2):606-611

[4] Statnikov K, Grzyb J, Heinemann B, Pfeiffer UR. 160-GHz to 1-THz multi-color active imaging with a Lens-coupled SiGe HBT Chip-set. IEEE Transactions on Microwave Theory Technology. 2015; **63**(2):520-532

[5] Puerta R, Yu J, Li X, Xu Y, Vegas Olmos JJ, Tafur Monroy I. Single-carrier dual-polarization 328-Gb/s wireless transmission in a D-band millimeter wave 2 × 2 Mu-MIMO radio-over-fiber system. Journal of Lightwave Technology. 2018;**36**(2):587-593

[6] Zhang J, Yu J, Chi N, Dong Z, Li X, Chang G. Multichannel 120-Gb/s data transmission over 2 × 2 MIMO fiber-wireless link at W-band. IEEE Photonics Technology Letters. 2013; **25**(8):780-783

[7] Yang B, Yu Z, Lan J, Zhang R, Zhou J, Hong W. Digital beamforming-based massive MIMO transceiver for 5G millimeter-wave communications. IEEE Transactions on Microwave Theory Technogy. 2018;**66**(7):3403-3418

[8] Uwaechia AN, Mahyuddin NM, Ain MF, Abdul Latiff NM, Zabah NF. On the spectral-efficiency of low-complexity and resolution hybrid precoding and combining transceivers for mmwave MIMO systems. IEEE Access. 2019;**7**:109259-109277

[9] Yu C et al. Full-angle digital predistortion of 5G Millimeter-wave massive MIMO transmitters. IEEE Transactions on Microw. Theory Technology. 2019;**67**(7):2847-2860

[10] Darvazehban A, Ahdi Rezaeieh S, Zamani A, Abbosh AM. Pattern reconfigurable metasurface antenna for electromagnetic torso imaging. IEEE Transactions on Antennas and Propagation. 2019;**67**(8):5453-5462

[11] He Y, Ding N, Zhang L, Zhang W, Du B. Short-length and high-aperture-efficiency horn antenna using low-loss bulk anisotropic metamaterial. IEEE Antennas Wireless Propagation Letters. 2015;**14**:1642-1645

[12] Valentine J, Zhang S, Zentgraf T, Zhang X. Development of bulk optical negative index fishnet metamaterials: Achieving a low-loss and broadband response through coupling. Proceedings of the IEEE. 2011;**99**(10): 1682-1690

[13] Costa F, Genovesi S, Monorchio A, Manara G. Low-cost metamaterial absorbers for sub-GHz wireless systems. IEEE Antennas Wireless Propagation Letters. 2014;**13**:27-30

[14] Li H, Wang G, Xu H, Cai T, Liang J. X-band phase-gradient metasurface for

high-gain lens antenna application. IEEE Transactions on Antennas and Propagation. 2015;**63**(11):5144-5149

[15] Han J, Li L, Ma X, Feng Q, Zhao Y, Liao G. A holographic metasurface based on orthogonally discrete unit-cell for flexible beam formation and polarization control. IEEE Antennas Wireless Propagation Letters. 2021;**20**(10): 1893-1897

[16] Su Y, Chen ZN. A flat dual-polarized transformation-optics beamscanning Luneburg lens antenna using PCB-stacked gradient index metamaterials. IEEE Transactions on Antennas and Propagation. 2018;**66**(10):5088-5097

[17] Jia Y, Liu Y, Zhang W, Wang J, Gong S, Liao G. High-gain Fabry-Perot antennas with wideband low monostatic RCS using phase gradient metasurface. IEEE Access. 2019;**7**:4816-4824

[18] Xiao Y, Yang F, Xu S, Li M, Zhu K, Sun H. Design and implementation of a wideband 1-bit transmitarray based on a Yagi–Vivaldi unit cell. IEEE Transactions on Antennas and Propagation. 2021; **69**(7):4229-4234

[19] Aziz A, Yang F, Xu S, Li M. A low-profile quad-beam transmitarray. IEEE Antennas Wireless Propagation Letters. 2020;**19**(8):1340-1344

[20] Tao Z, Bao D, Xu HX, Ma HF, Jiang WX, Cui TJ. A Millimeter-wave system of antenna Array and metamaterial Lens. IEEE Antennas Wireless Propagation Letters. 2016;**15**: 370-373

[21] Jiang M, Chen ZN, Zhang Y, Hong W, Xuan X. Metamaterial-based thin planar lens antenna for spatial beamforming and multibeam massive MIMO. IEEE Transactions on Antennas Propagation. 2017;**65**(2):464-472

[22] Guo X, Luo Y, Chen ZN, Yan N, An W, Ma K. Interferometric phase Transmitarray for spatial power combining to enhance EIRP of Millimeter-wave transmitters. IEEE Transactions on Antennas and Propagation. 2022;**70**(11):10485-10493

Section 3

# Channel Modeling

Chapter 12

# Multi-Cluster-Based MIMO-OFDM Channel Modeling

*Xin Li and Kun Yang*


## Abstract

In this chapter, the physical propagation environment of radio waves is described in terms of scattering clusters, in which each cluster could include many scattering objects. We use each single distant scattering cluster to study the characteristics of channel second-order statistics (CSOS) and build the multiple-input and multiple-output (MIMO) radio channels in accordance with the correlation properties of the channel. In this approach, each distant scattering cluster contributes a portion to the Doppler spectrum and corresponds to a state-space single-input and single-output (SISO) channel model. A MIMO channel model is then constructed by connecting multiple SISO channel models in parallel, in which a coloring matrix is used to adjust the channel spatial correlation properties between the SISO channels. A MIMO-OFDM (orthogonal frequency-division multiplexing) channel model is obtained in the same manner. This time, however, another matrix is used to adjust the channel spectral correlation properties between the MIMO channels. This approach has three advantages: Simple, the entire Doppler power spectrum can be formed from multiple uncorrelated distant scattering clusters, and the channels contributed by these clusters can be obtained by summing up the individual channels. In this way, we can reassemble the radio wave propagation environment in a simulated manner.


**Keywords:** channel second-order statistics, Cauchy-Rayleigh cluster, Rayleigh cluster, AOA, AOD, TOA, AR model, phase-shift method, SISO, MIMO, MIMO-OFDM, state-space model

## 1. Introduction

IN radio communications, from the traditional voice telephony to the current communication multimedia, to the future augmented reality (AR), virtual reality (VR), mixed reality (MR), and Internet of everything (IoE), the historical process shows that the increasing demand for higher data rates is the fundamental factor driving the development of communication technologies and methods.

One of the biggest challenges in radio communications is how to model radio channels. A radio channel refers to the influence of the propagation medium of electromagnetic (EM) waves on the signal from a transmitter to a receiver.

**Figure 1** depicts a typical terrestrial radio propagation environment. The basic characteristic of radio channels is fading, large-scale, and small-scale fading[1], which can be classified as distance loss, shadowing, and multi-path fading. Among them, multi-path fading is known as small-scale fading, and its characteristics vary with time, and change over frequency and space. That is, due to the multi-path propagation of EM waves, power-limited transmitted signals will be distorted at a receiver over time, frequency, and space simultaneously.

To better understand the mechanism behind the distortion, this physical phenomenon needs to be studied. Geometry-based stochastic multiple-input and multiple-output (MIMO) radio channel modeling was a hot topic [1–4]. The idea of this approach is to map the spatial location of scatterers in a cluster to an angular distribution of power through the trigonometric relationship among the scatterers, cluster center, and receiver or transmitter. Furthermore, the angular distribution of power has certain statistical properties if the cluster obeys a specific probability distribution [1, 2]. Hence, from an intuitive point of view, this approach is simple and straightforward.

A distant scattering cluster results in small variation in the angle-of-departure/angle of arrival (AOD/AOA) and produces a narrowband Doppler spectrum both at the base station (BS) and at the mobile station (MS) [5]. This can be used to explore the computation of the channel second-order statistics (CSOS) with a small angular approach, and this approach is suitable for the decomposition of the Doppler power spectrum into small uncorrelated portions.

Radio wave measurements indicate that the power azimuth spectrum typically has sharp, narrow peaks over a small range of angles [3, 4, 6, 7]. Each measurement has been modeled as a Laplace angular distribution [3, 4, 8–10]. Some measurements display smooth peaks [6, 11], which could be modeled by other distributions.

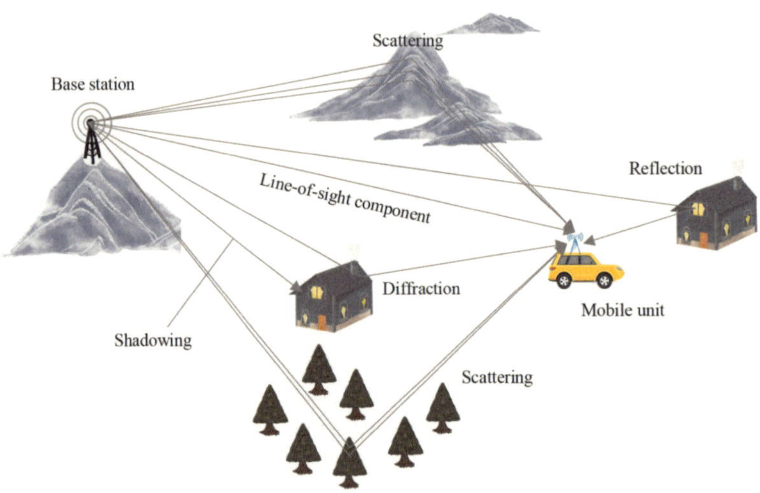


**Figure 1.**
*A typical mobile radio scenario for multi-path propagation in a terrestrial radio propagation environment.*

[1] It refers to the concept of distance described in terms of wavelengths.

In this chapter, the sharp peaks in the angular spectrum are modeled as Cauchy angular distributions of power and the smooth peaks are modeled as Gaussian angular power distributions. Other distributions can be approximated as weighted sums of Gaussian angular distributions of power or the combination of Gaussian and Cauchy angular distributions of power [5].

Although the Cauchy power distribution function (PDF) has fat tails as compared to the Laplace PDF, it could be used to achieve our goal if most of the power (such as 90% or more) is concentrated in a smaller angular range. Geometrically, it can be interpreted as that most of the scattering objects are located around the center of a cluster, while the rest contributes a much smaller amount of power to the antennas, which can be ignored. This idea can be used for truncated Gaussian angular power distributions as well.

It has been identified that, for a Cauchy angular power distribution function (APDF), the corresponding cluster has the following property: The distance between the cluster center and the scattering objects should obey the Cauchy-Rayleigh distribution [12], and for a Gaussian APDF, the corresponding cluster has the property that the distance between the cluster center and the scatterers should follow the Rayleigh distribution [13]. They are named as the Cauchy-Rayleigh cluster and Rayleigh cluster, respectively (**Figure 2**).

Based on the trigonometric relationship among transmitter, scatterers, and receiver, the APDFs of these two types of scattering clusters can be derived. In addition, the spatial-temporal correlation function is integrable according to the obtained APDF. The analytical solution, or closed-form solution, will be associated with a distant scattering cluster, i.e., the solution will depend on the characteristics of a given geometrical cluster. Furthermore, to be able to model a state-space MIMO channel, the correlation function needs to be separated into disjoint two parts, a temporal term over the movement and a spatial term over the antenna.

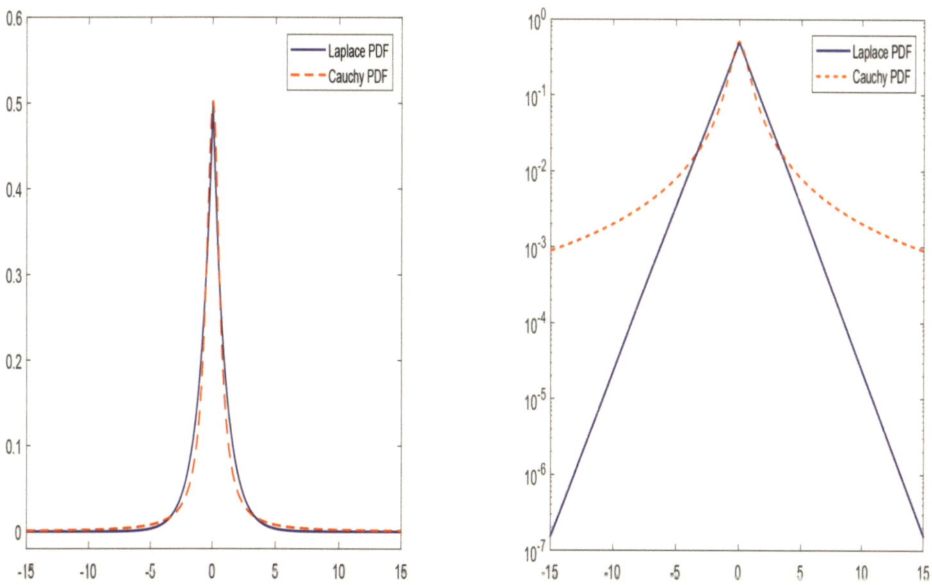


**Figure 2.**
*Laplace power distribution function (PDF)* $g_x(x) = \frac{1}{2}e^{-|x|}$ *and Cauchy PDF* $f_x(x) = \frac{1}{\pi}\frac{\eta}{\eta^2+x^2}$, *where* $\eta = 0.634$, *special case of parametrization.*

The beauty of this approach is that the expression of the channel second-order statistics can eventually be integrable. This analytical solution can be broken down into the temporal dynamics and spatial correlation parts of the channel. Depending on the type of cluster, the temporal dynamics part will be approximately modeled as an autoregressive order one (AR(1)) or an autoregressive order three (AR(3)) model, and the spatial correlation part will be described using the Kronecker matrix. An autoregressive order two (AR(2)) model can also be used if the requirement for approximation is acceptable. Therefore, one can construct the state-space MIMO/massive MIMO channel models, as well as the multi-cluster state-space MIMO-OFDM (orthogonal frequency-division multiplexing) channel models.

## 2. MIMO system

**Figure 3** depicts a $M_r \times M_t$ MIMO system, where $M_r$ and $M_t$ denote the numbers of receiving and transmitting antennas, respectively. This system has a total of $M_t M_r$ links between the BS and MS, in which each link is referred to as a radio channel.

Without loss of generality, a narrow-band, time-invariant channel model is used to compute the spatial correlation matrices. In this case, the channel matrix can be represented by [14].

$$\mathbf{H} = \begin{bmatrix} h_{11} & h_{12} & \cdots & h_{1M_t} \\ h_{21} & h_{22} & \cdots & h_{2M_t} \\ \vdots & \vdots & \ddots & \vdots \\ h_{M_r1} & h_{M_r2} & \cdots & h_{M_rM_t} \end{bmatrix} \quad (1)$$

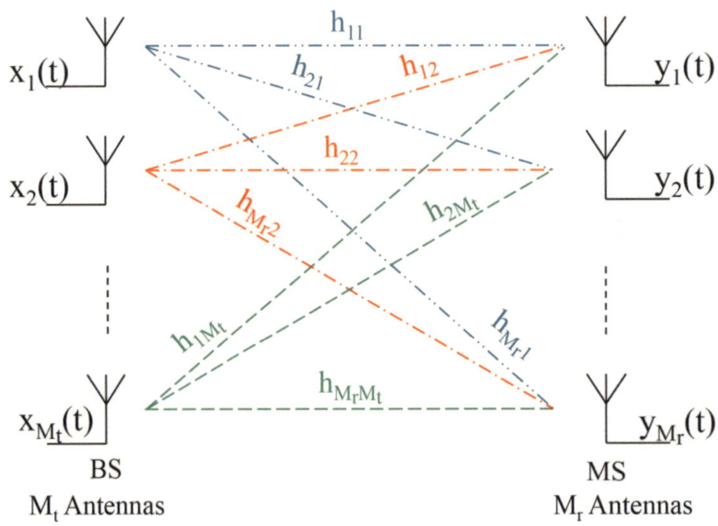


**Figure 3.**
*A MIMO system, $M_t$ antenna elements at the BS and $M_r$ antenna elements at the MS.*

where the elements $h_{ij}$ are the amplitude and phase change over the link between the *ith* MS antenna and the *jth* BS antenna.

To obtain the channel spatial correlation coefficients at the MS, we choose two arbitrary elements from a certain column of **H**, $h_{ik}, h_{jk}$, here and calculate the expectation value of the product of these two gains, i.e., $E\left[h_{ik}h_{jk}^*\right]$.

Usually, the distance between a transmitter and a receiver is quite large, so both transmitter and receiver will only be affected by scatterers in their vicinity. Therefore, the scatterers around the transmitter are uncorrelated with the scatterers around the receiver. That is, the spatial correlation between two arbitrary antennas at the MS does not depend on the transmitter antennas at the BS, but only depends on the antenna pair. Hence, the value, $E\left[h_{ik}h_{jk}^*\right]$, can be assumed to be independent of $k$.

All coefficients are then defined by

$$r_{i,j}^{\text{MS}} = E\left[h_{ik}h_{jk}^*\right] \tag{2}$$

Obviously, $r_{i,j}^{\text{MS}} = r_{j,i}^{\text{MS}^*}$ by this definition. Therefore, the corresponding spatial correlation matrix, a square matrix of order $M_r$, is represented by [14].

$$\mathbf{R}_{\text{MS}} = \begin{bmatrix} r_{1,1}^{\text{MS}} & r_{1,2}^{\text{MS}} & \cdots & r_{1,M_r}^{\text{MS}} \\ r_{2,1}^{\text{MS}} & r_{2,2}^{\text{MS}} & \cdots & r_{2,M_r}^{\text{MS}} \\ \vdots & \vdots & \ddots & \vdots \\ r_{M_r,1}^{\text{MS}} & r_{M_r,2}^{\text{MS}} & \cdots & r_{M_r,M_r}^{\text{MS}} \end{bmatrix} \tag{3}$$

Similarly, the channel spatial correlation coefficients at the BS can be given by

$$\mathbf{R}_{\text{BS}} = \begin{bmatrix} r_{1,1}^{\text{BS}} & r_{1,2}^{\text{BS}} & \cdots & r_{1,M_t}^{\text{BS}} \\ r_{2,1}^{\text{BS}} & r_{2,2}^{\text{BS}} & \cdots & r_{2,M_t}^{\text{BS}} \\ \vdots & \vdots & \ddots & \vdots \\ r_{M_t,1}^{\text{BS}} & r_{M_t,2}^{\text{BS}} & \cdots & r_{M_t,M_t}^{\text{BS}} \end{bmatrix} \tag{4}$$

where all elements $r_{m,n}^{\text{BS}}$ are defined by

$$r_{m,n}^{\text{BS}} = E\left[h_{lm}h_{ln}^*\right] \tag{5}$$

in terms of the channel gains $h_{lm}$ and $h_{ln}$, which are selected from a certain row of **H**, here $m, n \in \{1, 2, \cdots, M_t\}, l \in \{1, 2, \cdots, M_r\}$.

Finally, based on the assumptions that Eqs. (2) and (5) are independent of k and l, respectively, the spatial correlation matrix of the MIMO channels is given by [14].

$$\mathbf{R}_{\text{MIMO}} = E\left[\text{vec}(\mathbf{H})\,\text{vec}(\mathbf{H})^H\right] = \mathbf{R}_{\text{BS}} \otimes \mathbf{R}_{\text{MS}} \tag{6}$$

where $\otimes$ denotes the Kronecker product, vec $[\cdot]$ represents the vectorization of a matrix, which converts all elements of the matrix into a column vector.

## 3. Channel second-order statistics

Consider a MIMO system in which the BS has $M_t$ antennas, the MS has $M_r$ antennas, and the BS is fixed, while the MS is moving. Then, the spatial-temporal-spectral correlation function of a MIMO-OFDM channel can be expressed as [15, 16].

$$C_h\left(\Delta t, d_t, d_r, d_f\right) = \int_{\alpha,\beta,\tau} f_{\alpha,\beta,\tau}(\alpha,\beta,\tau)e^{j2\pi d_t \cos(\beta+\beta_0)} \times e^{j2\pi\left(d_r\cos(\alpha+\alpha_0)+f_D\Delta t\cos(\alpha+\alpha_0-\gamma)\right)}e^{-j2\pi d_f\tau}d\alpha d\beta d\tau \quad (7)$$

where $f_D$ is the Doppler frequency, $\Delta t$ is total time separation, and $f_D\Delta t$ is the MS moving distance. $\beta$ is the AOD and $\beta_0$ is its mean, $\alpha$ is the AOA and $\alpha_0$ is its mean, and $\gamma$ is the angle between the moving direction and the antenna array, as shown in **Figure 4**. $d_t = m_t\Delta d_t$ is the antenna spacing at the BS, $m_t \in \{0, 1, \cdots, M_t - 1\}$, $\Delta d_t$ is the spacing between two adjacent antenna sensors. $d_r = m_r\Delta d_r$ is the antenna spacing at the MS, $m_r \in \{0, 1, \cdots, M_r - 1\}$, $\Delta d_r$ is the spacing between two adjacent antenna sensors. $d_f = m_f\Delta f$ is frequency separation, $\Delta f$ denotes the frequency difference between two adjacent sub-carriers, and the sub-carrier frequencies are defined by $f_i = f_c + i\Delta f$, for all $i \in \{0,1,2,\cdots,M_f - 1\}$, here $f_c$ is the frequency range and $M_f$ denotes the number of sub-carrier frequencies required for transmission. The difference between two frequencies $f_i$ and $f_j$ is denoted by $m_f = j - i$. Hence, $m_f \in \{0,1,2,\cdots,M_f - 1\}$. When $m_f = 0$, it represents a single-carrier modulation system, the so-called MIMO system. $f_{\alpha,\beta,\tau}(\alpha,\beta,\tau)$ denotes the joint angular-delay power distribution function, here, $\tau$ denotes the time delay.

By assuming the independence of $\alpha, \beta$, and $\tau$, the joint angular-delay PDF $f_{\alpha,\beta,\tau}(\alpha,\beta,\tau)$ is separated into $f_{\alpha,\beta,\tau}(\alpha,\beta,\tau) = f_\alpha(\alpha)f_\beta(\beta)f_\tau(\tau)$, here $f_\alpha(\alpha)$ is the APDF of

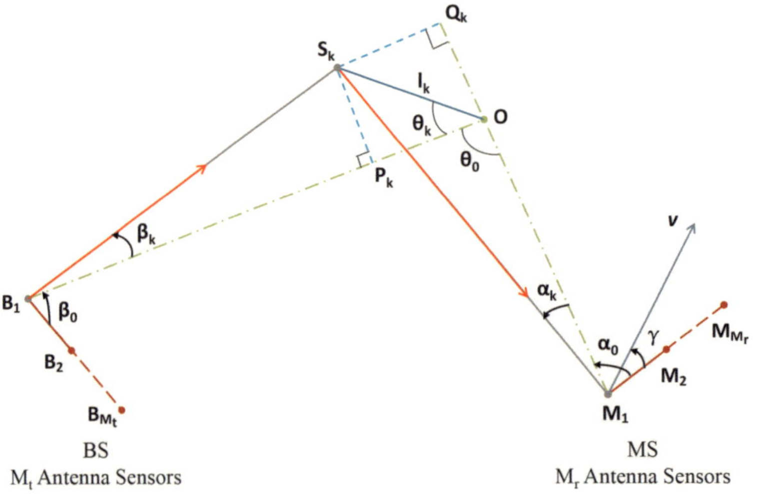


**Figure 4.**
*A distant cluster with an $M_r \times M_t$ MIMO antenna array, $l_k$ is the distance between the scattering object $S_k$ and the cluster center O.*

the AOA, $f_\beta(\beta)$ denotes the APDF of the AOD, and $f_\tau(\tau)$ is the delay power distribution function (DPDF) of the time-of-arrival (TOA).

This assumption is reasonable because usually radio signals pass through more than one scatterer in a cluster from a transmitter to a receiver, which means that $\alpha, \beta$ are independent. $\tau$ denotes the TOA, which is independent of $\alpha$ and $\beta$. Therefore, Eq. (7) becomes,

$$\begin{aligned} C_h\left(\Delta t, d_t, d_r, d_f\right) = & \int_{\alpha,\beta,\tau} f_\alpha(\alpha) f_\beta(\beta) f_\tau(\tau) e^{j2\pi d_t \cos(\beta+\beta_0)} \\ & \times e^{j2\pi\left(d_r \cos(\alpha+\alpha_0) + f_D \Delta t \cos(\alpha+\alpha_0-\gamma)\right)} e^{-j2\pi d_f \tau} d\alpha d\beta d\tau \end{aligned} \tag{8}$$

which two special cases are highlighted below,

- the channel temporal dynamic function is denoted by $R_h(\Delta t)$

$$R_h(\Delta t) = C_h(\Delta t, 0, 0, 0) = \int_\alpha f_\alpha(\alpha) e^{j2\pi f_D \Delta t \cos(\alpha+\alpha_0-\gamma)} d\alpha \tag{9}$$

- the spatial-temporal correlation function is denoted by $C_h(\Delta t, d_t, d_r, 0)$

$$C_h(\Delta t, d_t, d_r, 0) = \int_{\alpha,\beta} f_\alpha(\alpha) f_\beta(\beta) e^{j2\pi d_t \cos(\beta+\beta_0)} e^{j2\pi\left(d_r \cos(\alpha+\alpha_0) + f_D \Delta t \cos(\alpha+\alpha_0-\gamma)\right)} d\alpha d\beta \tag{10}$$

A distant scattering cluster causes the AOA and AOD to vary over a small angular range. This motivates us to approximate the CSOS in Eq. (8) and allows us to study its characteristics in a small angular range. Using the Taylor expansion[2], for all angles $\alpha$ and $\beta$ close to zero, the following approximate trigonometric identities are obtained,

$$\begin{aligned} \cos(\alpha+\alpha_0) &= \cos(\alpha)\cos(\alpha_0) - \sin(\alpha)\sin(\alpha_0) \\ &\approx \cos(\alpha_0) - \alpha \sin(\alpha_0) \\ \cos(\beta+\beta_0) &\approx \cos(\beta_0) - \beta \sin(\beta_0) \\ \cos(\alpha+\alpha_0-\gamma) &\approx \cos(\alpha_0-\gamma) - \alpha \sin(\alpha_0-\gamma) \end{aligned} \tag{11}$$

Substituting Eq. (11) into Eq. (8), an approximate channel spatial-temporal-spectral correlation function is obtained. This is the first time to approximate this expression. The notation $\overline{C}_h\left(\Delta t, d_t, d_r, d_f\right)$ is used to represent this approximation,

$$\begin{aligned} C_h\left(\Delta t, d_t, d_r, d_f\right) &\approx \overline{C}_h\left(\Delta t, d_t, d_r, d_f\right) \\ &= \int_{\alpha,\beta,\tau} f_\alpha(\alpha) f_\beta(\beta) f_\tau(\tau) e^{j2\pi d_t \cos(\beta_0)} \\ &\quad \times e^{-j2\pi d_t \sin(\beta_0)\beta} e^{j2\pi\left(d_r \cos(\alpha_0) + f_D \Delta t \cos(\alpha_0-\gamma)\right)} \\ &\quad \times e^{-j2\pi\left(d_r \sin(\alpha_0) + f_D \Delta t \sin(\alpha_0-\gamma)\right)\alpha} e^{-j2\pi d_f \tau} d\alpha d\beta d\tau \end{aligned} \tag{12}$$

---

[2] For small $\epsilon$, $\cos(\epsilon) = 1 + O(\epsilon^2)$ and $\sin(\epsilon) = \epsilon + O(\epsilon^3)$.

Considering the channel spatial-temporal correlation function $\overline{C}_h(\Delta t, d_t, d_r, 0)$, the separation of antenna spacing and motion into disjoint parts is an essential step to model a MIMO channel with a state-space representation.

However, the Cauchy angular distribution-based analytical solution contains an absolute sum of terms in the exponent related to antenna spacing and movement, in which the sign of the absolute value needs to be removed, while the Gaussian angular distribution-based solution has a cross-term that is related to the antenna spacing and motion, which can neither be classified as channel temporal dynamics nor as spatial correlation [12, 13].

To separate antenna spacing and motion (channel dynamics) while avoiding errors caused by unnecessary further approximations [12, 13], a linear transformation is introduced to handle this separation. Since this linear transformation eventually affects the phase of the CSOS, it is called the phase-shift method.

## 4. Linear transformation

Mathematically, the linear transformation approach implies converting the current Cartesian system to another system. In the new system, the antenna spacing and movement can be separated into error-free disjoint parts, and the channel characteristics can then be modeled using a state-space representation. Finally, an inverse linear transformation is performed to convert the channel properties back to and represent them in the original system.

To study the channel correlation properties caused by distant scattering clusters, the correlation related to the MS was approximated by the AOA near zero degrees around the angles $\alpha_0$ and $\alpha_0 - \gamma$, as expressed in Eq. (12).

As depicted in **Figure 5**, this approximation means decomposing the movement and antenna spacing into a phase change on OA and a damping change

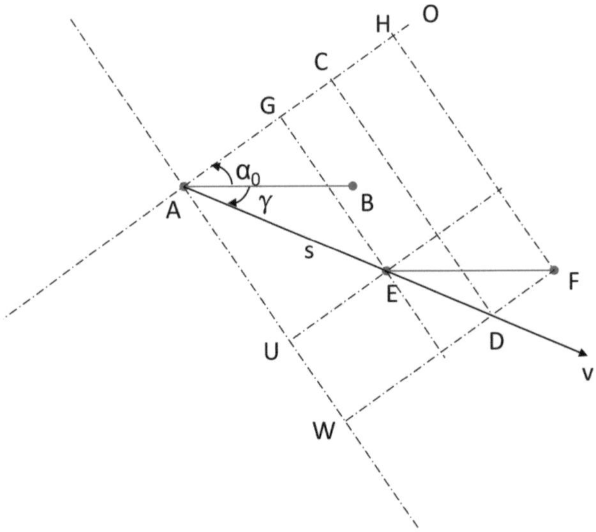


**Figure 5.**
*The motion vector is decomposed into a phase change on OA and a damping change on AW, where AB and EF denote the antenna arrays, and a, B, E, F are antenna sensors.*

on AW in accordance with the moving direction, in which the lines OA and AW are orthogonal.

Let AG $= d_{\mathrm{AG}}$, GH $= d_{\mathrm{GH}}$, AU $= d_{\mathrm{AU}}$, and UW $= d_{\mathrm{UW}}$, then,

$$\begin{aligned} d_{\mathrm{AG}} + d_{\mathrm{GH}} &= s\cos(\alpha_0 - \gamma) + d_r\cos(\alpha_0) \\ d_{\mathrm{AU}} + d_{\mathrm{UW}} &= s\sin(\alpha_0 - \gamma) + d_r\sin(\alpha_0) \end{aligned} \tag{13}$$

Geometrically, Eq. (13) interprets the meaning of the approximate expression in Eq. (12). Alternatively, AW can be considered as the result of AD projection.

Let AD $= d_{\mathrm{AD}} = \kappa$, then the right triangle relationship shows that,

$$\kappa = \frac{s\sin(\alpha_0 - \gamma) + d_r\sin(\alpha_0)}{\cos(90^o - \alpha_0 + \gamma)} = s + d_r\frac{\sin(\alpha_0)}{\sin(\alpha_0 - \gamma)} \tag{14}$$

This can be regarded as that antenna A moves to D, but its real position is at E. Hence, the changed phase will cause Eq. (12) to become

$$\begin{aligned} \overline{C}_h^{\kappa}(\Delta t_\kappa, d_t, d_r, 0) = &\ e^{-j2\pi d_r \sin(\gamma)/\sin(\alpha_0 - \gamma)} e^{j2\pi d_t \cos(\beta_0)} e^{j2\pi f_D \Delta t_\kappa \cos(\alpha_0 - \gamma)} \\ &\times \int_{\alpha,\beta} f_\alpha(\alpha) f_\beta(\beta) e^{-j2\pi d_t \sin(\beta_0)\beta} e^{-j2\pi f_D \Delta t_\kappa \sin(\alpha_0 - \gamma)\alpha} d\alpha d\beta \end{aligned} \tag{15}$$

in the new system.

Obviously, in the new system, the spatial correlation of the MS-related channels is represented by a phase rotation. In this way, we do separate the movement and antenna spacing into disjoint parts.

Moreover, this phase rotation is not related to the antennas at the BS. Thus, the Kronecker product can be used to construct the state-space MIMO channels [17].

The phase-shift approach provides an alternative way to study the same problem. By changing variables, an error-free and simple method is found to separate the movement and antenna spacing, in which the channel spatial-temporal correlation function can be regarded as the product of the phase rotation and channel temporal dynamics shifted by a value along the moving direction.[3]

## 5. Geometry-based approach

In this section, two APDFs are presented. They are obtained based on Cauchy-Rayleigh and Rayleigh clusters. This geometric approach provides an intuitive way to map scattering objects in a cluster as an angular distribution of power.

### 5.1 Cauchy-Rayleigh cluster

Given a distant Cauchy-Rayleigh cluster, the Cauchy APDFs are obtained according to the geometric relations shown in **Figure 4** [12],

$$f_\alpha(\alpha) \approx f_\alpha^c(\alpha) = \frac{1}{\pi}\frac{\eta_r}{\eta_r^2 + \alpha^2},\ f_\beta(\beta) \approx f_\beta^c(\beta) = \frac{1}{\pi}\frac{\eta_t}{\eta_t^2 + \beta^2} \tag{16}$$

[3] It can also be considered as the time delay of the channel dynamics.

where $[\cdot]^c$ indicates that the APDF is obtained in terms of the Cauchy-Rayleigh cluster, the parameters, $\eta_t = \zeta/d_{OB_1}$ and $\eta_r = \zeta/d_{OM_1}, d_{OB_1} = OB_1, d_{OM_1} = OM_1$, are used to control the angular width of these two distributions, respectively. $\zeta > 0$ is the dispersion of the Cauchy-Rayleigh distribution.

Obviously, both $f^c_\alpha(\alpha)$ and $f^c_\beta(\beta)$ in Eq. (16) are defined on $[-\pi, \ \pi]$, and they are not proper Cauchy angular power density functions because the Cauchy probability density function is defined over the infinite interval $(-\infty, \ \infty)$. Hence, it is necessary to extend the integral from $[-\pi, \ \pi]$ to $(-\infty, \ \infty)$ to use them to participate in the calculation of the integral.

Since $f^c_\alpha(\alpha)$ and $f^c_\beta(\beta)$ in Eq. (16) represent the angular powers of the MS and BS, which are similar, the discussion will focus only on the AOD. The same conclusions can be obtained for AOA.

Clearly, $f^c_\beta(\beta)$ is truncated tails, which lead to

$$\int_{-\pi}^{\pi} f^c_\beta(\beta) d\beta \lesssim 1 \tag{17}$$

However, if the intervals $(-\infty, \ -\pi)$ and $(\pi, \ \infty)$ contain much less power, then $f_\beta(\beta)$ in Eq. (16) is a suitable approximation.

Assuming that in the interval $\left[-\beta_{y\%}, \ \beta_{y\%}\right], P^c_\beta$ contains $y\%$ power, then the following equation describes the relationship among the critical angle, the power, and the width of the distribution,

$$P^c_\beta = \int_{-\beta_{y\%}}^{\beta_{y\%}} f^c_\beta(\beta) d\beta = \frac{2}{\pi} \arctan\left(\frac{\beta_{y\%}}{\eta_t}\right) \tag{18}$$

i.e., $\beta_{y\%} = \tan\left(\pi P^c_\beta/2\right)\eta_t$. Assuming $P^c_\beta = 90\%$, then $\beta_{90\%} = 6.3138\eta_t$. Similarly, $\alpha_{90\%} = 6.3138\eta_r$ for $P^c_\alpha = 90\%$.

Moreover, 90% of power within the angular interval $2\beta_{90\%}$ means that the intervals $(-\infty, -\beta_{90\%})$ and $(\beta_{90\%}, \infty)$ contain at most 10% of the transmitted power. With this in mind, the possibility of extending the angular interval from $[-\pi, \ \pi]$ to $(-\infty, \ \infty)$ is explored next.

Based on the formula,

$$P^c(\eta_t) = \int_{-\pi}^{\pi} f^c_\beta(\beta) d\beta = \frac{2}{\pi} \arctan\left(\frac{\pi}{\eta_t}\right) \tag{19}$$

and $\beta_{90\%} = 6.3138\eta_t$, the following table is obtained.

**Table 1** indicates that for each critical angle $\beta_{90\%}$, the interval $[-\pi, \pi]$ contains almost all the power contributed from the cluster. Therefore,

$$\frac{2}{\pi} \arctan\left(\frac{\pi}{\eta_t}\right) = \int_{-\pi}^{\pi} f^c_\beta(\beta) d\beta \approx \int_{-\infty}^{\infty} f^c_\beta(\beta) d\beta = 1 \tag{20}$$

Moreover, the assumption of a small angular range with most of the power will help one to obtain the following characteristic function as well,

$$\Phi^c_\beta(\omega) = \int_{-\infty}^{\infty} f^c_\beta(\beta) e^{-j\omega\beta} d\beta \approx \int_{-\pi}^{\pi} f^c_\beta(\beta) e^{-j\omega\beta} d\beta \tag{21}$$

| $\beta_{90\%}$ | $1^o$ | $5^o$ | $10^o$ | $15^o$ | $20^o$ | $30^o$ |
|---|---|---|---|---|---|---|
| $\eta_t$ | 0.003 | 0.0134 | 0.028 | 0.042 | 0.055 | 0.083 |
| $P^c(\eta_t)$ | 0.999 | 0.997 | 0.994 | 0.991 | 0.989 | 0.983 |
| Residue | 0.1% | 0.3% | 0.6% | 0.9% | 1.1% | 1.7% |

**Table 1.**
*The widths $\eta_t$ and the corresponding powers.*

Eq. (21) indicates that if some power is left out in one domain, then the same small amount will be missing in the other.

Therefore, Eq. (21) can be used to solve the integrals in Eq. (15) as

$$\Phi^c_\beta(2\pi d_t \sin(\beta_0)) \approx \int_\beta f^c_\beta(\beta) e^{-j2\pi d_t \sin(\beta_0)\beta} d\beta \tag{22}$$

which is known,

$$\begin{aligned} \tilde{C}_h^{c_\kappa}(\Delta t_\kappa, d_t, d_r, 0) = \; & e^{-2\pi\eta_t d_t |\sin(\beta_0)|} e^{j2\pi d_t \cos(\beta_0)} e^{-j2\pi d_r \sin(\gamma)/\sin(\alpha_0-\gamma)} \\ & \times e^{-2\pi\eta_r f_D \Delta t_\kappa |\sin(\alpha_0-\gamma)|} e^{j2\pi f_D \Delta t_\kappa \cos(\alpha_0-\gamma)} \end{aligned} \tag{23}$$

and the channel dynamic function,

$$\tilde{R}_h^{c_\kappa}(\Delta t_\kappa) = \tilde{C}_h^{c_\kappa}(\Delta t_\kappa, 0,0,0) = e^{-2\pi\eta_r f_D \Delta t_\kappa |\sin(\alpha_0-\gamma)|} e^{j2\pi f_D \Delta t_\kappa \cos(\alpha_0-\gamma)} \tag{24}$$

According to Eq. (23), a specific expression of each element of $\mathbf{R}_{\mathrm{BS}}$ in Eq. (4) is assigned,

$$r^{c,\mathrm{BS}}_{m,n}(d_t) = e^{-2\pi\eta_t d_t |\sin(\beta_0)|} e^{j2\pi d_t \cos(\beta_0)} \tag{25}$$

and the notation $\mathbf{R}^c_{\mathrm{BS}}$ is to replace $\mathbf{R}_{\mathrm{BS}}$. Furthermore, all elements of $\mathbf{R}_{\mathrm{MS}}$ in Eq. (3) will have the following specific expression,

$$r^{\mathrm{MS}}_{i,j}(d_r) = e^{-j2\pi d_r \sin(\gamma)/\sin(\alpha_0-\gamma)} \tag{26}$$

Eq. (26) indicates that the spatial correlation between MS channels will depend only on the antenna spacing $d_r$ but not on the cluster type. Thus, the notation $\mathbf{R}_{\mathrm{MS}}$ will be kept.

### 5.2 Rayleigh cluster

Similarly, given a distant Rayleigh cluster, the following approximate Gaussian APDFs are derived [13],

$$f^r_\alpha(\alpha) = \frac{1}{\sqrt{2\pi}\sigma_r} e^{-\frac{\alpha^2}{2\sigma_r^2}}, \quad f^r_\beta(\beta) = \frac{1}{\sqrt{2\pi}\sigma_t} e^{-\frac{\beta^2}{2\sigma_t^2}} \tag{27}$$

where $[\cdot]^r$ indicates that the APDF is obtained from the Rayleigh cluster, $\sigma_t = \sigma/d_{\mathrm{OB}_1}, \sigma_r = \sigma/d_{\mathrm{OM}_1},$ and $\sigma$ is obtained from the Rayleigh distribution.

As described in the previous section, the truncated Gaussian APDFs can also be extended from $\pi$ to infinity, and the analytical solution of the CSOS, denoted by the notation $\tilde{C}_h^{r_\kappa}(\Delta t_\kappa, d_t, d_r, 0)$, is obtained by substituting Eq. (27) into Eq. (15) [13],

$$\begin{aligned}\tilde{C}_h^{r_\kappa}(\Delta t_\kappa, d_t, d_r, 0) = \; & e^{-2\pi^2\sigma_t^2 d_t^2 \sin^2(\beta_0)} e^{j2\pi d_t \cos(\beta_0)} e^{-j2\pi d_r \sin(\gamma)/\sin(\alpha_0-\gamma)} \\ & \times e^{-2\pi^2\sigma_r^2 \sin^2(\alpha_0-\gamma) f_D^2 \Delta t_\kappa^2} e^{j2\pi \cos(\alpha_0-\gamma) f_D \Delta t_\kappa}\end{aligned} \tag{28}$$

and the channel temporal dynamic function is

$$\tilde{R}_h^{r_\kappa}(\Delta t_\kappa) = e^{-2\pi^2\sigma_r^2 \sin^2(\alpha_0-\gamma) f_D^2 \Delta t_\kappa^2} e^{j2\pi \cos(\alpha_0-\gamma) f_D \Delta t_\kappa} \tag{29}$$

Thus, each element of $\mathbf{R}_{\mathrm{BS}}^r$ can be given by

$$r_{m,n}^{r,\mathrm{BS}}(d_t) = e^{-2\pi^2\sigma_t^2 d_t^2 \sin^2(\beta_0)} e^{j2\pi d_t \cos(\beta_0)} \tag{30}$$

and all elements of $\mathbf{R}_{\mathrm{MS}}^r$ are also given by Eq. (26).

## 6. AR-based state-space channel model

In the previous sections, two types of scattering clusters were introduced to obtain the analytical solutions of the CSOS. These two analytical solutions were decomposed into the product of channel temporal dynamics and spatial correlation. In this section, the channel temporal dynamics will be approximated as an AR(p) model, by which, a state-space MIMO channel model can be constructed for fitting the channel spatial-temporal correlation function.

A state-space model describes a dynamic system associated with the input, state variables, and output. The system input and output are linked by a state vector which is determined by a state transition matrix, and the last variable in the state vector will be the contribution from the cluster to the channel.

That is, for each scattering cluster, an AR(p) model is used to describe the MIMO radio channel temporal dynamics, and the Kronecker correlation matrix is employed to characterize the channel spatial correlation.

A coloring matrix is used to drive input Gaussian noise innovations to create channel spatial correlations, and the coloring matrix is determined by the channel correlation properties of the BS and MS described by the Kronecker product.

### 6.1 AR(p) model

An AR(p) model specifies that the output variable depends linearly on its previous values. It is a very ordinary model and has a wide variety of applications in time series. One of the significant features of an AR(p) model is that it can be transformed into a state-space representation. Therefore, a large number of approaches in the control domain can potentially be applied to MIMO channel modeling and be used to study radio channels.

An AR(p) model can be represented by [18].

$$x_k = \sum_{i=1}^{p} \phi_i x_{k-i} + w_k \tag{31}$$

where $\phi_1, \cdots, \phi_p \left(\phi_p \neq 0\right)$ are complex coefficients and $w_k$ is a complex Gaussian sequence $\mathbb{CN}(0; \sigma_w^2)$. That is, the stochastic variable $x_k$ is defined as a linear combination of its previous p values of the series plus an innovation noise.

In this section, Eq. (24) will be described by an AR(1) model, while Eq. (29) will be approximated by an AR(3) model. Since Eq. (24) is itself an AR(1) model, its coefficient $\phi$ and standard variance $\sigma_{\text{AR}(1)}$ can be obtained directly from equations [12, 19]. The coefficient $\phi_i$ and standard variance $\sigma_{\text{AR}(3)}$ of an AR(3) model can be estimated using the least-squares (LS) method [20] or computed using the spectral-equivalent (SE) method [13].

In this way, a single peak on the Doppler spectrum corresponding to the contribution from a distant scattering cluster is modeled by an AR(p) model.[4] The advantage of using an AR(p) model is that it can be directly parameterized by the properties of the cluster, and it allows changing the angles in the simulation, which corresponds to changing the directions of the mobile receiver.

### 6.2 SISO channel model

The AR(p) model given by Eq. (31) can be transformed into the controllable canonical form [18, 21, 22] to obtain a state-space representation,

$$\begin{aligned} \mathbf{x}_{k+1} &= \mathbf{A}\mathbf{x}_k + \mathbf{B}w_k \\ h_k &= \mathbf{C}\mathbf{x}_k \end{aligned} \tag{32}$$

where $\mathbf{B} = [0 \;\; 0 \cdots \;\; 0 \;\; 1]^T$ is a $p \times 1$, $\mathbf{C} = [0 \;\; 0 \;\; \cdots \;\; 0 \;\; 1]$ is a $1 \times p$ vector, the output $h_k = x_k$ is a scalar, the channel, and

$$\begin{aligned} \mathbf{x}_{k+1} &= \begin{bmatrix} x_{k-p+1} \\ x_{k-p+2} \\ \vdots \\ x_k \\ x_{k+1} \end{bmatrix}, \mathbf{A} = \begin{bmatrix} 0 & 1 & 0 & \cdots & 0 \\ 0 & 0 & 1 & \cdots & 0 \\ \vdots & \vdots & \vdots & \ddots & \vdots \\ 0 & 0 & 0 & \cdots & 1 \\ \phi_p & \phi_{p-1} & \phi_{p-2} & \cdots & \phi_1 \end{bmatrix} \\ \mathbf{x}_k &= \begin{bmatrix} x_{k-p} & x_{k-p+1} & \cdots & x_{k-1} & x_k \end{bmatrix}^T \end{aligned} \tag{33}$$

The input vector $\mathbf{B} = \begin{bmatrix} 0 & 0 & \cdots & 0 & \sigma_{\text{AR(p)}} \end{bmatrix}^T$ is redefined, i.e., the noise input will be scaled by $\sigma_{\text{AR(p)}}$, then the variance of $x_k$ is 1, i.e., $\sigma_x^2 = 1$ [13]. It makes a lot of sense to let $x_k$ have unit variance before $\mathbf{C}$ and let $\mathbf{C}$ scale be the contribution from a cluster, including path loss to $h_k$.

It must be noted that, in reality, the matrices $\mathbf{A}$ and $\mathbf{B}$ are time-variant because both of them have angle-dependent elements. The angle $\alpha_0 - \gamma$ is used to describe the moving direction to the cluster center, which may change all the time during the movement.

However, these matrices are assumed to be time-invariant due to small movements compared with the distance from the MS to the center of a scattering cluster. That is, within some time slots, all matrices are approximately constant. This implies that

---

4 The channel dynamics due to the Cauchy-Rayleigh clusters is modeled as an AR(1) model and it approximately represents the channel dynamics due to the Rayleigh clusters by an AR(3) model.

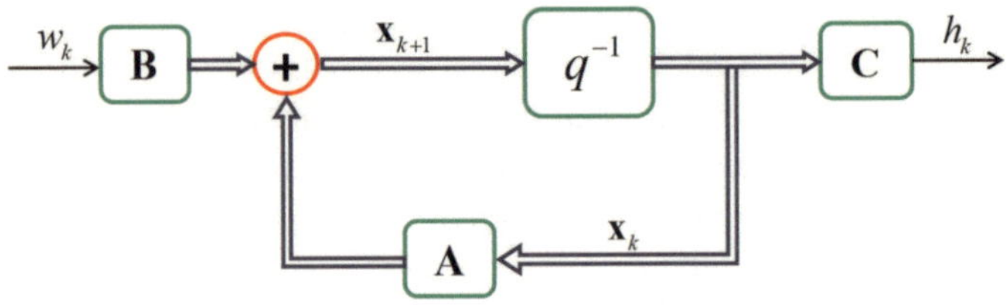


**Figure 6.**
*Block diagram of the AR(p)-based state-space SISO channel model.*

constant angles toward clusters, constant speed during the movement, and hence a time-invariant environment is satisfied. This assumption is related to the stationarity of $\mathbf{x}_k$ and $\mathbf{h}_k$ sequences as well.

The block diagram corresponding to the singe-input and single-output (SISO) channel model in Eq. (32) is shown below,

**Figure 6** is also known as the AR(p)-based state-space SISO channel model block. In the block diagram, the inputs and outputs are scalars, described by a single line, and double lines are used to represent vectors.

This is the simplest state-space model used to describe the channel temporal dynamics and will be employed to construct state-space single-input and multiple-output (SIMO) and MIMO channel models.

### 6.3 SIMO channel model

Based on the SISO channel model block shown in **Figure 6**, a state-space SIMO channel model is constructed by connecting multiple SISO channel model blocks in parallel, in which a correlated innovation process is employed to adjust the spatial correlation between these SISO channel blocks, the SIMO channels, as shown in **Figure 7**. This can be done by introducing $\mathbf{\Phi}_{\mathrm{M_r}}$, an $M_r \times M_r$ coloring matrix. The number of SISO channel model blocks required for the SIMO channels will depend on the number of receiving antenna elements $M_r$.

Mathematically, this parallel connection can be interpreted as the following state-space representation,

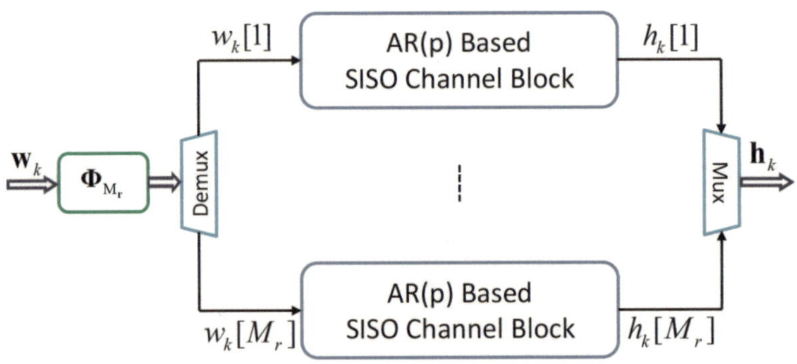


**Figure 7.**
*Block diagram of the AR(p)-based state-space SIMO channel model.*

$$\begin{array}{rcl} \mathbf{x}_{k+1} & = & \boldsymbol{\Gamma}^{simo}\mathbf{x}_k + \boldsymbol{\Psi}^{simo}\mathbf{w}_k \\ \mathbf{h}_k^{simo} & = & \boldsymbol{\Omega}^{simo}\mathbf{x}_k \end{array} \tag{34}$$

where the state vector $\mathbf{x}_k \in \mathbb{C}^{\mathrm{pM_r}}$, $\mathbf{w}_k \sim \mathcal{CN}\left(\mathbf{0}; \sigma_w^2 \mathbf{I}_{\mathrm{M_r}}\right) \in \mathbb{C}^{\mathrm{M_r}}$ are independent and identically distributed (i.i.d), the channel vector and the driving noise vector are expressed as

$$\begin{array}{rcl} \mathbf{h}_k^{simo} & = & [h_k[1] \;\; h_k[2] \;\; \cdots \;\; h_k[M_r]]^T \in \mathbb{C}^{\mathrm{M_r}} \\ \mathbf{w}_k & = & [w_k[1] \;\; w_k[2] \;\; \cdots \;\; w_k[M_r]]^T \in \mathbb{C}^{\mathrm{M_r}} \end{array} \tag{35}$$

Moreover, the matrices $\boldsymbol{\Gamma}^{simo}$, $\boldsymbol{\Psi}^{simo}$, and $\boldsymbol{\Omega}^{simo}$ are defined by

$$\begin{array}{rcl} \boldsymbol{\Gamma}^{simo} & = & \mathbf{I}_{\mathrm{M_r}} \otimes \mathbf{A} = \begin{bmatrix} \mathbf{A} & 0 & \cdots & 0 \\ 0 & \mathbf{A} & \cdots & 0 \\ \vdots & \vdots & \ddots & \vdots \\ 0 & 0 & \cdots & \mathbf{A} \end{bmatrix}_{\mathrm{pM_r \times pM_r}} \\ \boldsymbol{\Psi}^{simo} & = & (\mathbf{I}_{\mathrm{M_r}} \otimes \mathbf{B})\boldsymbol{\Phi}_{\mathrm{M_r}} = \begin{bmatrix} \mathbf{B} & 0 & \cdots & 0 \\ 0 & \mathbf{B} & \cdots & 0 \\ \vdots & \vdots & \ddots & \vdots \\ 0 & 0 & \cdots & \mathbf{B} \end{bmatrix}_{\mathrm{pM_r \times M_r}} \boldsymbol{\Phi}_{\mathrm{M_r}} \\ \boldsymbol{\Omega}^{simo} & = & \mathbf{I}_{\mathrm{M_r}} \otimes \mathbf{C} = \begin{bmatrix} \mathbf{C} & 0 & \cdots & 0 \\ 0 & \mathbf{C} & \cdots & 0 \\ \vdots & \vdots & \ddots & \vdots \\ 0 & 0 & \cdots & \mathbf{C} \end{bmatrix}_{\mathrm{M_r \times pM_r}} \end{array} \tag{36}$$

where $\mathbf{I}_{\mathrm{M_r}}$ is the identity matrix of size $M_r$, $\boldsymbol{\Psi}^{simo}$ is a $pM_r \times M_r$ matrix, and $\boldsymbol{\Phi}_{\mathrm{M_r}}$ is the coloring matrix employed to control the spatial correlation properties between the channels.

To acquire $\boldsymbol{\Phi}_{\mathrm{M_r}}$, we need to study the system output in Eq. (34). By definition, the auto-covariance matrix of channels $\mathbf{R}_h$ equals $E\left[\mathbf{h}_k^{simo}\,\mathbf{h}_k^{simo^H}\right]$. Simple algebra gives $\mathbf{R}_h = \boldsymbol{\Phi}_{\mathrm{M_r}}\boldsymbol{\Phi}_{\mathrm{M_r}}^H = \mathbf{R}_{\mathrm{MS}}$.

The Cholesky decomposition method can be employed to solve the equation $\boldsymbol{\Phi}_{\mathrm{M_r}}\boldsymbol{\Phi}_{\mathrm{M_r}}^H = \mathbf{R}_{\mathrm{MS}}$ numerically. This results in a lower triangular matrix with strictly positive diagonal entries. For small $M_r$, however, a lower triangular matrix $\boldsymbol{\Phi}_{\mathrm{M_r}}$ can be found analytically [12]. Therefore, it significantly reduces the computational complexity of getting all the elements in a closed-form representation.

The key idea of modeling the SIMO channel using the state-space representation is the modular approach, i.e., just add the required number of SISO channel blocks to form another bigger block called an AR(p)-based state-space SIMO channel model block. This constructed block has an i.i.d. Gaussian noise input vector and a correlated output vector, i.e., the SIMO channels.

## 6.4 MIMO channel model

To build an AR(p)-based state-space MIMO channel model, the spatial correlation properties at the BS will be added. Thus, based on the Kronecker matrix given in Eq. (6), a correlated innovation matrix, a coloring matrix, is employed to characterize the spatial correlation of the channels.

Similar to modeling the state-space SIMO channel model, a state-space MIMO channel model is constructed by connecting multiple SIMO channel blocks in parallel, as **Figure 8** illustrates, in which $\mathbf{\Phi}_{\mathrm{M_rM_t}}$ is the coloring matrix, and the number of SIMO channel blocks needed for the MIMO channels will depend on the number of transmitting antenna elements $M_t$.

Mathematically, this block diagram can be implemented as the following state-space representation,

$$\begin{aligned} \mathbf{x}_{k+1} &= \mathbf{\Gamma}^{mimo}\mathbf{x}_k + \mathbf{\Psi}^{mimo}\mathbf{w}_k \\ \mathbf{h}_k^{mimo} &= \mathbf{\Omega}^{mimo}\mathbf{x}_k \end{aligned} \tag{37}$$

where $\mathbf{x}_k \in \mathbb{C}^{\mathrm{pM_rM_t}}$, $\mathbf{w}_k \sim \mathcal{CN}(0;1) \in \mathbb{C}^{\mathrm{M_rM_t}}$ and $\mathbf{h}_k^{mimo} = \mathrm{vec}(\mathbf{H}_k)$, here

$$\mathbf{H}_k = \begin{bmatrix} h_k[1,1] & h_k[1,2] & \cdots & h_k[1,M_t] \\ h_k[2,1] & h_k[2,2] & \cdots & h_k[2,M_t] \\ \vdots & \vdots & \ddots & \vdots \\ h_k[M_r,1] & h_k[M_r,2] & \cdots & h_k[M_r,M_t] \end{bmatrix} \tag{38}$$

and $\mathbf{\Psi}^{mimo}, \mathbf{\Omega}^{mimo}, \mathbf{\Gamma}^{mimo}$ are defined by

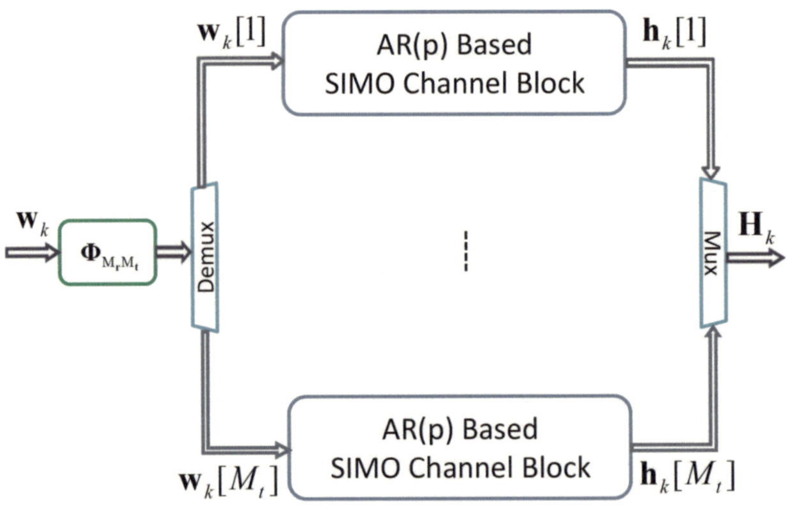


**Figure 8.**
*Block diagram of the AR(p)-based state-space MIMO channel model.*

$$\boldsymbol{\Gamma}^{mimo} = \mathbf{I}_{\mathrm{M_rM_t}} \otimes \mathbf{A} = \begin{bmatrix} \mathbf{A} & 0 & \cdots & 0 \\ 0 & \mathbf{A} & \cdots & 0 \\ \vdots & \vdots & \ddots & \vdots \\ 0 & 0 & \cdots & \mathbf{A} \end{bmatrix}_{\mathrm{pM_rM_t \times pM_rM_t}}$$

$$\boldsymbol{\Psi}^{mimo} = (\mathbf{I}_{\mathrm{M_rM_t}} \otimes \mathbf{B})\boldsymbol{\Phi}_{\mathrm{M_rM_t}} = \begin{bmatrix} \mathbf{B} & 0 & \cdots & 0 \\ 0 & \mathbf{B} & \cdots & 0 \\ \vdots & \vdots & \ddots & \vdots \\ 0 & 0 & \cdots & \mathbf{B} \end{bmatrix} \boldsymbol{\Phi}_{\mathrm{M_rM_t}} \tag{39}$$

$$\boldsymbol{\Omega}^{mimo} = \mathbf{I}_{\mathrm{M_rM_t}} \otimes \mathbf{C} = \begin{bmatrix} \mathbf{C} & 0 & \cdots & 0 \\ 0 & \mathbf{C} & \cdots & 0 \\ \vdots & \vdots & \ddots & \vdots \\ 0 & 0 & \cdots & \mathbf{C} \end{bmatrix}_{\mathrm{M_rM_t \times pM_rM_t}}$$

where $\boldsymbol{\Phi}_{\mathrm{M_rM_t}}$ is defined as a lower triangular matrix, which fulfills the condition,

$$\boldsymbol{\Phi}_{\mathrm{M_rM_t}}\boldsymbol{\Phi}_{\mathrm{M_rM_t}}^{H} = \mathbf{R}_{\mathrm{MIMO}} = \mathbf{R}_{\mathrm{BS}} \otimes \mathbf{R}_{\mathrm{MS}} \tag{40}$$

Similarly, the Cholesky decomposition method can be used to solve Eq. (40) numerically. However, for a small size matrix $\boldsymbol{\Phi}_{\mathrm{M_rM_t}}$, like a $2 \times 2$ MIMO channel model, an analytical solution of a lower triangular matrix $\boldsymbol{\Phi}_4$ is obtained [13, 19].

## 7. MIMO-OFDM channel model

The demand for multimedia services requires high data rates for communications. However, in a single-carrier modulation system, this is limited by inter-symbol interference, which occurs due to time dispersion of channel caused by multi-path propagation [23, 24]. A multi-carrier modulation technique, OFDM, is proposed to overcome this problem. That is, OFDM is employed to the channels that exhibit a time delay spread, or equivalently, have the characteristic of frequency selectivity.

Notice that the MIMO channel model presented earlier is used for narrow-band and single-carrier frequency. In this section, as a promising strategy, a combination of MIMO and OFDM technology is proposed to deal with the frequency-selective fading channels, i.e., a wide-band MIMO channel model and a MIMO-OFDM channel model.

To this end, the so-called time delay factor is introduced to describe the delay spread due to the two-dimensional (2D) scattering clusters, which will focus on the spatial-temporal-spectral correlation properties of the channel and not only on the spatial-temporal correlation characteristics of the channel.

### 7.1 Spectral correlation matrix

Let us define the elements of the channel spectral correlation matrix below,

$$r_f\left(d_f\right) = \int_{\tau} f_{\tau}(\tau) e^{-j2\pi d_f \tau} d\tau \tag{41}$$

then the spectral correlation matrix of size $M_f \times M_f$ can be represented by the sequence $r_f[m_f]$,

$$\mathbf{C}_f = \begin{bmatrix} r_f[0] & r_f[1] & \cdots & r_f[M_f-1] \\ r_f^*[1] & r_f[0] & \cdots & r_f[M_f-2] \\ \vdots & \vdots & \ddots & \vdots \\ r_f^*[M_f-1] & r_f^*[M_f-2] & \cdots & r_f[0] \end{bmatrix} \quad (42)$$

where the diagonal element $r_f[0] = \int_\tau f_\tau(\tau)d\tau = 1$. The above matrix will be used to derive the coloring matrix for the MIMO-OFDM channels.

## 7.2 Building a MIMO-OFDM channel model

Similarly, based on the MIMO channel model block, a MIMO-OFDM channel model can be constructed. This time, however, a colored input noise vector for the MIMO-OFDM channels is generated using the spectral correlation matrix $\mathbf{C}_f$.

The vector $\left[\mathbf{h}(t,f_0)^T \ \mathbf{h}(t,f_1)^T \ \cdots \ \mathbf{h}\left(t,f_{M_f-1}\right)^T\right]^T$ is used to represent all of the MIMO-OFDM channels. This results in the channels that are characterized by a spatial-temporal-spectral correlation function.

In **Figure 9**, $\mathbf{h}[k, i]$ is a discretized representation of the continuous-time channel vector $\mathbf{h}(k\Delta t,f_i)$, each dotted box represents a MIMO channel model, which includes a total of $M_rM_t$ state-space SISO channel blocks and one spatial correlation matrix $\mathbf{\Phi}_{\mathrm{M_rM_t}}$. Moreover, each block involves a single-carrier frequency, and this parallel

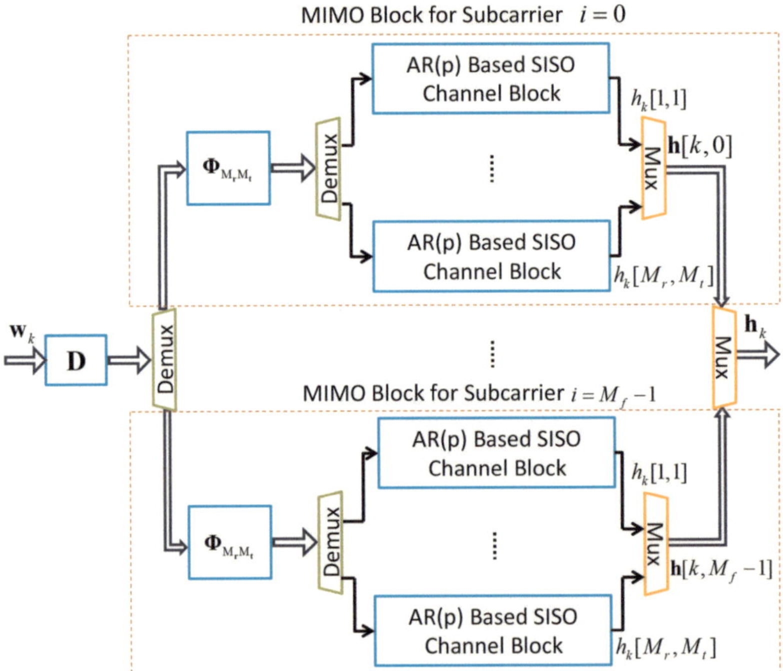


**Figure 9.**
*Block diagram of the MIMO-OFDM channel model.*

connection will generate $M_f$ frequency-selective channels. In addition, the block diagram $\mathbf{D}$ is a square matrix of order $M_f$ obtained from the spectral correlation matrix $\mathbf{C}_f$ in Eq. (42). This matrix is employed to adjust the spectral correlation properties between the MIMO channel blocks.

Mathematically, this state-space MIMO-OFDM channel model can be represented by

$$\begin{aligned} \mathbf{x}_{k+1} &= \mathbf{\Gamma}\mathbf{x}_k + \mathbf{\Psi}\mathbf{w}_k \\ \mathbf{h}_k &= \mathbf{\Omega}\mathbf{x}_k \end{aligned} \tag{43}$$

where $\left[\mathbf{h}[k,0]^T \ \ \mathbf{h}[k,1]^T \ \cdots \ \mathbf{h}\left[k, M_f-1\right]^T\right]^T \in \mathbb{C}^{\mathrm{M_f M_r M_t}}$ is denoted by $\mathbf{h}_k$, $\mathbf{x}_k \in \mathbb{C}^{\mathrm{pM_f M_r M_t}}$, $\mathbf{w}_k \in \mathbb{C}^{\mathrm{M_f M_r M_t}}$, $\mathbf{\Gamma}$ is a complex square matrix of order $pM_fM_rM_t$, $\mathbf{\Psi}$ is a $pM_fM_rM_t \times M_fM_rM_t$ complex matrix, and $\mathbf{\Omega}$ is a $M_fM_rM_t$ by $pM_fM_rM_t$ real matrix. The matrices $\mathbf{\Gamma}$, $\mathbf{\Psi}$, and $\mathbf{\Omega}$ are given by

$$\mathbf{\Gamma} = \mathbf{I}_{M_f} \otimes \mathbf{\Gamma}^{mimo}, \mathbf{\Psi} = \mathbf{D} \otimes \mathbf{\Psi}^{mimo}, \mathbf{\Omega} = \mathbf{I}_{M_f} \otimes \mathbf{\Omega}^{mimo} \tag{44}$$

where $p$ is the order of the AR model, $\mathbf{\Gamma}^{mimo}$, $\mathbf{\Psi}^{mimo}$, and $\mathbf{\Omega}^{mimo}$ are given by Eq. (39), and $\mathbf{D}$ is defined as a lower triangular matrix that satisfies,

$$\mathbf{D}\mathbf{D}^H = \mathbf{C}_f \tag{45}$$

As mentioned earlier, the Cholesky decomposition method can also be used to obtain all of the elements of the lower triangular matrix $\mathbf{D}$. However, in the case of two sub-carriers, simple algebra will result in the following closed-form solution.

$$\mathbf{D} = \begin{bmatrix} 1 & 0 \\ r_f^*\left[m_f\right] & \sqrt{1 - \left|r_f\left[m_f\right]\right|^2} \end{bmatrix} \tag{46}$$

Therefore, given a DPDF, the corresponding spatial-temporal-spectral correlation function can be obtained. This will be presented next.

### 7.3 Cauchy delay PDF

Similarly, given a distant Cauchy-Rayleigh cluster, the delay PDF of TOA is approximately equal to Cauchy [25],

$$f_\tau^c(\tau) = f_\tau(\tau) \approx \frac{1}{\pi} \frac{\eta}{\eta^2 + (\tau - \overline{\tau})^2} \tag{47}$$

where $\overline{\tau}$ denotes the average time delay,

$$\overline{\tau} = \frac{d_{\mathrm{OB_1}} + d_{\mathrm{OM_1}}}{v_c}, \qquad \eta = \frac{\zeta\sqrt{2 + 2\cos(\theta_0)}}{v_c}, \qquad \cos(\theta_0) = \frac{d_{\mathrm{OB_1}}^2 + d_{\mathrm{OM_1}}^2 - d_{\mathrm{B_1M_1}}^2}{2 d_{\mathrm{OB_1}} d_{\mathrm{OM_1}}} \tag{48}$$

and $v_c$ is the speed of light, $\theta_0$ is the angle between the two edges $\mathrm{B_1O}$ and $\mathrm{OM_1}$, as illustrated in **Figure 4**. Notice that the time delay $\tau$ is a non-negative variable. Hence, Eq. (47) is valid only if the main area under the curve is in the positive direction of the delay axis. In other words, the area under the tail in the negative direction of the delay axis is small and can be ignored.

Since $\zeta = \alpha_{90\%} d_{\mathrm{OM}_1}/6.3138$, from Eq. (48), we get,

$$\eta = \frac{\alpha_{90\%}\sqrt{2+2\cos(\theta_0)}}{6.3138}\frac{d_{\mathrm{OM}_1}}{v_c} \tag{49}$$

Notice that the ratio of $d_{\mathrm{OM}_1}$ and $v_c$ is very small, the width of the delay power distribution function $\eta$ will be a very small value. Therefore, the integration of Eq. (47) will be approximately equal to 1 over the interval $[0,\ \tau_\epsilon]$, and it can thereby be extended to $[0,\ \infty)$. Here, $\tau_\epsilon \gg \tau_{\max}$ is a number and $\tau_{\max}$ denotes the maximum delay.

Adding the spectrum $d_f$ to the expression, the following equation is obtained by substituting $f_\alpha^c(\alpha)$,$f_\beta^c(\beta)$. and $f_\tau^c(\tau)$ into Eq. (15),

$$\begin{aligned} \overline{C}_h^{\kappa}(\Delta t_\kappa, d_t, d_r, d_f) &\approx \tilde{C}_h^{c_\kappa}(\Delta t_\kappa, d_t, d_r, d_f) \\ &= \tilde{R}_h^{c_\kappa}(\Delta t_\kappa) r_{m,n}^{c,\mathrm{BS}}(d_t) r_{i,j}^{\mathrm{MS}}(d_r) r_f^c(d_f) \end{aligned} \tag{50}$$

where $\tilde{R}_h^{c_\kappa}(\Delta t_\kappa)$ is the channel dynamics given in Eq. (24), $r_{m,n}^{c,\mathrm{BS}}(d_t)$, $r_{i,j}^{\mathrm{MS}}(d_r)$ are spacing correlations given in Eqs. (25) and (26), respectively, and $r_f^c(d_f)$ is the spectral correlation given by

$$r_f^c(d_f) = e^{-2\pi\eta d_f} e^{-j2\pi d_f \bar{\tau}} \tag{51}$$

### 7.4 Gaussian delay PDF

Given a distant Rayleigh cluster, the approximate Gaussian DPDF of TOA is obtained [26],

$$f_\tau^r(\tau) = f_\tau(\tau) \approx \frac{1}{\sqrt{2\pi}\sigma_0} e^{-\frac{(\tau-\bar{\tau})^2}{2\sigma_0^2}} \tag{52}$$

where

$$\sigma_0 = \frac{\sigma\sqrt{2+2\cos(\theta_0)}}{v_c} \tag{53}$$

and $\cos(\theta_0)$ and $\bar{\tau}$ are defined in Eq. (48). Therefore, for Gaussian distributed TOA, we have,

$$\begin{aligned} \overline{C}_h^{\kappa}(\Delta t_\kappa, d_t, d_r, d_f) &\approx \tilde{C}_h^{r_\kappa}(\Delta t_\kappa, d_t, d_r, d_f) \\ &= \tilde{R}_h^{r_\kappa}(\Delta t_\kappa) r_{m,n}^{r,\mathrm{BS}}(d_t) r_{i,j}^{\mathrm{MS}}(d_r) r_f^r(d_f) \end{aligned} \tag{54}$$

where $\tilde{R}_h^{r_\kappa}(\Delta t_\kappa)$ is defined in Eq. (29), $r_{m,n}^{r,\mathrm{BS}}(d_t)$, $r_{i,j}^{\mathrm{MS}}(d_r)$ are spacing correlations defined in Eqs. (30) and (26), respectively, and $r_f^r(d_f)$ is the spectral correlation given by

$$r_f^r(d_f) = e^{-2\pi^2\sigma_0^2 d_f^2} e^{-j2\pi\bar{\tau} d_f} \tag{55}$$

Thus, an AR(p)-based state-space MIMO-OFDM channel model has been constructed. However, this approach is only applicable to a single scattering cluster. Next, the method for constructing a multi-cluster MIMO-OFDM channel model is described.

## 8. Multi-cluster MIMO-OFDM channel model

According to previous studies, Eqs. (50) and (54) are two key functions for building the MIMO-OFDM channel model based on a single scattering cluster. Combining these two types of channel models, a multi-cluster MIMO-OFDM channel model is constructed. In this way, a physical propagation environment of radio waves is reconstructed by simulations.

Considering a radio wave propagation environment with $K$ distant Cauchy-Rayleigh and Rayleigh scattering clusters, as shown in **Figure 10**, it is assumed that the BS is fixed while the MS is moving with speed $v$, and there is noline of sight (LOS) between the BS and MS, all of the signals transmitted and received are via these $K$ uncorrelated scattering clusters. Each cluster is grouped into resolvable multi-path components. Besides, within a cluster, the trigonometric relationship among the BS, scatterers, and MS has been introduced, as shown in **Figure 4**.

For this model, only a single scattering event along each path between the transmit and receive antenna arrays is considered. That is, it is assumed that the contribution to the power due to multiple scattering events is much lower and will be ignored.

In addition, the radio waves contributed from different scattering clusters can be added to obtain the contributions of all.

The power contributed from each cluster is dedicated in a portion to the Doppler power spectrum. From this point of view, under the assumption of uncorrelated scattering clusters, the summation of the radio waves can be regarded as adding up each individual portion of power. These contributions will result in a $K$-cluster MIMO-OFDM channel model if the delay factor is taken into account.

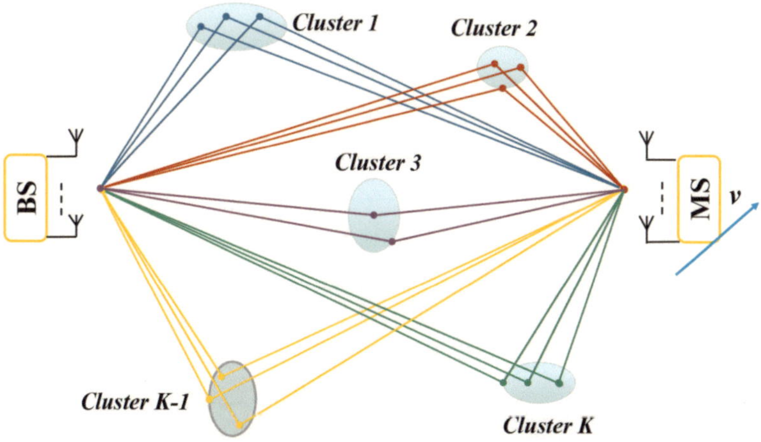


**Figure 10.**
*Multiple distant scattering clusters, cluster no. 1 to cluster no. K, in a radio wave propagation environment, in which each cluster is grouped into resolvable multi-path components.*

## 8.1 Multi-cluster angular-delay Spectrum

The joint angular-delay spectrum associated with $K$ scattering clusters can be written as

$$\begin{aligned} f_{\alpha,\tau}(\alpha,\tau) &= \frac{\sum_{k=1}^{K} P_k f_{\alpha_k,\tau_k}(\alpha_k,\tau_k)}{\sum_{k=1}^{K} P_k} \\ &= \frac{\sum_{k=1}^{K} P_k f_{\alpha_k}(\alpha_k) f_{\tau_k}(\tau_k)}{\sum_{k=1}^{K} P_k} \end{aligned} \tag{56}$$

where $P_k$ denotes the power contributed from the *kth* cluster. Taking summation over the angles, the marginal distribution represents the PDP, $f_{\tau}(\tau)$, of the clusters. The sum over the delays stands for the angular power distribution, $f_{\alpha}(\alpha)$, of the clusters.

## 8.2 Building a multi-cluster MIMO-OFDM Channel model

Connecting multiple MIMO-OFDM channel model blocks in parallel, a multi-cluster MIMO-OFDM channel model is constructed, as shown in **Figure 11**, and the number of blocks required depends on $K$.

The connection illustrated in **Figure 11** can be transformed into the following mathematical representation,

$$\begin{aligned} \mathbf{x}_{k+1} &= \mathbf{\Gamma}\mathbf{x}_k + \mathbf{\Psi}\mathbf{w}_k \\ \mathbf{h}_k &= \mathbf{\Omega}\mathbf{x}_k \end{aligned} \tag{57}$$

where $\mathbf{\Gamma}$, $\mathbf{\Psi}$, and $\mathbf{\Omega}$ are given below,

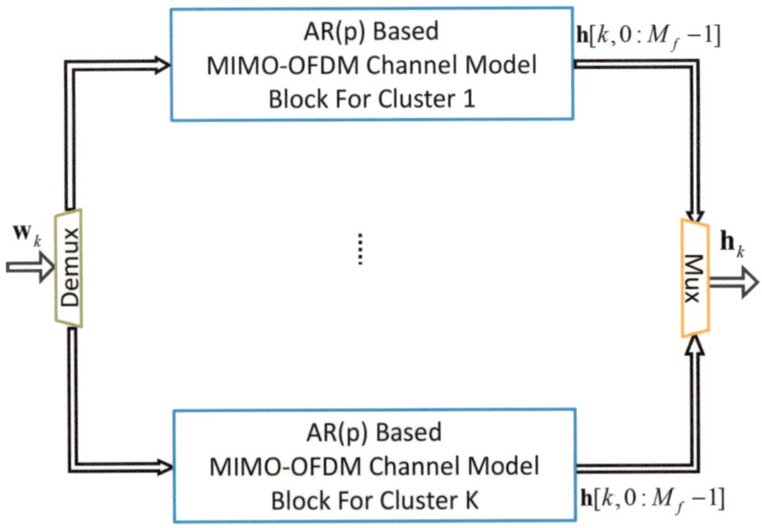


**Figure 11.**
*Block diagram of a K-cluster MIMO-OFDM channel model, where the input noise vector* $\mathbf{w}_k \in \mathbb{C}^{KM_fM_rM_t}$*, the output channel vector* $\mathbf{h}[k, 0 : M_f - 1]$ *means that there are* $M_f$ *sub-carriers from* $m_f = 0$ *to* $m_f = M_f - 1$*, and the AR(p)-based MIMO-OFDM channel model block is shown in* ***Figure 9****.*

$$\boldsymbol{\Gamma} = \begin{bmatrix} \boldsymbol{\Gamma}_1 & 0 & \cdots & 0 \\ 0 & \boldsymbol{\Gamma}_2 & \cdots & 0 \\ \vdots & \vdots & \ddots & \vdots \\ 0 & 0 & \cdots & \boldsymbol{\Gamma}_K \end{bmatrix}, \boldsymbol{\Psi} = \begin{bmatrix} \boldsymbol{\Psi}_1 & 0 & \cdots & 0 \\ 0 & \boldsymbol{\Psi}_2 & \cdots & 0 \\ \vdots & \vdots & \ddots & \vdots \\ 0 & 0 & \cdots & \boldsymbol{\Psi}_K \end{bmatrix} \\ \boldsymbol{\Omega} = \begin{bmatrix} \boldsymbol{\Omega}_1 & 0 & \cdots & 0 \\ 0 & \boldsymbol{\Omega}_2 & \cdots & 0 \\ \vdots & \vdots & \ddots & \vdots \\ 0 & 0 & \cdots & \boldsymbol{\Omega}_K \end{bmatrix} \quad (58)$$

where $\boldsymbol{\Gamma}_i, \boldsymbol{\Psi}_i,$ and $\boldsymbol{\Omega}_i$ are defined in Eq. (44), which represent the matrices either from the AR(1)-based MIMO-OFDM channel model or from the AR(3)-based MIMO-OFDM channel model.

## 9. Conclusions

This chapter presents a state-space-based simulation model for MIMO-OFDM channels. Based on this model, a physical propagation environment of radio waves can be reconstructed by simulations.

In this approach, for each distant scattering cluster, the received power renders a narrow peak, which contributes a portion to the Doppler power spectrum. The entire Doppler power spectrum is obtained by summing the contributions of all these uncorrelated scattering clusters.

One of the fundamental assumptions in this chapter is the probability distribution of scattering clusters. The AOD, AOA, and TOA due to distant Cauchy-Rayleigh scattering clusters can be approximately modeled as the Cauchy angular and delay power distribution functions, while distant Rayleigh clusters result in the Gaussian angular and delay power distribution functions.

Another underlying assumption is that more than 90% of the power is within a small angular spread. The narrow distribution enables us to study the CSOS using approximations for small angles. This implies that both the upper and lower limits of the integral of the channel spatial-temporal correlation function can be extended from $[-\pi, \ \pi]$ to $(-\infty, \ \infty)$ without losing its main features. Meanwhile, the assumption of independence of the AOD and AOA makes the channel correlation function integrable.

One of the main results is the decomposition of the spatial-temporal correlation function caused by a single cluster. The CSOS can be decomposed into disjoint antenna spacing and movement parts using the phase-shift method. Thus, an AR(p) model can be employed to describe the temporal dynamics of the channel.

A major result is that the radio channels can be built modularly. A state-space-based MIMO-OFDM channel model is another major result. A distant scattering cluster contributed to each antenna at a mobile receiver is associated with an AR(1)- or AR(3)-based state-space SISO channel model block. The beauty of using state-space representation is that a MIMO-OFDM channel model can be constructed using multiple SISO channel blocks. Meanwhile, a correlated innovation process is employed to adjust the channel spatial correlation within each MIMO block and spectral correlation between the MIMO blocks. Following the same process, it is easy to extend this model to the multi-cluster case.

Therefore, the spatial-temporal-spectral correlation characteristics of the channel are achievable in the simulated channels.

## 10. Future work

Future work may include:

- AOD/AOA Measurement

  The angular-delay spectrum is an important parameter in the modeling of the state-space-based MIMO-OFDM channels. In practice, how to measure the directional information will directly affect the results of a realistic channel correlation accuracy. On the other hand, the effective channel modeling largely relies on well-defined correlation functions.

- AOD/AOA Estimation

  Extracting or estimating AOD/AOA from measurements is another issue. This is a hot research topic that attracts people. Many results have been published in the literature, for example, the multiple signal classification (MUSIC) algorithm [27, 28], the estimation of signal parameters via rotational invariance techniques (ESPRIT) algorithm [28, 29], the expectation-maximization (EM) algorithm [30, 31], and the space-alternating generalized expectation-maximization (SAGE) algorithm [31, 32].

  Several issues related to these algorithms need to be addressed, for example, how to estimate the number of signal sources and estimate the arbitrariness of the DOA. In addition, these algorithms do not work when the number of signal sources is larger than the number of antennas. The recurrent neural network (RNN) and convolutional neural network (CNN) may be suitable to solve this problem.

- Reduce Simulation Complexity

  In simulations, the computation complexity depends on the size of the antenna arrays $M_r \times M_t$, the number of sub-carriers $M_f$, and the number of uncorrelated scattering clusters $K$.

  For each $M_r \times M_t$ block, we may assign a small number to $M_f$ and use the interpolation technique to increase the size of the channels. This idea makes sense because the contributing channels have high coherence bandwidth, which renders close to flat fading.

  In this way, the size of a spectral correlation matrix and the computational complexity in a simulation will be highly reduced. Hence, the problem of decomposition of the spectral correlation matrix using the Cholesky decomposition method may be avoided. For the large size of the matrix, the Cholesky decomposition method may lead to numerical problems.

- Massive MIMO

  The massive MIMO technology uses a large number of antennas at the BS to serve multiple users simultaneously. It is proposed to improve the performance of wireless communication systems, such as higher data rates, improved spectral efficiency, and better link reliability. Due to the large number of antennas, the propagating wave will no longer be a plane wave. That is, the spherical wave model for near-field should be taken into account. In this case, a mathematical model describing the radio channel characteristics is needed.

- Channel Generators

  The spatial channel model (SCM) [33] and the WINNER II [34] are channel models used in wireless communication systems. They are designed to simulate the propagation of radio waves in different environments and are used for evaluating and testing the performance of wireless communication systems.

  They are good channel models and have been used in many radio propagation scenarios [35–37]. However, the scatterers in both SCM and WINNER II are limited and they cannot be used to describe situations such as the propagation of a large number of signal sources, i.e., the presence of a large number of scatterers in the propagation environments.

  The channel model presented in this chapter can be employed to describe the situations of a large number of scattering objects in the radio wave propagation environment and to evaluate the performance of the designed wireless communication systems.

## Author details

Xin Li and Kun Yang*
School of Information Engineering, Zhejiang Ocean University, Zhoushan, China

*Address all correspondence to: yangkun@zjou.edu.cn

## References

[1] Ertel RB, Reed JH. Angle and time of arrival statistics for circular and elliptical scattering models. IEEE Journal on Selected Areas in Communications. 1999;**17**(11):1829-1840

[2] Janaswamy R. Angle and time of arrival statistics for the Gaussian scatter density model. IEEE Transactions on Wireless Communications. 2002;**1**(3):488-497

[3] Pedersen KI, Mogensen PE, Fleury BH. Power azimuth Spectrum in outdoor environments. IEEE Transactions on Wireless Communications. 1997;**33**(18):1583–1584

[4] Martin U. Spatio-temporal Radio Channel characteristics in urban macrocells. IEE Proceedings - Radar, Sonar and Navigation. 1998;**145**(1): 42-49

[5] Ekman T. Prediction of Mobile Radio Channels. Sweden: Uppsala University; 2002

[6] Chong C, Tan C, Laurenson DI, McLaughlin S, Beach MA, Nix AR. A new statistical wideband Spatio-Temporal Channel model for 5-GHz band WLAN systems. IEEE Journal on Selected Areas in Communications. 2003;**21**(2):139-150

[7] Spencer QH, Jeffs BD, Jensen MA, Swindlehurst AL. Modeling the statistical time and angle of arrival characteristics of an indoor Multipath Channel. IEEE Journal on Selected Areas in Communications. 2000;**18**(3):347-360

[8] Michael Buehrer R. The impact of angular energy distribution on spatial correlation. In: Proceedings of IEEE 56th Vehicular Technology Conference, 24–28 September 2002, Vancouver, BC, Canada; 2022. pp. 1173-1177

[9] Dong L, Ma J, Zhou J, Kikuchi H. Performance of MIMO with UCA and Laplacian Angular Distribution using correlation matrix. In: Proceedings of 2007 International Conference on Wireless Communications, Networking and Mobile Computing, 21-25 September 2007, Shanghai, China; 2007

[10] Forenza A, Love DJ, Heath RW. Simplified spatial correlation models for clustered MIMO channels with different Array configurations. IEEE Transactions on Vehicular Technology. 2007;**56**(4): 1924-1934

[11] Kalliola K, Sulonen K, Laitinen H, Kivekäs O, Krogerus J, Vainikainen P. Angular power distribution and mean effective gain of mobile antenna in different propagation environments. IEEE Transactions on Vehicular Technology. 2002;**51**(5):823-838

[12] Li X, Ekman T. Cauchy angular distribution for clustered radio propagation SIMO Channel model. In: IEEE the 71st Vehicular Technology Conference (VTC) 2010 Spring, May 16–19. Taiwan: Taipei; 2010

[13] Li X, Ekman T. Gaussian angular distributed MIMO Channel model. In: IEEE the 74th Vehicular Technology Conference (VTC) 2011 Fall, September 5–8. San Francisco, USA: IEEE; 2011

[14] Costa N, Haykin S. Multiple-Input Multiple-Output Channel Models Theory and Practice. Hoboken, New Jersey: John Wiley & Sons, Inc.; 2010

[15] Kunnari E, Linatti J. Stochastic Modeling of Rice fading channels with temporal, spatial and spectral correlation. IET Communications. 2007;**1**(2):215-224

[16] Lamahewa TA, Abhayapala TD, Iqbal R, Athaudage C. A framework to calculate space-frequency correlation in multi-carrier systems. IEEE Transactions on Wireless Communications. 2010;**9**(6):1825-1831

[17] Oestges C. Validity of the Kronecker model for MIMO correlated channels. In: 2006 IEEE 63rd Vehicular Technology Conference. Vol. 6, No. 3. Melbourne, VIC, Australia; 07-10 May 2006. pp. 2818-2822

[18] Shumway RH, Stoffer DS. Time Series Analysis and its Applications. 2nd ed. New York, USA: Springer; 2006

[19] Li X, Ekman T. Cauchy power azimuth Spectrum for clustered radio propagation MIMO Channel model. In: IEEE the 72st Vehicular Technology Conference (VTC) 2010 Fall, September 6–9. Ottawa: Canada; 2010

[20] Stoica P. Introduction to Spectral Analysis. Upper Saddle River, New Jersey: Prentice Hall; 2005

[21] Ogata K. Modern Control Engineering. 3rd ed. Upper Saddle River, New Jersey: Prentice-Hall Inc.; 1996

[22] Brockwell PJ, Davis RA. Introduction to Time Series and Forecasting. 2nd ed. New York, Inc.: Springer-Verlag; 2002

[23] Proakis JG. Digital Communications. 3rd ed. New York, USA: McGraw-Hill, Inc.; 1995

[24] Goldsmith A. Wireless Communications. Cambridge, United Kingdom: Cambridge University Press; 2005

[25] Li X, Ekman T. Cauchy-Rayleigh scattering cluster based spatial-temporal-spectral correlation properties with MIMO-OFDM Channel model. In: IEEE the International Conference on Wireless Communications and Signal Processing (WCSP), November 9–11. Nanjing, China: IEEE; 2011

[26] Li X, Ekman T. Rayleigh scattering cluster based spatial-temporal-spectral correlation properties with MIMO-OFDM Channel model. In: IEEE the 76st Vehicular Technology Conference (VTC) 2012 Fall. Canada: Québec City; September 3 – 6, 2012

[27] Schmidt RO. Multiple emitter location and signal parameter estimation. IEEE Transactions on Antennas and Propagation. 1986;**34**(3):276-280

[28] Krim H, Viberg M. Two decades of Array signal processing research. IEEE Signal Processing Magazine. 1996;**13**(4): 67-69

[29] Roy R, Kailath T. ESPRIT—Estimation of signal parameters via rotational invariance techniques. IEEE Transactions on Acoustics, Speech, and Signal Processing. 1989;**37**(7):984-995

[30] Dempster AP, Laird NM, Rubin DB. Maximum likelihood from incomplete data via the EM algorithm. Journal of the Royal Statistical Society: Series B. 1977; **39**(1):1-38

[31] Chung PJ, Bohme JF. DOA estimation using fast EM and SAGE algorithms. IEEE Signal Processing. 2002;**82**:1753-1762

[32] Xiong K, Liu Z, Jiang W. SAGE-based algorithm for direction-of-arrival estimation and array calibration. Journal of Systems Engineering and Electronics. 2014;**30**(6):1074-1080

[33] Matlab Implementation of the 3GPP Spatial Channel Model (3GPP TR 25.996). v1.2, Nov. 1. 2005

[34] ST-4-027756 WINNER II D1.1.2 V1.2. WINNER II channel models. 2008, Feb. [Online]. Available from: http:// www.ist-winner.org

[35] Narandzic M, Schneider C, Thoma R, Jamsa T, Kyosti P, Zhao X. Comparison of SCM, SCME, and WINNER Channel models. In: IEEE 65st Vehicular Technology Conference, Apr. 22–25. Dublin, Ireland: IEEE; 2007

[36] Narandzic M, Kaske M, Schneider C, Milojevic M, Landmann M, Sommerkorn G, et al. 3D-antenna Array model for IST-WINNER Channel simulations. In: IEEE 65st Vehicular Technology Conference, Apr. 22–25. Dublin, Ireland: IEEE; 2007

[37] Schneider C, Narandzic M, Kaske M, Sommerkorn G, Thoma RS. Large scale parameter for the WINNER II Channel model at 2.53 GHz in urban macro cell. In: IEEE 71st Vehicular Technology Conference. Taipei, Taiwan: IEEE; May 16-19 2010

Section 4

# Autonomous Driving and Radars

Chapter 13

# Autonomous Driving and Cybersecurity by Design

*Cecil Bruce-Boye, Thomas Eisenbarth, Moritz Krebbel, Andreas Fechner, Robert Luyken and Telse David*

## Abstract

So far, real-time requirements for the overall autonomous driving (AD) have been addressed only in a few cases. Cybersecurity and real-time capability are usually addressed separately. However, with regard to a justifiable mobility quality, these requirements are in direct interaction with each other. Therefore, as suggested here, it makes sense to consider the provision of a suitable IT infrastructure with cybersecurity, QoS (Quality of Service) and simultaneous real-time IoT capabilities. The early integration of security and real-time by design, as well as the architecture concepts mentioned, are measures that limit development costs, make the solution modular, scalable and thus sustainable. We introduce the adaptive-real-time-manager (ARM), an innovative concept for continuous assessment and optimization of the real-time capability of autonomous driving systems. The paper also proposes a cloud-broker-concept and simulation as essential building blocks to accelerate the integration of the ARM into an autonomous driving system (ADS). Furthermore, we discuss aspects of multisensory data acquisition and processing, addressing the integration of various data sources and their qualities. Finally, we highlight the importance of driveability for autonomous vehicles, emphasizing its role in comfort, safety, and user acceptance.

**Keywords:** autonomous driving, cybersecurity, real-time, adaptive real-time, real-time management multisensory data, driveability

## 1. Introduction

### 1.1 General thoughts about autonomous driving

The transition from self-driving individual transport to driverless on-demand mobility with ADS is a major challenge for both people and technology. Innovative mobility concepts such as Mobility-as-a-Service (MaaS) and Transport-as-a-Service (TaaS) are being developed to bring safe, environmentally friendly, cost-effective and convenient solutions to the market [1].

As the mobility market evolves, companies will need to offer diverse hardware, software, and services portfolios to meet the expectations of their customers. Among the top priorities is Level 4 safety: Today's vehicles are already equipped with numerous safety and assistance systems, making driving very safe. On average, a human driver causes a fatal accident every 600 million kilometers [2]. Self-driving systems are expected to further reduce the number of accidents. To achieve this, the system needs to be extremely robust, which is not only challenging in terms of design, but also in terms of verification.

## 1.2 A new approach: real-time IoT and cybersecurity

ADS-based mobility requires a secure, uninterrupted connection between all traffic participants. It therefore relies on a smooth and fast flow of data for each individual information chain between all relevant participants. This also means that these chains must be protected from attack. All possible attack vectors must be secured, regardless of the point of attack. At the same time, it must be ensured that the acquisition and response times of all data in the relevant information chains are reliable, deterministic and predictable. A suitable computing infrastructure with cybersecurity—QoS (Quality of Service) and real-time IoT capabilities—is therefore required [3].

The ADS system must ensure both cybersecurity and real-time IoT capabilities across all information chains of the entire system. While these requirements may seem contradictory at first, it is essential to perform the necessary analysis during the design phase to develop concepts, architectures and strategies that resolve this contradiction. By doing so, we can avoid the costly and often unattainable process of implementing security and real-time capabilities in an ADS after the fact.

## 1.3 Cybersecurity for autonomous driving system

With the increase in connectivity and communication between vehicles, traffic management systems and other elements of the transport infrastructure, the attack surface and potential vulnerabilities are also increasing. One of the main challenges in implementing cybersecurity in autonomous systems is that security mechanisms such as encryption, authentication and integrity checks require time and computing resources. These additional requirements can potentially impact the real-time capabilities of the systems by increasing latency and slowing the response time of autonomous vehicles. However, with careful planning and innovative solutions, it is possible to achieve both cybersecurity and real-time performance without compromising the safety and reliability of autonomous driving systems.

The importance of cybersecurity has been acknowledged by lawmakers, leading to the introduction of UNECE Regulations R.155 and R.156 [4, 5]. These regulations establish requirements for the cybersecurity of vehicles and their systems, and require the automotive industry to take appropriate security measures to ensure the cyber resilience of their vehicles.

The combination of cybersecurity and real-time capability requires close collaboration between the various disciplines involved in the development of autonomous driving systems, such as vehicle engineering, software development and IT security. An appropriate IT infrastructure that provides both cybersecurity QoS and real-time IoT capabilities is crucial for the safety and reliability of autonomous vehicles. To achieve this, the following concepts and ideas are presented, which enable an efficient combination of cybersecurity and real-time capability to ensure the safety and functionality of autonomous driving systems.

### 1.4 Multisensory input information

In addition to vehicle data, a variety of external sensor-generated data, server-based environmental data and even satellite-based positioning information are used as input variables in the ADS. External sensor-generated data includes Car2Car communication. This ensures that the speed and distance of autonomous road users in the vicinity are monitored.

The real-time requirements in the immediate vicinity of autonomous vehicles are obviously higher than those in the superimposed environments, from which, for example, spatial or environmental data are obtained. Decentralization (edge computing) in the IoT network allows the next action decision to be made as close as possible to the distributed sensors. This decision is then made available to higher-level intelligent instances for further coordination and regulation of the overall process. As a result, there are multiple levels of interaction in the IoT network. During the software development process, it is important to consider the transitions between the different interaction levels.

In Section 6, we will consider velocity and position control. It is important to note that the time to acquire data, calculate the next action and provide instructions must be at least twice as fast as the process speed or constant to control the current process in real-time [6]. In addition, certain safety requirements for the ADS can only be ensured by guaranteeing real-time conditions in the information chain. It is obvious that there is some interplay between cybersecurity and safety in terms of real-time requirements. However, this issue is not addressed in this article.

We assume that both cascaded and cross-layer control loops are likely to become necessary to meet the varying requirements of the different layers of the hierarchical model, e.g. hardware, operating system, software, Car2Car, server and cloud. In order to calculate the continuous autonomous driving speed for all collision-free positions, the information chains require the processing of multisensory input information, resulting in a MIMO (multiple input, multiple output) system [7, 8]. We consider the multiple antenna approach on the transport level as given. We also want to evaluate the driving behavior of the AD, and for this purpose we introduce the term "ADS driveability" in Section 7. We want to encourage an objective evaluation of the driving experience of an AD, as this can ultimately be a decisive factor in the competitive use of ADS services.

## 2. Information chain according to the shell model

In order to achieve real-time control, it is essential that the data acquisition, computation, and provision of the next instruction occur at a speed that is at least twice as fast as the process being controlled [6].

Accordingly, all interaction levels in IoT must be measured for their QoS (Quality of Service) in addition to Round-Trip Times (RTT). Only then can a reliable decision be made as to which processes can be controlled in real-time. Alternatively, the process speeds can be adjusted to the determined real-time characteristics (or real-time limits) of the respective interaction levels. The speed at which the autonomous vehicle performs over the measured interaction level should not exceed half of its real-time capability.

**Figure 1** gives a rough overview of the information flow to the IoT interaction levels and back.

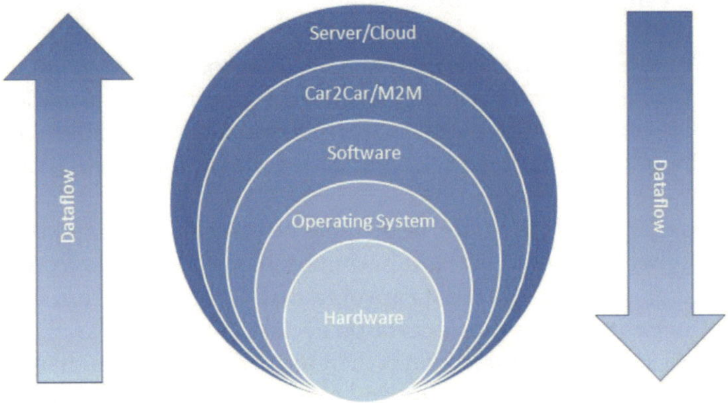


**Figure 1.**
*Shell model for the IoT interaction levels [9].*

Our objective is to propose a software development methodology for real-time IoT interactions. The term "propose" implies that this is a sketch that does not claim to be complete, but rather represents one of many possible solutions. Given the enormous complexity of the subject, it cannot be fully represented within the scope of this framework.

## 3. Cybersecurity and real-time

In the context of autonomous driving, ensuring cybersecurity and real-time capability is crucial. With the increasing networking and automation of vehicles, new challenges and questions arise that will be discussed in this section.

A central problem is ensuring end-to-end cybersecurity under real-time conditions. To do this, security measures must be implemented at all levels of the system, starting with the sensors and extending to communications and the cloud.

Examples include authentication and key exchange under real-time conditions. The typical use of asymmetric crypto methods is problematic for key renewal during

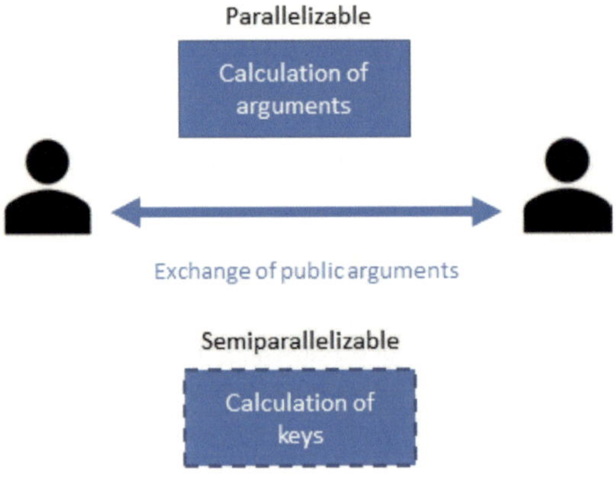


**Figure 2.**
*Challenges of parallelizability in key exchange with asymmetric cryptography.*

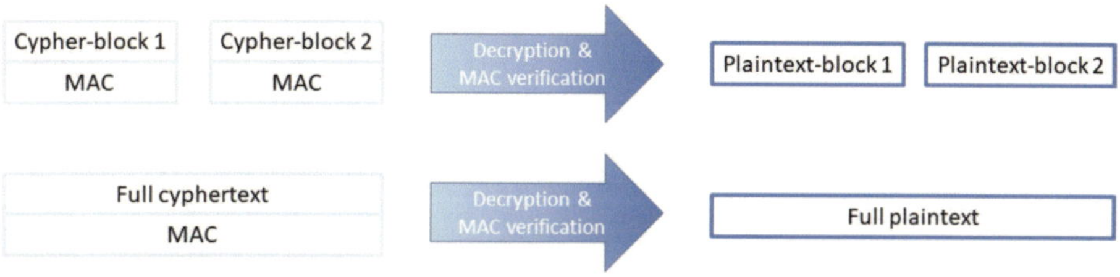


**Figure 3.**
*Example of parallelizing a MAC calculation.*

runtime due to their slow runtime. So, if you want to still have real-time capability, you must consider parallel key renewal during runtime (**Figure 2**). In addition, the use of parallelizable crypto algorithms can be an important building block; for example, authentication procedures, such as the Message Authentication Code (MAC) procedure, can be parallelized to guarantee real-time capability (**Figure 3**).

Another component is edge computing, where data processing and analysis take place in the vehicle instead of in the cloud, which can help optimize latency and data rates. This supports real-time guarantees by reducing the amount of data transmitted over the network and increasing the speed of response to events.

A major challenge arises from the fact that vehicles are in the field for a long period of time, so future systems should be prepared for changing crypto computing power and key length requirements by considering or balancing newer crypto techniques such as post-quantum cryptography etc. For example, the ongoing development of quantum computers poses a particular challenge by challenging the security of traditional asymmetric key exchange methods [10].

In summary, ensuring cybersecurity and real-time capability in autonomous driving is a challenging task that requires a combination of different technologies and concepts. The integration of edge computing, parallel key renewal and authentication, as well as the adaptation to future crypto requirements are key elements to ensure the security and performance of autonomous vehicles in the connected world.

## 4. Real-time management

Our adaptive-real-time-manager (ARM) is an innovative concept that aims continuously assessing and optimizing the real-time capability of autonomous driving systems. This section discusses the basic design of the ARM and its advantages compared to existing solutions.

Factors such as vehicle environment, traffic conditions, visibility, and network connection quality influence the real-time capability of autonomous driving systems. The ARM constantly evaluates these factors and adjusts driving speed and strategy accordingly (**Figure 4**).

A crucial aspect of the ARM concept is the Round Trip Time (RTT) of the closed information chain from the vehicle's sensors and actuators to the cloud and back. The RTT varies depending on the preferred cybersecurity mechanisms, which can be selectively integrated at different security levels.

The ARM assesses the real-time capability of the respective closed information chain by considering the RTT and, if necessary, other system parameters. This enables optimal adjustment of driving speed and strategy to the respective conditions.

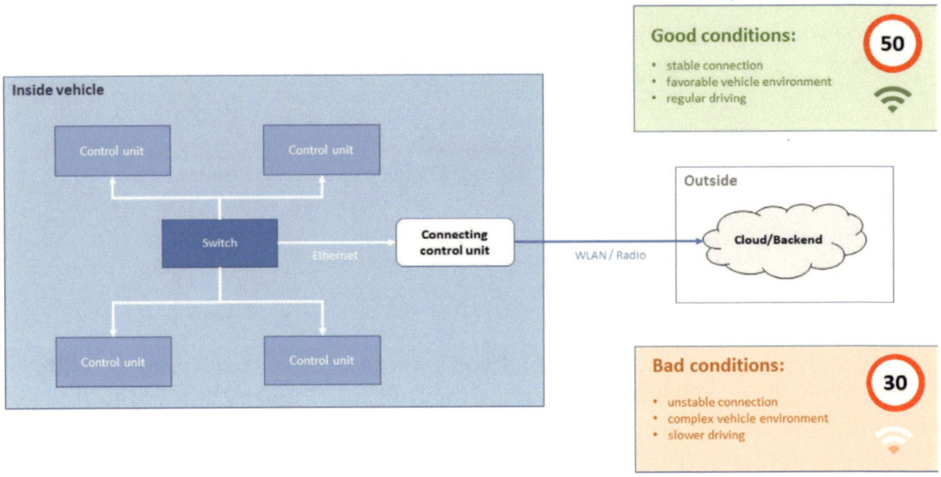


**Figure 4.**
*The ARM might suggest a speed of 50 km/h when the connection quality is good and the vehicle environment is favorable, but only 30 km/h when the connection quality is poor or the vehicle environment is more complex.*

The ARM can reduce the impact of traffic control systems on the real-time capability of autonomous driving systems. This is achieved by continuously adapting driving strategies and speeds to the current conditions and, if necessary, to the information provided by traffic control systems.

A real-world scenario illustrates this benefit of ARM: An autonomous vehicle stops before a green light at an intersection. One possible explanation for this behavior is that the intelligent traffic light has informed the autonomous driving system of the time remaining in the green phase. However, the ARM has suggested a driving speed that is not sufficient to cross the intersection without a collision, so in this case the vehicle waits for the next full green phase.

Compared to existing solutions, the ARM offers a more dynamic approach to real-time assessment and optimization of autonomous driving systems. The continuous analysis of influencing factors and the adaptation of driving speed and strategy increase the safety, efficiency and flexibility of these systems.

Another advantage of the ARM is the ability to selectively incorporate cybersecurity mechanisms at different security levels. This ensures data security and system integrity without unnecessarily compromising the real-time capability of the autonomous driving system.

In summary, the automITe Adaptive Real-time Manager offers a promising approach for addressing the challenges related to autonomous driving and cybersecurity. Continuous real-time assessment and optimization, selective incorporation of cybersecurity mechanisms, and enhanced interaction with infrastructure make the ARM a unique and forward-looking solution in this field. It remains to be seen how the ARM will prove itself in real application scenarios and what further developments and optimizations are possible in the future (**Figure 5**).

Together with the ARM there are two essential building blocks that can accelerate the integration into an ADS system:

- Cloud-Broker-Concept: This ensures independence from the cloud provider and a uniform interface on the ADS side to the cloud. An essential step here is the

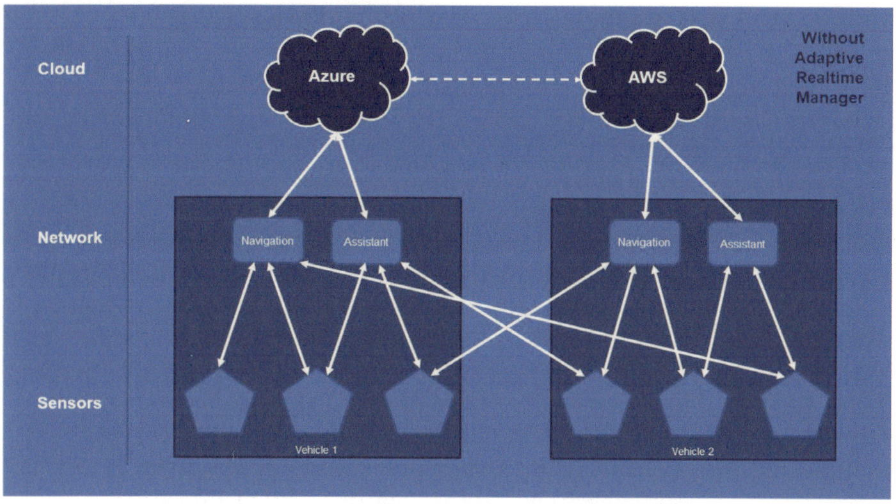


**Figure 5.**
*Typical vehicle architecture without middleware and adaptive-real-time-manager.*

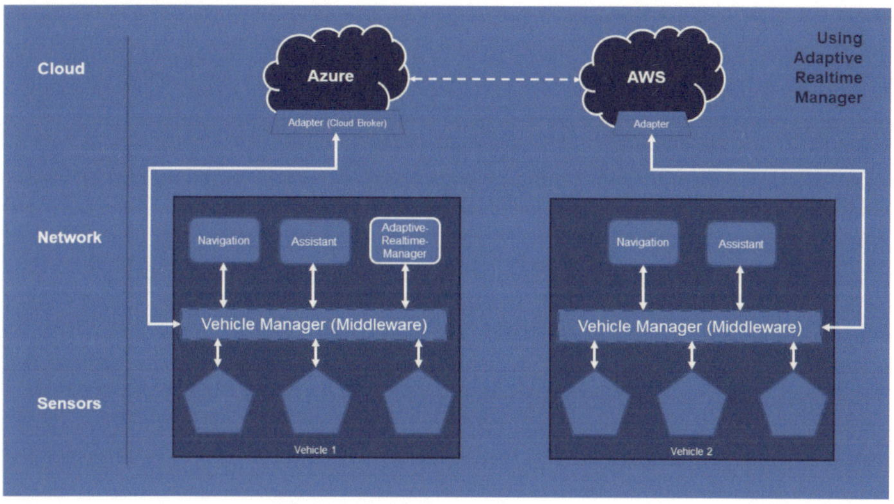


**Figure 6.**
*Vehicle architecture with middleware and adaptive-real-time-manager in place.*

integration and the interface management of the cloud broker into the system of the ARM (see **Figure 6**).

- Simulation of the Adaptive Real-time Manager and the Cloud Broker: For this purpose, we are currently designing a driving simulator that can be used and extended to simulate the driving of a vehicle in a city, with all driving information obtained from the cloud. By using the simulator, the effort required for testing in the field can be reduced, as many shortcomings are already revealed by simulations.

## 5. Aspects of multisensory data acquisition and processing

Data necessary for driving a car comprised of various sources:

- Physical measurements such as location and speed;
- Events like states of traffic lights
- Linguistic variables like human descriptions of traffic congestion. A linguistic variable gives an imprecise description of some perceived value like high, medium, low.

Physical measurements can be direct and indirect. The direct measurements are performed by various sensors while indirect measurements estimate values from other measurements, events, linguistic variables. The same physical value can be measured by different ways each characterized by different qualities of:

- Accuracy, how close a given set of measurements (observations or readings) are to their true value;
- Precision, how fine measured values can be specified;
- Confidence, the level of trust in the measurement source quantified in some measure like probability or possibility;
- Availability of the measurement source, e.g. in the cases of remote services like satellites, cloud servers, neighbor traffic participants;
- Latency, the time needed for the measurement to become accessible;
- Time span and spatial location of the measurement point.

The events and linguistic variables are characterized by:

- Confidence
- Availability
- Latency
- Time span and spatial location.

Thus the same physical value can be measured by different sensors and estimated indirectly with a great variety in accuracy, confidence, availability. For example, the car speed can be estimated using the car wheels with high availability and low precision; or from the GPS system with low availability, precision and high latency; or from a radar with high precision; or queried from other traffic participants with low confidence etc.

The wide variety of sources must be integrated using plausibility checks and inference based on the reliability and availability of the sources.

Since the sources may contradict each other, the inference model must support conditional reasoning. This is necessary when the confidence measure of a measurement is conditional on some event or other measurements, such as in the case of computed values. At the same time driving has mission critical aspect. Therefore it should allow reasoning under contradiction when different events and measurements contradict each other since so that contradictions be resolved using information from other sources available.

The confidence measure may describe either or both kind of uncertainty:

- Probabilistic resulting from a stochastic measurement process;
- Fuzzy incoming from human estimations and processes of ill-defined nature.

In addition to uncertainty the confidence measure also needs to describe contradiction allowing combination of erroneous sources.

Furthermore, the inference process is a subject of real-time constraints. Therefore the choice of must consider:

- Support of fine-grained parallelism, e.g. when walking down a decision tree of alternatives;
- Gradual refinement of the estimation in order to be able to get an answer even if the deadline was prematurely reached at the cost of accuracy and certainty loss;
- Using conditionals in reasoning and decision making.

The latencies imposed on the measurement process consideration of the time aspect, such as the time stamps and time intervals of the values, events and linguistic variables.

## 6. AD-velocity and position control

The performance of WLAN communication of multiple antennas is an important aspect in this context, especially as a MIMO system, to improve the channel capacities [11]. However, it essentially concerns the transport level. We consider it as a given [12]. And on the other hand we focus on the MIMO concepts for the control of the driving behavior of an ADS via the information chain [9]. For the present ADS with MIMO (multiple input, multiple output) characteristics [7, 8], we define the multi-sensory input information in a simplified way as follows:

- Vehicle board data
- External sensor-generated near-field data
- Server and satellite-based information.

The following should be considered as output variables:

- Driving speed and the collision-free
- Current position of the autonomously driving vehicle.

To control the driving behavior of the ADS, the RTT of the information chain plays a crucial role.

Suitable methods for controller synthesis are available according to (Ackermann). For digitization, the choice of sampling time is

$$T = \frac{RTT}{2} \tag{1}$$

or sampling angular frequency is

$$\omega_T = \frac{2\pi}{T} \tag{2}$$

where $\omega_T$, according to the Shannon theorem, is the largest angular frequency occurring in the information chain. However, the angular frequencies of the disturbance signals, the multisensory input variables, and the bandwidths of the controls must also be considered in this context. These considerations apply to both control variables, ADS velocities and the continuous determination of collision-free positions.

## 7. AD-driveability

Initially, driveability refers to a vehicle's driving dynamics, particularly in terms of power, throttle response, engine, transmission, braking and steering control. It is an important aspect of the overall ride quality of a vehicle and has a significant impact on driver experience and customer satisfaction.

Good driveability means that the vehicle responds smoothly and predictably in all driving situations. Driveability is particularly important in modern vehicles with electronic controls, as it ensures precise and responsive control of the engine and other systems.

Driveability is of high importance for both comfort and safety. For example, in critical situations such as emergency braking or quick evasive maneuvers, good driveability can help the vehicle remain stable and the driver to maintain control.

Autonomous vehicles are not driven by human drivers. Therefore, the term driveability should be redefined as AD-driveability. This creates a basis for objectively evaluating different MaaS and TaaS concepts in terms of driving style and experience.

As far as comfort is concerned, passengers should not be impaired in their activities (working, reading, sleeping...) during the journey. For example, by braking too hard, accelerating too fast or driving in a jerky manner.

Good and safe driving behavior "AD-driveability" will become a competitive factor for autonomous vehicles, as the purchase decision will essentially depend on it. It is expected that the MaaS, TaaS concept, which reaches the destination faster with smooth driving comfort, will achieve a higher acceptance in the MaaS and TaaS service market.

The solutions outlined here, for the correlation of real-time and cybersecurity and adaptive real-time managers can make a decisive contribution to this.

## 8. Conclusion

This paper has presented a comprehensive overview of various challenges and potential solutions related to autonomous driving and cybersecurity by design.

Ensuring real-time control, end-to-end cybersecurity, and driveability are critical aspects of developing successful autonomous driving systems. The proposed adaptive-real-time-manager (ARM) concept is a promising approach to addressing these challenges by continuously assessing and optimizing the real-time capability of autonomous driving systems while considering various influencing factors and selectively integrating cybersecurity mechanisms.

The integration of edge computing, parallel key renewal, and authentication, as well as the adaptation to future crypto requirements, are essential elements for ensuring the security and performance of autonomous vehicles in the connected world. The Cloud-Broker-Concept and simulation of the Adaptive Real-time Manager and the Cloud Broker further support these efforts by facilitating the integration into an ADS system and allowing for more effective testing and optimization.

Aspects of multisensory data acquisition and processing have also been explored, emphasizing the importance of integrating a variety of data sources and managing uncertainties and contradictions in the inference process. Speed and position control have been addressed as crucial aspects of autonomous driving, highlighting the significance of considering the round trip time of the information chain in controller synthesis.

Finally, the concept of driveability has been discussed in the context of autonomous vehicles, underlining its importance for passenger comfort, safety, and overall user experience. As the field of autonomous driving continues to evolve, the strategies and concepts presented in this paper serve as valuable building blocks for developing secure, efficient, and adaptable autonomous driving systems that meet the demands of an increasingly connected world. Future research and development efforts will undoubtedly reveal new challenges and opportunities for further enhancing the safety, performance, and acceptance of these innovative transportation solutions.

## Conflict of interest

The authors declare no conflict of interest.

## Author details

Cecil Bruce-Boye[1*], Thomas Eisenbarth[2], Moritz Krebbel[3], Andreas Fechner[4], Robert Luyken[5] and Telse David[6]

1 Prof. Dr.-Ing., International Project Consultant (iPcon), Lübeck, Germany

2 Prof. Dr.-Ing., Director of the Institute for IT Security, University of Lübeck, Lübeck, Germany

3 M. Sc., AutomITe-Engineering GmbH, Lübeck, Germany

4 Dipl.-Ing., AutomITe-Engineering GmbH, Lübeck, Germany

5 cand. B. Sc., Hochschule Flensburg, Flensburg, Germany

6 Dr.-Ing., Technische Hochschule Lübeck, Lübeck, Germany

*Address all correspondence to: cecil@bruce-boye.com

## References

[1] The Transformation has Begun. Available from: https://www.moia.io/de-DE/innovation. [Accessed: May 9, 2023]

[2] Volkswagen Plans to Make Autonomous Driving Market-ready, 2019. Available from: https://www.volkswagenag.com/en/news/2019/10/autonomous_driving.html. [Accessed: May 9, 2023]

[3] Bruce-Boye C, Kazakov DA. Quality of uni- and multicast services in a middleware LabMap study case? In: Innovative Algorithms and Techniques in Automation, Industrial Electronics and Telecommunications, Dordrecht: Springer; 2007. pp. 89-94

[4] UN Regulation No. 155 - Cyber Security and Cyber Security Management System. Available from: https://unece.org/transport/documents/2021/03/standards/un-regulation-no-155-cyber-security-and-cyber-security. [Accessed: May 9, 2023]

[5] UN Regulation No. 156 - Software Update and Software Update Management System. Available from: https://unece.org/transport/documents/2021/03/standards/un-regulation-no-156-software-update-and-software-update. [Accessed: May 9, 2023]

[6] Shannon RV. A model of safe levels for electrical stimulation. IEEE Transactions on Biomedical Engineering. 1992;**39**:424-426

[7] O. Nelles, Regelungstechnik. Siegen: University of Siegen. Available from: https://www.mb.uni-siegen.de/mrt/lehre/rt/rt_skript.pdf. [Accessed: May 9, 2023]

[8] Weiss GHM. Repetitive control of MIMO systems using H∞ design. Automatica. 1999;**35**(7):1185-1199

[9] Bruce-Boye CLDRM. Echtzeit-IoT im 5G-Umfeld. Vol. 39. Wiesbaden: Springer Fachmedien Wiesbaden; DOI: 10.1007/978-3-658-28307-0_14

[10] Shor PW. Polynomial-Time Algorithms for Prime Factorization and Discrete Logarithms on a Quantum Computer. Washington, DC Annual Symposium on Foundations of Computer Science, Washington, DC; 1996

[11] Ghayoula ABRG-YE. Capacity and performance of MIMO systems for wireless communications. Journal of Engineering Science and Technology Review. 2014;7(3):108-111

[12] Ackermann JPTMAHGFRWT. A robust digital predistortion algorithm for 5G MIMO: Modeling a MIMO scenario with two nonlinear MIMO transmitters including a cross-coupling effect. IEEE Microwave Magazine. 2020;**21**(7):54-62

Chapter 14

# MIMO Radar

*Motoyuki Sato*


## Abstract

We show the concept of multiple-input multiple-output (MIMO) radar and introduce practical applications, which include ground based synthetic aperture radar (GB-SAR) and ground penetrating radar (GPR). As an example, a 17 GHz MIMO GB-SAR system to be used for landslide monitoring and infrastructure measurement is described. We also show that a MIMO GPR system "Yakumo" can achieve dense three-dimensional (3D) subsurface imaging compared to conventional GPR. We also explain that MIMO GPR can be used for common midpoint (CMP) measurement, which can be used for the estimation of the vertical profile of EM velocity, which is related to soil moisture.


**Keywords:** GPR, GB-SAR, MIMO radar, multi-static radar, DInSAR

## 1. Introduction

Ground based synthetic aperture radar (GB-SAR) has been used for the observation of the displacement of ground surface and can be applied, for example, to remote landslide monitoring. GPR is a useful method for shallow subsurface imaging and widely used for the detection of buried pipes. Conventional GB-SAR systems and GPR systems are equipped with a pair of a transmitting antenna and receiving antenna, and synthetic aperture radar (SAR) processing is applied to the data sets acquired by moving the pair of antennas.

Instead of moving antennas for radar imaging, we introduce MIMO technique, where we use fixed multiple antennas for equivalent SAR imaging. In both GB-SAR and GPR systems, we use multiple transmitting and receiving antennas equivalent to multiple-input and multiple-output (MIMO). This radar configuration is referred as multi-static radar. However, we acquire all the combination of transmitting and receiving antennas, which is not common in the conventional multi-static radar. This is the reason why we call it MIMO radar, and we show that it expands the potential of radar drastically. The targets of MIMO GB-SAR and MIMO GPR such as land slope and buried pipes are stational, and we can acquire radar signal from these targets by switching all the transmitting and receiving antenna combinations. We do not need orthogonal signal transmission for the identification of the transmitted signal by receiver, because signals can be separated by the time sequence.

## 2. GB-SAR

Differential interferometric synthetic aperture radar (DInSAR) by GB-SAR is used to measure the displacement of the target surface [1]. This method has been used for monitoring landslide slopes [2–4], volcanic lava domes [4, 5], and inspection of large-scale infrastructure facilities such as dams and bridges [6, 7]. However, by conventional GB-SAR, the data for SAR processing is acquired by physically moving a radar unit equipped with a pair of transmitting and receiving antennas on a rail. The size of the rail determines the synthetic aperture length, which is typically about 2 m for 17 GHz GB-SAR. The data acquisition takes several tens of seconds to several minutes for one SAR image. Recently, MIMO radar [8–11], which does not have to move a radar unit, has been proposed to use for GB-SAR applications.

MIMO radar is a multi-static array type radar that has multiple transmitting and receiving antennas. However, MIMO radar transmits electromagnetic wave from one of the transmitting antennas, and the reflected signal is received by all the receiving antennas. Consequently, for a radar system with M transmitting and N receiving antennas, M × N independent radar signals can be measured. This is equivalent to acquire radar signal by using M × N independent antenna pairs. This concept is called virtual array.

Compared to conventional GB-SAR, MIMO radar can acquire data in a short time by using electronic switches for multiple transmitting and receiving antennas. Since MIMO radar has no mechanical moving parts, it can improve the reliability of long-term operation.

## 3. MIMO GB-SAR

MIMO radar uses multiple transmitting and receiving antennas independently to form a single SAR image, and a virtual array replaces the physical transmitting and receiving array which is equivalent to an array composed of monostatic radar capable of transmitting and receiving.

**Figure 1** shows the relationship between a physical bistatic radar consisting of a pair of transmitting and receiving antennas and a monostatic radar with a virtual

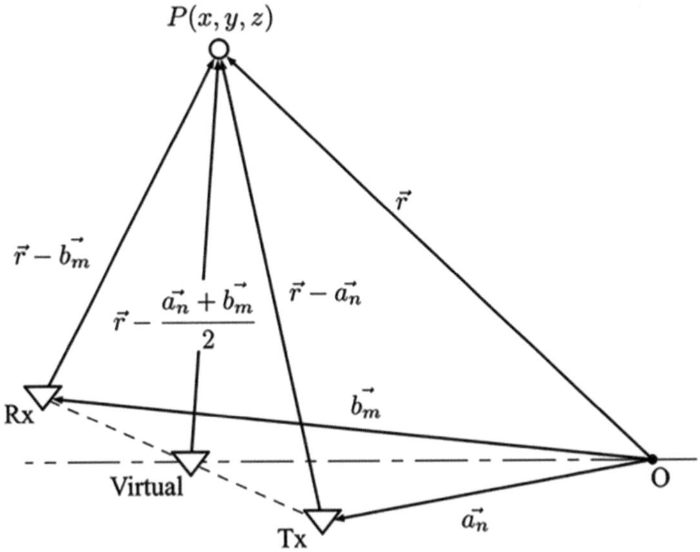


**Figure 1.**
*The relationship between a physical bistatic radar consisting of a pair of transmitting and receiving antennas and a monostatic radar with a virtual array.*

array. Here, $O$ is the coordinate origin, $P$ is the target position, Tx and Rx are the transmitting and receiving antenna positions, $\vec{a}_n$ and $\vec{b}_m$ are the position vectors of the transmitting and receiving antennas, and $\vec{r}$ is the target position vector. The path length $R_{n,m}$ is that of the EM wave propagating from the $n$-th Tx antenna to the target and to the $m$-th Rx antenna is given as:

$$R_{n,m} = \left|\vec{r} - \vec{a}_n\right| + \left|\vec{r} - \vec{b}_m\right| \tag{1}$$

When the target is far enough from the origin compared to the wavelength, it can be approximated by

$$R_{n,m} = 2\left|\vec{r} - \frac{\vec{a}_n + \vec{b}_m}{2}\right| \tag{2}$$

The condition for this approximation [11] is determined by the total length of the transmitting and receiving array $L_{\mathrm{Tx}}$, $L_{\mathrm{Rx}}$ as shown in (3). If (3) is satisfied, the array factor generated by the virtual array will be given by the product of the physical transmitting array factor (4) and the receiving array factor (5), where $\lambda$ is the wavelength, $k$ is the wavenumber, and $\vec{l}$ is the directional $\vec{r}$ vector given by (6).

$$|\vec{r}| \geq 1.24\sqrt{\frac{L_{\mathrm{Tx}}^3 + L_{\mathrm{Rx}}^3}{\lambda}} \tag{3}$$

$$F_{\mathrm{Tx}}(\theta, \phi) = \frac{1}{N}\mathrm{e}^{-jk|\vec{r}|}\sum_{n=1}^{N}\mathrm{e}^{-jk\vec{a}_n\cdot\vec{l}} \tag{4}$$

$$F_{\mathrm{Rx}}(\theta, \phi) = \frac{1}{M}\mathrm{e}^{-jk|\vec{r}|}\sum_{n=1}^{M}\mathrm{e}^{-jk\vec{b}_m\cdot\vec{l}} \tag{5}$$

$$\vec{l} = \begin{pmatrix} \sin\theta\cos\phi \\ \sin\theta\sin\phi \\ \cos\theta \end{pmatrix} \tag{6}$$

To prevent the generation of grating lobes in a basic concept for designing array antenna, and the antenna spacing $d$ must satisfy the condition $d < \lambda/2$. In MIMO radar, we consider this condition for the virtual array, but not for the physical antenna positions.

Back-projection algorithm is used to reconstruct the SAR image from data acquired by MIMO GB-SAR. The SAR image $I\left(\vec{r}\right)$ is obtained by (7), where, $s_{n,m}$ is the radar waveform (range profile) measured by the combination of the $n$-th transmit antenna and the $m$-th receive antenna.

$$I\left(\vec{r}\right) = \sum_{m=1}^{M}\sum_{n=1}^{N} s_{n,m}(t)e^{j4\pi R_{n,m}/c} \tag{7}$$

To estimate the surface displacement of the imaged objects, DInSAR is performed using the phase difference of a pair of SAR images acquired at different times.

Assuming two SAR images acquired at different time as master and slave images, the phase difference $\Delta\phi$ between the master image I M and the slave image I S is given by (8).

$$\Delta\phi\left(\vec{r}\right) = \arctan\left(\frac{\mathrm{Im}\left(I_M\left(\vec{r}\right)I_S^*\left(\vec{r}\right)\right)}{\mathrm{Re}\left(I_M\left(\vec{r}\right)I_S^*\left(\vec{r}\right)\right)}\right) \tag{8}$$

The actual displacement $\Delta d$ is obtained by (9), where $\lambda_c$ is the wavelength at the center frequency.

$$\Delta d\left(\vec{r}\right) = \frac{\lambda_c}{4\pi}\Delta\phi\left(\vec{r}\right) \tag{9}$$

## 4. 17 GHz MIMO radar design

By the recommendation of ITU, 17 GHz is one of the standard frequencies used for GB-SAR all over the world, and it is suitable for the measurement of bare soil ground surface. We use 17 GHz for our system, and the specifications of the MIMO radar that we designed are shown in **Table 1**. The designed antenna arrangement is shown in **Figure 2**. By using these technical specifications, the antenna array factors are simulated and shown in **Figure 3**. **Figure 3** shows the array factors of the transmitting antenna array and the receiving antenna array and the virtual array. The separation of adjacent transmitting antennas is 17.5 mm, which is one wavelength at 17 GHz, and the separation of the adjacent receiving antennas is 131.3 mm, which corresponds to 7.5 wavelengths, and the separation of the adjacent virtual antennas is 4.4 mm, which

| | | |
|---|---|---|
| Center frequency | $f_c$ | 17.1 GHz |
| Frequency bandwidth | $B$ | 200 MHz |
| FM-CW sweep time | $T$ | 100 μs |
| Number of transmitting antennas | $N$ | 15 |
| Number of receiving antennas | $M$ | 15 |

**Table 1.**
*The technical specification of the 17 GHz MIMO radar.*

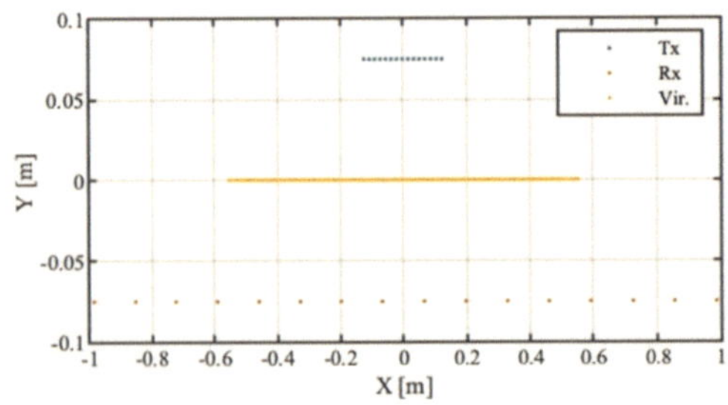


**Figure 2.**
*The antenna arrangement of the 17 GHz MIMO radar. 15Tx, 16Rx and 240 virtual antennas.*

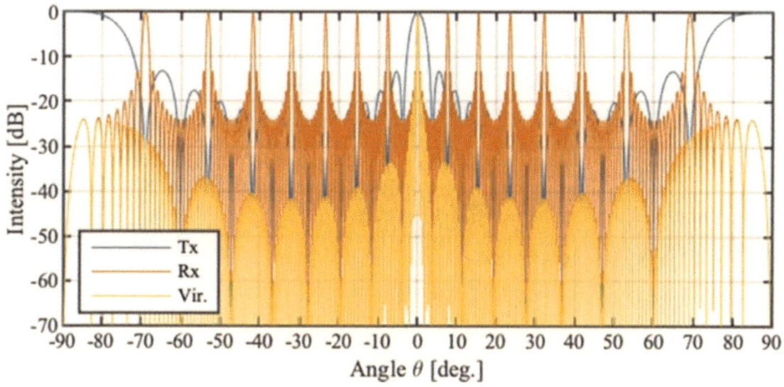


**Figure 3.**
*The antenna factor of the 15 × 16 17 GHz MIMO radar.*

is the 1/4 wavelength. The transmitting and receiving equidistant arrays are separated by 150 mm vertically.

In **Figure 3**, we find that the grating lobes are generated in the physical receiving antenna array. However, since the null points of the transmitting antenna array overlap it and cancel in the virtual array and the radar system has no grating lobes. We should note that the number of physical antennas can drastically be reduced from $M \times N$ to $M + N$ by MIMO GB-SAR.

## 5. Evaluation of 17 GHz MIMO radar

A prototype MIMO radar based on the above design was built and evaluated. We used a patch antenna for the array antenna element [12], which has a wide beam in the horizontal direction and sharp beam in the vertical direction, to avoid the ground surface clutter. We adopt FMCW radar system, and the antennas were connected with coaxial cables through a 16ch semiconductor switch.

Experiments were conducted to evaluate the MIMO radar system. **Figure 4** shows the MIMO radar facing the targets. A 15 cm trihedral square metal corner reflector is placed at 5 m from the center of the radar. In addition to the corner reflectors, there are also some targets. **Figure 5** shows the reconstructed SAR intensity image. In **Figure 5**, we can find the image of the corner reflector at $[X, Z] = [0, 5]$. The images of other targets are also formed accurately; we think the system works properly.

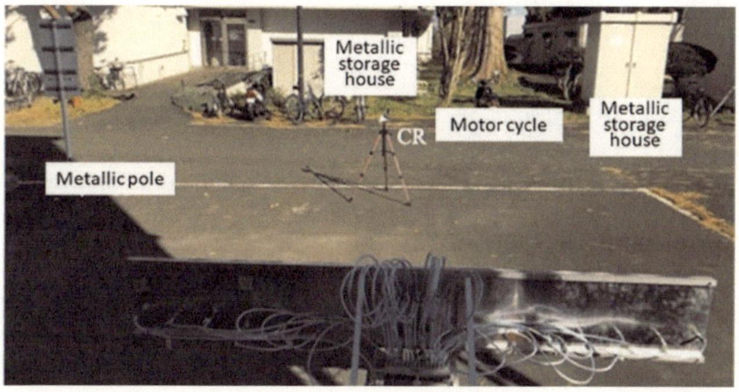


**Figure 4.**
*The 17 GHz MIMO radar and targets.*

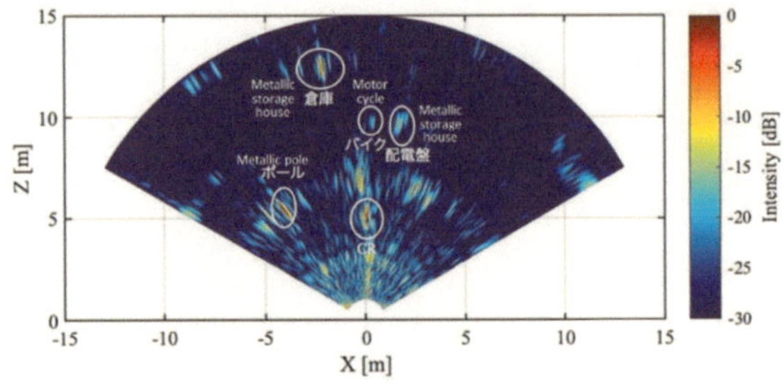


**Figure 5.**
*The reconstructed SAR intensity image of the 17 GHz MIMO radar.*

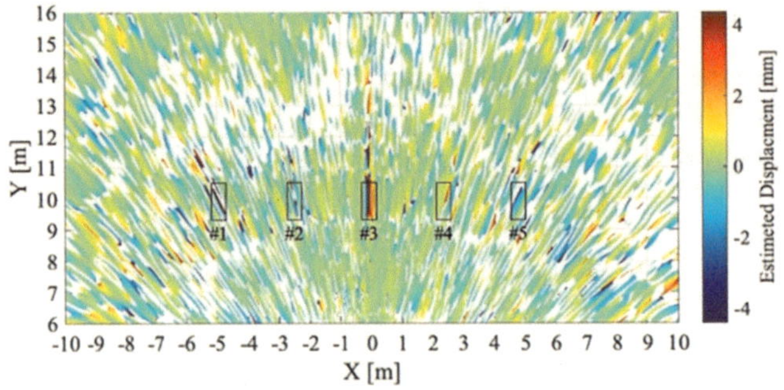


**Figure 6.**
*SAR interferograms when the displacement of all five patches is –4 mm. The positions of the displacement are also shown.*

We will use the prototype MIMO GB-SAR for ground surface displacement measurement. In order to evaluate the capability of DInSAR, we made a wooden wall having 10 m width and 2 m height, with 20 cm × 20 patches, which will be displaced from the flat surface. This wall has five displacement patches at 2.5 m intervals. The wall has a rough surface to suppress the specular reflection. In this experiment, the distance from the radar to the wall was 10 m. **Figure 6** shows SAR interferograms when the displacement of all five patches is –4 mm. At this time, pixels below −35 dB were masked in the SAR intensity image in order to extract the displacement on the wall surface. Also, the squares in **Figure 6** indicate the position of each displacement plane.

We can confirm that the displacement was detected at the position of each displacement plane in **Figure 6**; **Figure 7** shows a comparison of displacement and

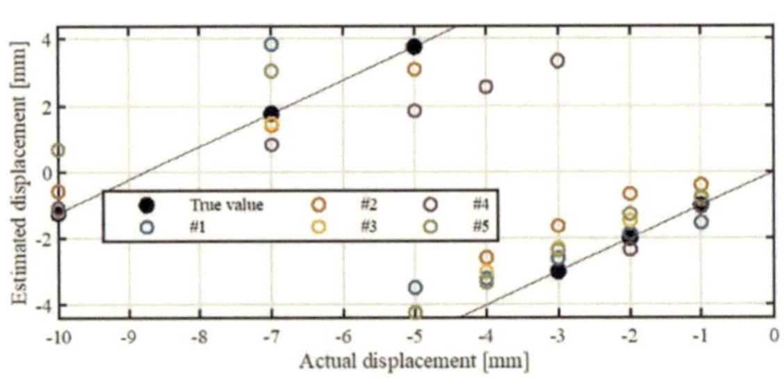


**Figure 7.**
*A comparison of displacement and estimated displacement in each displacement plane.*

estimated displacement in each displacement plan, and we can see that the displacement is correctly estimated.

## 6. GPR

Ground penetrating radar (GPR) is a useful method for shallow geophysical exploration and is widely used for the detection of buried objects such as pipes and cables and voids under pavement. GPR basically has a pair of transmit and receive antennas. By scanning the GPR unit, GPR profiles along the survey line can be obtained. In order to extend the swath width in the direction perpendicular to the survey line, we can set multiple radar units and measure simultaneously. If the multiple radar devices are synchronized, it is a multi-static radar and can greatly improve the quality of radar data. And if we use all the combinations of the transmit and receive antennas, we can configurate MIMO GPR.

## 7. MIMO-GPR "Yakumo"

We developed a MIMO GPR system "Yakumo" shown in **Figure 8**, for scanning a large area [13–15]. Yakumo was developed for surveying 1–2 m in depth, which is relatively deep compared to the similar multi-static GPR systems. Yakumo is a SF-CW radar that uses 50 MHz–1.5 GHz, which is a relatively low frequency compared to MIMO-GPR for pavement inspection. Since this device operates in a wide frequency bandwidth, it can select optimal frequency.

**Figure 8.**
*MIMO GPR system "Yakumo".*

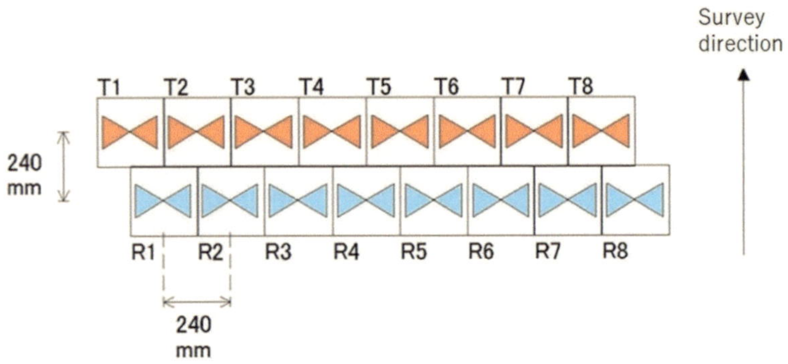


**Figure 9.**
*The antenna arrangement of the MIMO GPR system "Yakumo".*

| Frequency | 50 MHz–1.5 GHz |
|---|---|
| Radar system | SF-CW |
| Antenna element | Bowtie antenna |
| Number of antenna element | Tx 8, Rx 8 |
| Data acquisition interval | 1 cm |
| Data acquisition speed | 7 km/h (1 cm interval) |

**Table 2.**
*The technical specifications of the MIMO GPR.*

**Figure 9** shows the antenna arrangement of this system, which is equipped with eight transmitting and receiving antennas, and **Table 2** shows the technical specifications. Antenna feeding point separation in the same row is 240 mm but a minimum of 120 mm in the transverse direction between transmit and receive antenna by the staggered position.

Yakumo is a multi-static radar, but by measuring the radio waves transmitted from one transmitting antenna with all receiving antennas, it is possible to acquire complete three-dimensional (3D) subsurface information by looking at the target from different angles. This leads to advanced 3D imaging.

## 8. Measurement example

An example of C-scan imaging by Yakumo is shown in **Figure 10**, which was acquired in a rice paddy field in winter time [14]. The radar was scanned in the horizontal direction of the figure, and six images of 2 m swarth width are superimposed vertically. The two white lines that can be seen in the C-scan image are agricultural drainage. Due to the high accuracy of the position control, the water pipes for drainage are correctly visualized in a straight line.

## 9. CMP measurement

Common midpoint (CMP) technique is used for estimating vertical profile of the velocity of electromagnetic wave in subsurface geological layers. In order to acquire

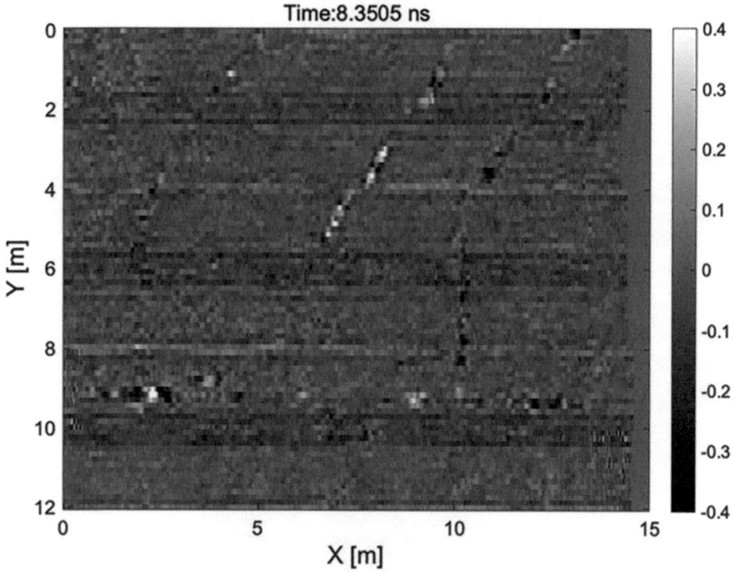


**Figure 10.**
*C-scan imaging by Yakumo. The two white lines are agricultural drainage.*

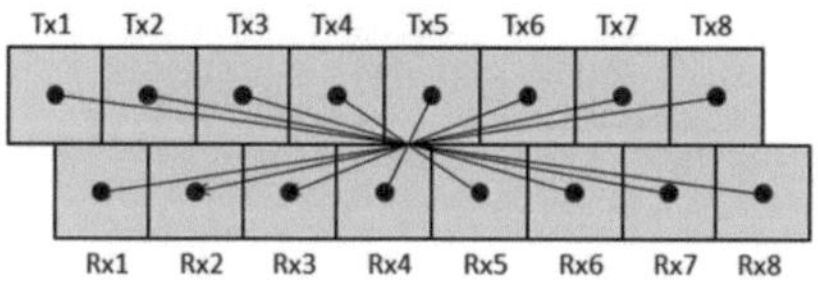


**Figure 11.**
*Combinations of the transmitting and receiving antennas to acquire CMP data sets.*

the CMP data by using a conventional GPR system, we move the transmit and receive antenna simultaneously to the opposite direction so that the reflection from the CMP point stays at one position. We fit theoretical arrival time of the reflected wave from the target at the midpoint position and estimate the velocity and the depth of the reflecting layer simultaneously by the use of a velocity spectrum. MIMO GPR can achieve CMP measurement by selecting a combination of antennas so that the center of the array is the midpoint (midpoint) of the transmitting and receiving antennas, as shown in **Figure 11**. CMP measurement can be performed without moving antennas by MIMO GPR [15, 16].

We show an example of simultaneous CMP and profile measurements performed by Yakumo near Sendai Airport, which was damaged by the tsunami of the Great East Japan Earthquake in 2011. This site was a rice paddy field, but the tsunami invaded, and then, the surface soil was releveled. **Figure 12** is the CMP data, and **Figure 13** is the velocity spectrum obtained by the CMP analysis. Spectral peaks are seen at four different depths, detecting four stratified geological boundaries. **Figure 14** shows a continuous display of the velocity obtained by the CMP analysis along the survey line. Under the assumption that homogeneous soil moisture is almost uniform, the distribution of geological boundaries can be detected. These are considered to contain information from geological deposited by the Great East Japan Earthquake in 2011 to past tsunami deposits from more than 1000 years ago.

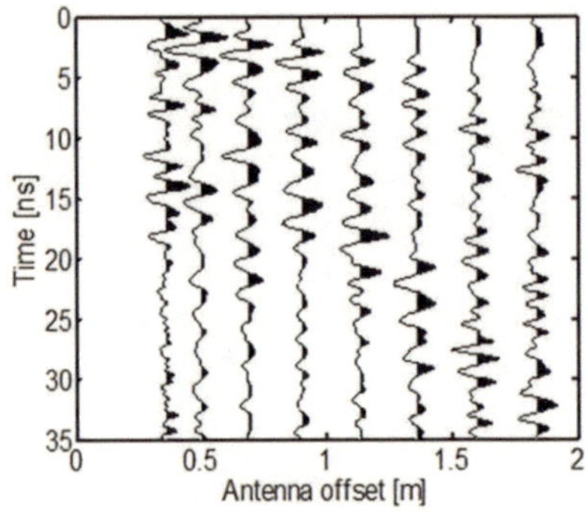


**Figure 12.**
*CMP profile measured by Yakumo near Sendai Airport.*

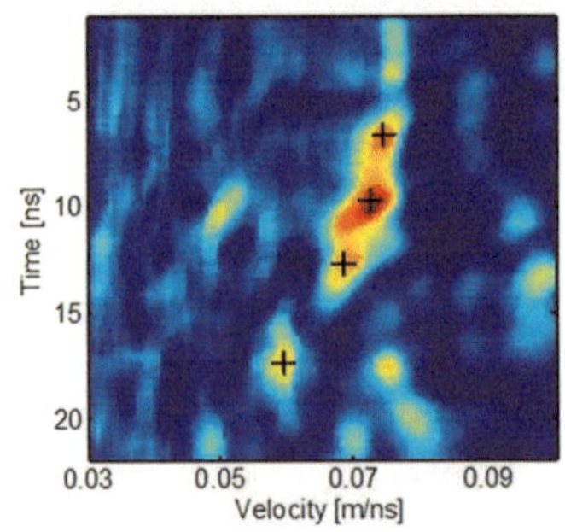


**Figure 13.**
*The velocity spectrum of the CMP profile in **Figure 12a**.*

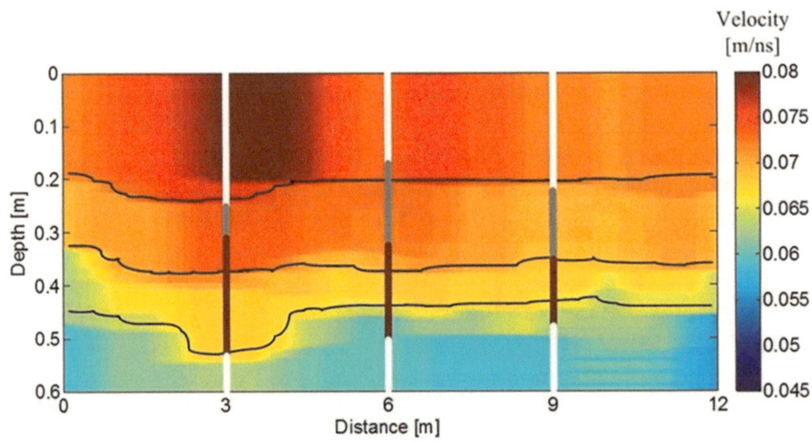


**Figure 14.**
*Continuous display of the velocity obtained by the CMP analysis along the survey line.*

## 10. Conclusion

The design and prototype MIMO GB-SAR was shown in this chapter. Higher pulse repetition frequency (PRF) of MIMO GB-SAR can easily be achieved, and it can be used for vibration measurement. Compared to the conventional GB-SAR, MIMO GB-SAR has advantage in maintenance, because there is no mechanical moving component.

By using MIMO-GPR, it is possible to measure a wide area with a wide swarth width for one scan. However, MIMO-GPR is not limited to wide-area measurement, but it can be used for simultaneous measurement of the wave velocity by CMP and common offset profiling [17, 18].

## Acknowledgements

MIMO GB-SAR and MINO GPR have been conducted by graduate students of Tohoku University. I acknowledge the contribution especially by Yuya Akiyama and Amarsaikhan Tsogtbaatar. Part of this work was supported by JSPS KAKENHI Grant Number 20K20990.

## Author details

Motoyuki Sato
Tohoku University, Sendai, Japan

*Address all correspondence to: motoyuki.sato.b3@tohoku.ac.jp

## References

[1] Takahashi K, Matsumoto M, Sato M. Continuous observation of natural-disaster-affected areas using ground-based SAR interferometry. IEEE Journal of Selected Topics in Applied Earth Observations and Remote Sensing. 2013;**6**(6):1286-1294

[2] Leva D, Nico G, Tarchi D, Fortuny-Guasch J, Sieber A. Temporal analysis of a landslide by means of a ground-based SAR interferometer. IEEE Transactions on Geoscience and Remote Sensing. 2003;**41**(4):745-752

[3] Noferini L, Pieraccini M, Mecatti D, Macaluso G, Atzeni C, Mantovani M, et al. Using GB-SAR technique to monitor slow moving landslide. Engineering Geology. 2007;**95**:88-98

[4] Kuraoka S, Nakashima Y, Doke R, Mannen K. Monitoring ground deformation of eruption center by ground-based interferometric synthetic aperture radar (GB-InSAR): A case study during the 2015 phreatic eruption of Hakone volcano. Earth Planets and Space. 2018;**70**(181). DOI: 10.1186/s40623-018-0951-0

[5] Schaefera LN, Tragliab FD, Chaussardc E, Lud Z, Nolesinie T, Casaglib A. Monitoring volcano slope instability with synthetic aperture radar: A review and new data from Pacaya (Guatemala) and Stromboli (Italy) volcanoes. Earth-Science Reviews. 2019; **192**:236-257

[6] Xiang X, Chen J, Wang H, Pei L, Wu Z. PS selection method for and application to GB-SAR monitoring of dam deformation. Advances in Civil Engineering. 2019. DOI: 10.1155/8320351

[7] Hu J, Guo J, Xu Y, Zhou L, Zhang S, Fan K. Differential ground-based radar interferometry for slope and civil structures monitoring: Two case studies of landslide and bridge. Remote Sensing. 2019;**11**(24):2887. DOI: 10.3390/rs11242887

[8] Tarchi D, Oliveri F, Sammartino PF. MIMO radar and ground-based SAR imaging systems: Equivalent approaches for remote sensing. IEEE Transactions on Geoscience and Remote Sensing. 2013;**51**(1):425-435

[9] Feng W, Zou L, Sato M. 2D imaging by sparse array radar system. IEICE Technical Report. 2016;**116**(309):65-70 EMT2016-49

[10] Ender J, Klare J. System Architectures and Algorithms for Radar Imaging By MIMO-SAR. Proc. IEEE Radar Conf., 2009. pp. 1-6

[11] Hu C, Wang J, Tian W, Zeng T, Wang R. Design and imaging of ground-based multiple-input multiple-output synthetic aperture radar (MIMO SAR) with non-collinear arrays. Sensors. 2017; **17**(3):598

[12] ASA E-H, Akiyama Y, Sato M. MIMO antenna array for GB-SAR. IEICE Technical Report. 2019;**119**(120):95-100 AP2019-39

[13] Sato M, Yi L, Iitsuka Y, Zou L, Takahashi K. Optimization of antenna polarization of the multistatic GPR system "Yakumo". In: Proceedings of the IEEE 16th International Conference on GPR; 13-16 June 2016; Hong Kong. DOI: 10.1109/ICGPR.2016.7572664

[14] Wong PTW, Lai WWL, Sato M. Time-frequency spectral analysis of step frequency continuous wave and impulse ground penetrating radar. In:

Proceedings of the IEEE 16th International Conference on GPR; 13-16 June 2016; Hong Kong. DOI: 10.1109/ ICGPR2016.7572694

[15] Li Y, Zou L, Sato M. Practical approach for high-resolution airport pavement inspection with the Yakumo multistatic array ground-penetrating radar system. Sensors. 2018;**18**:2684. DOI: 10.3390/s18082684

[16] Kikuta K, Li Y, Zou L, Sato M. Robust subsurface velocity change detection method with Yakumo multistatic GPR system. In: Proceedings of the IEEE International Geoscience and Remote Sensing Symposium; 28 July 2019-02 August; Yokokohama-Japan. DOI: 10.1109/IGARSS.2019.8900570

[17] Li Y, Takahashi K, Sato M. High-resolution velocity analysis method using the $\ell$-1 norm regularized least-squares method for pavement inspection. IEEE Journal of Selected Topics in Applied Earth Observations and Remote Sensing. 2018;**11**(3): 1005-1015

[18] Zou L, Li Y, Sato M. On the use of lateral wave for the interlayer debonding detecting in an asphalt airport pavement using a multi-static GPR system. Transactions on Geoscience and Remote Sensing. 2020;**58**(6):4215-4224

Chapter 15

# Localization Techniques in Multiple-Input Multiple-Output Communication: Fundamental Principles, Challenges, and Opportunities

*Katarina Vuckovic and Nazanin Rahanvard*


## Abstract

This chapter provides an overview of localization techniques in Multiple-Input Multiple-Output (MIMO) communication systems. The chapter mainly focuses on sub-6 GHz and mmWave bands. MIMO technology enables high-capacity wireless communication, but also presents challenges for localization due to the complexity of the signal propagation environment. Various methods have been developed to overcome these challenges, which utilize side information such as the map of the area, or techniques such as Compressive Sensing (CS), Deep Learning (DL), Gaussian Process Regression (GPR), or clustering. These techniques utilize wireless communication parameters such as Received Signal Strength Indicator (RSSI), Channel State Information (CSI), Angle-Delay-Profile (ADP), Angle-of-Departure (AoD), Angle-of-Arrival (AoA), or Time-of-Arrival (ToA) as inputs to estimate the user's location. The goal of this chapter is to offer a comprehensive understanding of MIMO localization techniques, along with an overview of the challenges and opportunities associated with them. Furthermore, it also aims to provide the theoretical background on channel models and wireless channel parameters required to understand the localization techniques.


**Keywords:** localization techniques, positioning system, channel model, channel parameters, machine learning

## 1. Introduction

The proliferation of smartphone devices has enabled the expansion of Location Based Services (LBS) [1]. With the increasing popularity of LBS applications, there is a growing demand for more accurate localization solutions. Wireless MIMO localization is an alternative solution to the widely accepted Global Positioning System (GPS) in environments where GPS falls short. Specifically, GPS faces a challenge in maintaining accuracy and availability with urban canyons and indoor environments

[2]. Wireless MIMO systems already exist in these environments for communication purposes. Therefore, the existing wireless communication infrastructure can also be leveraged to provide localization services without investing in additional equipment. In fact, many LBS applications are enabled by wireless MIMO localization. While compiling a comprehensive list of these applications would be difficult, the following subsections provide an overview of some interesting LBS applications.

## 1.1 Applications

### *1.1.1 Emergency services*

The purpose of emergency services is to identify a caller's location and provide this information to the emergency responders. Emergency service is the oldest LBS application. The need to position mobile users was first advocated back in 1996 when the Federal Communication Commission (FCC) announced its mandate to enhance emergency services. During that time, the main motivation was mostly centered around locating emergency calls [3]. Since then, both FCC Enhanced 911 (E911) and 3rd Generation Partnership Project (3GPP) requirements for localization accuracy have become more stringent [4, 5].

### *1.1.2 Autonomous vehicles and urban air mobility*

Precise positioning systems play a crucial role in autonomous vehicles and Unmanned Aerial Systems (UASs) [6]. The purpose of these positioning systems is to provide accurate estimations of the vehicle's location and orientation relative to the road and other vehicles (whether terrestrial or aerial). Moreover, the localization systems facilitate tracking of other vehicles, pedestrians, and obstacles in the surroundings. This information is utilized to plan safe and efficient routes, and to avoid collisions. The wireless MIMO system can provide primary location estimation or a backup in the event of GPS failure or loss of other proximity sensors [2]. Several studies have explored using MIMO localization for vehicles [7–11] and UASs [12–14].

### *1.1.3 Field surveying and mapping*

Field surveying and mapping has both civilian and military applications including creating detailed topographical maps, measuring land boundaries, and collecting data on natural resources. For example, in construction surveying, positioning and localization systems are used to ensure that buildings and infrastructures are positioned and aligned correctly. In military applications, these systems can be used for reconnaissance of enemy territory and targeting of enemy or enemy assets. Simultaneous Localization and Mapping (SLAM) is often employed in these types of applications. SLAM is an active area of research and over the past few years, various surveys have been published that summarize the state-of-the-art SLAM solutions [15–17].

### *1.1.4 Indoor tracking and localization*

Indoor tracking and localization technology have numerous practical applications across various industries. In healthcare, it can be used to track the location of medical equipment, staff and patients, ensuring efficient use of resources and timely delivery of care [18]. In the retail industry, it can help to optimize store layouts and improve the customer experience by providing personalized recommendations and targeted advertising. In industrial settings, it can improve warehouse logistics and inventory management by providing real-time tracking of goods and equipment [19]. Additionally, indoor tracking and localization can be used to enhance the safety of buildings and occupants by detecting and responding to emergencies, such as fires or security breaches. The technology also has potential applications in the field of smart architectures (smart homes [20], smart buildings [21], smart cities [22], and smart grids [23]) where it can be used to automate and optimize tasks and energy consumption.

#### *1.1.5 Agriculture*

Highly accurate localization systems have a wide range of applications in agriculture, including precision farming, autonomous equipment, livestock tracking, and soil mapping [24–28]. In precision farming, localization systems are used to collect data on soil conditions, crop growth, weather patterns, and other factors, which can then be analyzed to make informed decisions about crop management, including planting, fertilization, irrigation, and harvesting. Moreover, the accurate localization systems are also used to guide autonomous equipment to carry out tasks such as planting, spraying, and harvesting with greater precision and efficiency.

#### *1.1.6 Social networking*

LBS-enabled social networking applications aim to connect people who are located near each other and share similar interests. These applications use location data to recommend nearby events, activities, or groups that users might be interested in, and facilitate connections with others who are nearby. This approach offers benefits for both individuals and businesses. Some popular LBS-enabled social networking applications include Meetup, Foursquare, Yelp, and Facebook Places.

## 2. Wireless MIMO system

### 2.1 Sub-6 GHz and mmWave massive MIMO systems

Fifth-Generation and Beyond (5G&B) mobile networks offer the potential for significantly greater communication capacity and ultra high-speeds that exceed those of previous generations by several orders of magnitude [29]. The large number of antennas in massive MIMO allows for more precise control of the signals, leading to increased capacity, better coverage, improved energy efficiency and reliability [30, 31]. Specifically, massive MIMO antennas enable the generation of narrow and highly directional signal beams. A beam can be steered towards a user to provide a high-quality signal that is less susceptible to interference and fading.

Sub-6 GHz bands are typically between 1 and 6 GHz. This frequency range is commonly used for wireless communication technologies such as cellular networks (3G, 4G, and 5G), Wi-Fi, Bluetooth, and other wireless communication standards. Sub-6 GHz systems are typically implemented using small-scale MIMO antennas. Regarding the sub-6 GHz channel, several measurement campaigns have been carried out to characterize it [32–34]. The propagation that depends on path-loss and shadowing results in large-scale fading, and multi-path propagation, results in small-scale fading [35].

The massive increase in data traffic has made the sub-6 GHz spectrum congested. This results in limited bandwidth for users, causing slower and unreliable connections [36]. One solution to this problem is to move to a different frequency band such as milimeter-Wave (mmWave) frequency channels. The channels are called mmWave because their wavelength ranges between 1 mm and 10 mm, which is equivalent to a frequency range between 30 GHz and 300 GHz. The mmWave channels can provide significantly more bandwidth compared to sub-6 GHz, which will be required for next generation wireless communication systems. Therefore, mmWave frequency has been identified as a key technology-enabler in 5G&B [30, 35, 36]. However, there are some disadvantages in mmWave communication such as severe signal attenuation and blockage. The signals cannot penetrate obstacles and tend to get absorbed by rain [37, 38].

In an experimental study, a comprehensive channel measurement campaign was conducted in Europe in 2014–2016 in numerous indoor and outdoor scenarios. The study showed that geometry of the main propagation paths at sub-6 GHz and mmWave bands are almost similar [39]. However, the blockage at mmWave band causes higher losses, rendering the path completely blocked. This experimental outcome has motivated several recent studies to use sub-6 GHz channel information for mmWave applications [40–42].

## 2.2 Single-site system model

In wireless communication, the Base Station (BS) and User Equipment (UE) engage in point-to-point communication as shown in **Figure 1**. The BS may function

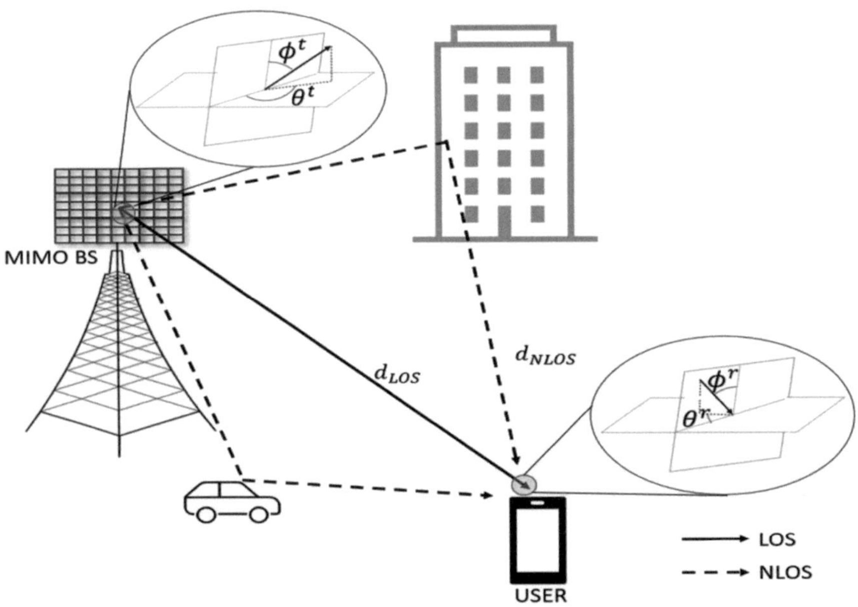


**Figure 1.**
*Single-site wireless MIMO channel model showing LOS and NLOS propagation paths between BS and UE.*

as an Access Point (AP) or as another device in device-to-device communication. Typically, the BS has multiple antenna array elements while the UE may have one or more antenna elements. A general assumption is that the BS and UE are located in the far-field zones of each other, and multiple propagation paths exist between them. Multipaths arise from either reflection off objects or scattering [43]. Typically, there is a Line-of-Sight (LOS) path and several Non-LOS (NLOS) paths. The LOS path can be blocked, in which case only NLOS paths may exist.

Regardless of which side transmits the signal, the propagation path geometry between the BS and UE remains the same. Each path is characterized by an Angle-of-Departure (AoD), an Angle-of-Arrival (AoA), a Time-of-Arrival (ToA), and a complex gain. Since the signal geometry is invariant, it is possible to use AoA and AoD interchangeably. The AoD and AoA are vectors that define the azimuth and elevation angles in 3D space, while ToA represents the time it takes for the propagating signal to travel from the transmitter to the receiver. The ToA is sometimes referred to as the propagation path *delay*. ToA is equal to the length of the path traveled ($d$) divided by the speed of light ($c$):

$$\tau = d/c. \tag{1}$$

The 2D multipath propagation geometry is illustrated in **Figure 2**. In the LOS case, the shortest distance between the BS and UE represents the path traveled by the LOS signal. Furthermore, **Figure 2(a)** shows the AoD from the BS $\theta^t_{LOS}$ and AoA at the UE $\theta^r_{LOS}$. On the other hand, the NLOS propagation path can be modeled using a *virtual BS* (BS') [43] as depicted in **Figure 2(b)**. A NLOS path can be thought of as direct path from a virtual node behind the reflecting surface. The virtual BS is on the opposite side of the reflecting surface, maintaining the same distance from it as the original BS, resulting in $d_{NLOS1} = d'_{NLOS1}$. The total path traveled by the NLOS signal is $d_{NLOS} = d_{NLOS1} + d_{NLOS2}$. Furthermore, the AoD from the virtual BS can be calculated as $\pi - \theta^t_{NLOS}$, where $\theta^t_{NLOS}$ is the AoD at the original BS.

## 2.3 Channel model

The wireless communication community has widely adopted the COST 2100 MIMO channel model [44] as the predominant geometric channel model. This model

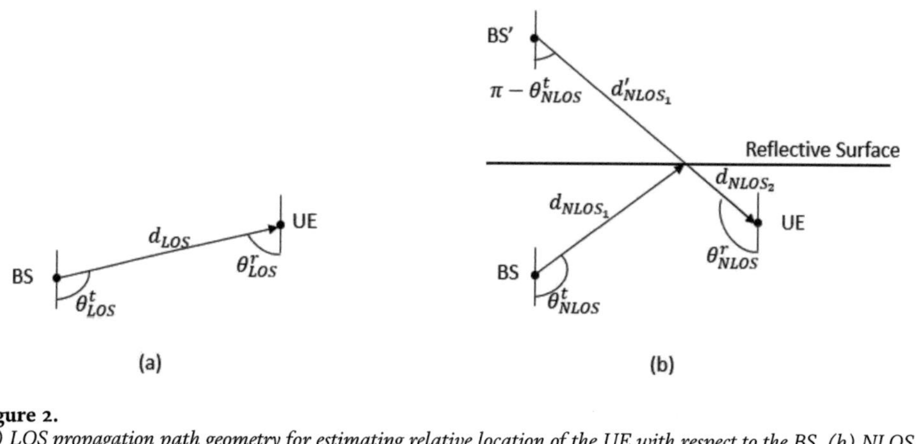


**Figure 2.**
*(a) LOS propagation path geometry for estimating relative location of the UE with respect to the BS. (b) NLOS propagation path and virtual BS (BS') geometry for estimating relative location of the UE with respect to the BS.*

expresses that a propagation environment can be defined by a set of scatterers that create clusters of multipath components. The model is applicable for both sub-6 GHz and mmWave band frequencies.

Consider a MIMO Orthogonal Frequency-Division Multiplexing (OFDM) wireless system, in which the BS and the UE are equipped with antenna arrays with $N_B$ and $N_U$ elements, respectively. The system uses OFDM signaling with $N_C$ subcarriers and the wideband channel has $L$ taps. The received signal at the $l^{th}$ subcarrier of the UE antenna array can be expressed as

$$\boldsymbol{y}[l] = \boldsymbol{H}[l]\boldsymbol{s}[l] + \boldsymbol{n}[l]. \tag{2}$$

Here, $\boldsymbol{y}[l] \in \mathbb{C}^{N_U \times 1}$ denotes the received signal, $\boldsymbol{H}[l] \in \mathbb{C}^{N_U \times N_B}$ represents the channel matrix, $\boldsymbol{s}[l] \in \mathbb{C}^{N_B \times 1}$ represents the transmitted signal, and $\boldsymbol{n}[l] \sim \mathcal{N}\mathbb{C}(\mathbf{0},, \sigma^2\boldsymbol{I})$ denotes the noise at the receiver.

The propagation paths between the BS and the UE can be split into $C$ distinguishable path clusters, with each cluster containing $R_C$ distinguishable paths. Each path cluster is characterized by a mean time delay $\tau_m^{(k)}, k \in 1, \dots, C, m \in 1, \dots, R_C$, a mean AoD $\theta_c^{tx}, \phi_c^{tx} \in [0, \pi)$, and a mean AoA $\theta_c^{rx}, \phi_c^{rx} \in [0, 2\pi)$. Each cluster contributes $R_C$ paths between the transmitter and the receiver, where each path has a relative time delay $\tau_{c_m}$ (relative with respect to mean), a relative AOD $\theta_{c_m}^{tx}, \phi_{c_m}^{tx}$, a relative AOA $\theta_{c_m}^{rx}, \phi_{c_m}^{rx}$, and a complex path gain $\alpha_{c_m}$. The mean and relative paths are illustrated in **Figure 3**.

## 2.4 Channel state information (CSI)

Assuming the channel model defined above, the complex baseband delay-$\ell$ MIMO channel matrix $\boldsymbol{H}[\ell] \in \mathbb{C}^{N_U \times N_B}$ can be written as [45, 46]

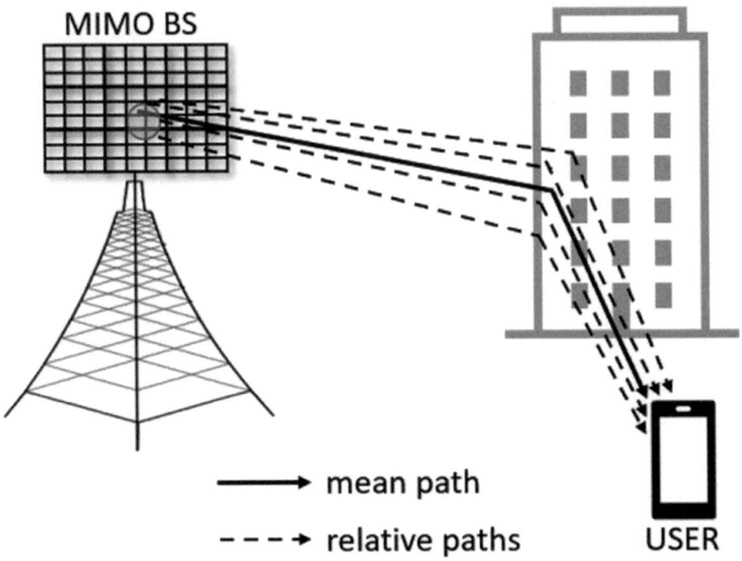


**Figure 3.**
*The mean and relative paths of a NLOS path.*

$$\boldsymbol{H}[\ell] = \sqrt{\frac{N_B N_U}{P_{pl}}} \sum_{c=1}^{C} \sum_{c_m=1}^{R_C} \alpha_{c_m} \boldsymbol{e}_{rx}\left(\theta_c^{rx} + \theta_{c_m}^{rx}, \phi_c^{rx} + \phi_{c_m}^{rx}\right) \boldsymbol{e}_{tx}^H\left(\theta_c^{tx} + \theta_{c_m}^{tx}, \phi_c^{tx} + \phi_{c_m}^{tx}\right) \delta(\ell T_s - n_{c_m} T_s), \tag{3}$$

where $\ell = 0, 1, \ldots, L-1$. Furthermore, $P_{pl}$ indicates the pathloss between the transmitter and the receiver, while $\boldsymbol{e}_{tx}(\theta,\phi) \in \mathbb{C}^{N_B \times 1}$ and $\boldsymbol{e}_{rx}(\theta,\phi) \in \mathbb{C}^{N_U \times 1}$ denote the antenna array response vectors of the transmitter and the receiver, respectively. $\delta(t)$ is the Dirac function, $T_s$ is the signaling time, and $n_{c_m} = \lfloor \frac{\tau_c + \tau_{c_m}}{T_s} \rfloor$.

The channel matrix at subcarrier $k$, denoted as $\mathcal{H}[k]$, can be written as $\mathcal{H}[k] = \sum_{\ell=0}^{L-1} \boldsymbol{H}[\ell] e^{-j\frac{2\pi k}{N_C}\ell}$. The overall Channel Frequency Response (CFR) matrix, denoted as $\mathbf{H}$, can be expressed as $\mathcal{H} = [\mathcal{H}[0], \mathcal{H}[1], \ldots, \mathcal{H}[N_C - 1]]$, where $Nc$ is the number of subcarriers. This matrix is also known as the Channel State Information (CSI) and its estimation is referred to as the channel estimation problem.

The direct measurement of CSI is possible using MIMO-OFDM systems with fully digital beamforming which is available at sub-6 GHz bands. However, in the mmWave band, only analog beamforming is available, making direct CSI measurement not feasible. Instead, estimation techniques are used to obtain the CSI indirectly [47]. Channel estimation in mmWave massive MIMO channel is under extensive research and several CSI estimation methods have been proposed to this end [48–50]. Accurate estimation of these parameters is crucial for effective localization.

### 2.5 Angle-delay-profile (ADP)

Assuming a single antenna at the UE and a uniform linear array antenna at the BS, the ADP is a linear transformation of the CSI computed by multiplying it with two Discrete Fourier Transform (DFT) matrices $\boldsymbol{V} \in \mathbb{C}^{N_B \times N_B}$ and $\boldsymbol{F} \in \mathbb{C}^{N_C \times N_C}$. The ADP matrix $\boldsymbol{G} \in \mathbb{C}^{N_B \times N_C}$ is defined as follows [51]

$$\boldsymbol{G} = \boldsymbol{V}^H \mathcal{H} \boldsymbol{F}, \tag{4}$$

where $\boldsymbol{V} \in \mathbb{C}^{N_B \times N_B}$ is defined as

$$[\boldsymbol{V}]_{i,k} \triangleq \frac{1}{\sqrt{N_B}} e^{-j2\pi \frac{\left(i\left(k - \frac{N_B}{2}\right)\right)}{N_B}}, \tag{5}$$

and $\boldsymbol{F} \in \mathbb{C}^{N_C \times N_C}$ as

$$[\boldsymbol{F}]_{i,k} \triangleq \frac{1}{\sqrt{N_C}} e^{-j2\pi \frac{ik}{N_C}}, \tag{6}$$

where $i = 0, \ldots, N_C - 1$ and $k = 0, \ldots, N_B - 1$.

This transformation has proven to be quite useful for various localization applications. **Figure 4** illustrates an example of the magnitude of the raw CSI $|\mathcal{H}|$ and its ADP transformation $|\boldsymbol{G}|$. The transformation converts the data into a sparse representation which has shown to improve the performance and generalizability of data-driven models [52]. Furthermore, in the visual representation of the raw CSI data, the scattering characteristics of the multipaths are ambiguous [53]. In contrast, the ADP

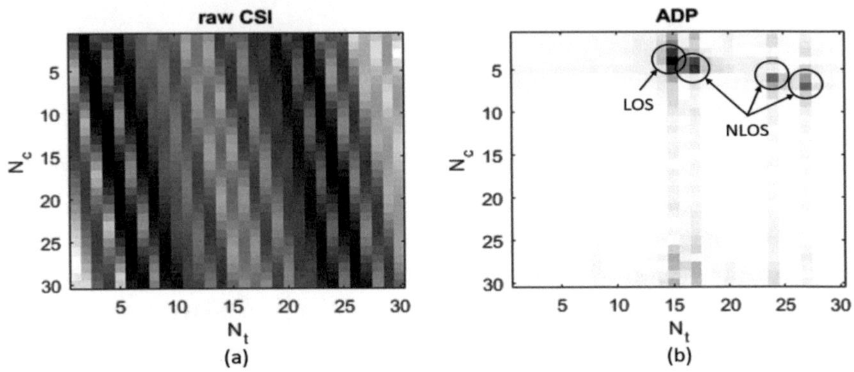


**Figure 4.**
*(a) Raw CSI data of a OFDM-MIMO system with 30 sub-carriers. The BS is equipped with a uniform linear array antenna with 30 antenna elements, and UE with single antenna. (b) the ADP transformation of the CSI in (a) with LOS and NLOS path clusters labeled.*

provides semantic visual interpretation of the channel multipath, where $[\mathbf{G}]_{i,k}$ denotes the power of path associated with the angle

$$\theta_k = \arccos\left(\frac{2k - N_B}{N_B}\right), \tag{7}$$

and delay

$$\tau_i = iT_s. \tag{8}$$

The semantic visual interpretation means that the path clusters can easily be identified visually in the ADP. Referring to **Figure 4(b)**, the strongest peak in the ADP is the LOS path cluster and the remaining peaks are NLOS path clusters. This information is not visually observable in the raw CSI in **Figure 4(a)**.

### 2.6 Received signal strength indicator (RSSI)

The RSSI is a metric used in wireless communication systems that measures the strength of a received signal. RSSI parameters are typically used in distributed (or cell free) MIMO localization systems. Cell-free MIMO uses a large number of distributed antennas and MIMO techniques to improve coverage, capacity, and reliability compared to single-site MIMO system shown in **Figure 1**. Specifically, it aims to improve the performance of single-site MIMO systems by dynamically assigning antennas to users based on their location and available resources.

An example of a distributed MIMO system is illustrated in **Figure 5**. In this example, there are multiple BSs distributed in the environment. The RSSI is measured for each BS to create a RSSI vector $\mathbf{p} = (p_1, p_2, \dots, p_M)$, where $M$ represents the number of BSs and $p_i$ is the RSSI from the $i^{th}$ BS. The RSSI vector should be unique for every location in the environment. To ensure the uniqueness of the RSSI vector, multiple BSs are necessary.

### 2.7 Channel parameters summary

The common channel parameters discussed above are summarized in **Table 1**. These parameters can be used individually or in combination to estimate the location

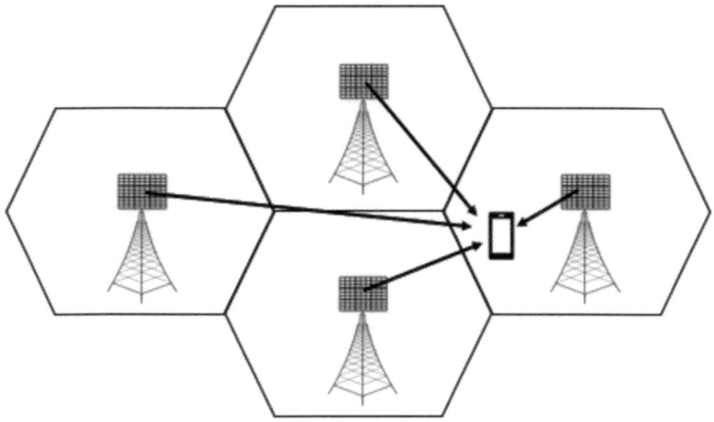

**Figure 5.**
*Illustration of a distributed MIMO system with four BSs and one UE.*

| Parameter | Description | Notation |
|---|---|---|
| AoD | Angle-of-departure | $(\theta^{(t)}, \phi^{(t)})$ |
| AoA | Angle-of-arrival | $(\theta^{(r)}, \phi^{(r)})$ |
| ToA | Time-of-arrival (delay) | $\tau$ |
| CSI | Channel state information | $\mathbf{H}$ |
| ADP | Angle delay profile | $G$ |
| RSSI | Received signal strength indicator | $\mathbf{p}$ |

**Table 1.**
*List of MIMO channel parameters utilized in localization techniques.*

of a UE. For instance, to define a propagation path, AoD or AoA is often used in conjunction with ToA.

## 3. Localization techniques

Localization is an extensive area of research in wireless MIMO communication and several different approaches have been proposed to solve this problem. This section provides an overview of the common localization techniques in sub-6 GHz and mmWave MIMO systems.

### 3.1 Map-assisted localization

Map-assisted localization techniques leverage 2D or 3D environment maps along with channel parameters to determine the location of UEs. The map provides information about the scattering surfaces and other obstacles in the environment. Then, by utilizing the AoD and delay of the signal path, multiple beam paths can be traced from the BS to the UE. This is illustrated in **Figure 6**. The paths are traced using the geometry defined in **Figure 2**. The point where these paths intersect is the UE's

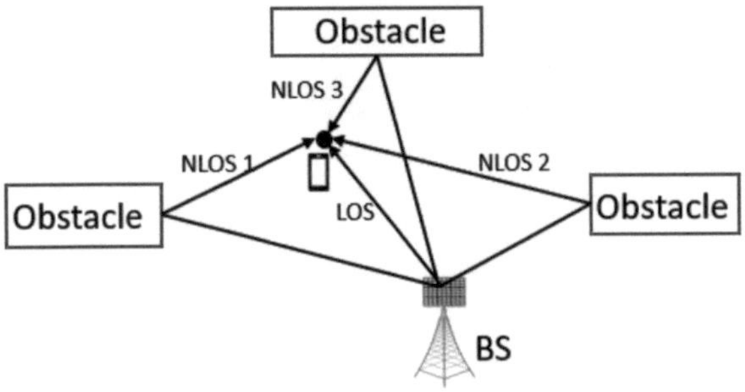


**Figure 6.**
*Map-assisted localization using propagation path tracing. The figure illustrates the LOS path and three NLOS paths between the BS and UE. The intersection of these four paths represents the UE's location.*

location. The minimum requirement to localize the UE is the AoD of two different paths. Alternatively, the UE can be localized if the angle and delay of a single path are known. The delay is used to estimate the length of the path by solving for $d$ in (1). However, more precise localization is achieved by utilizing multiple paths and incorporating both angle and delay information. Furthermore, since the communication is bi-directional, either AoD or AoA can be used to estimate the UE's location.

When analog beamforming is available, which is typically at lower frequency bands (i.e. sub-6 GHz), the angle and delay can be directly measured. However, the mmWave bands digital beam forming is still prevalent, which does not enable measuring angle and delay directly. Therefore, angle and delay parameters have to be estimated. One approach to this problem is to estimate CSI and convert it to ADP. Then, the angle and delay can be estimated using (7) and (8), respectively.

### 3.2 Localization using compressive sensing techniques

*Compressive Sensing* (CS), also known as compressed sensing or sparse sampling, is a signal processing technique that allows for the reconstruction of a sparse signal from a small number of measurements or samples. CS has found its way in many applications [54, 55]. *Sparsity* is the property of a signal or data representation whereby a small number of coefficients or elements carry most of the signal's energy or information content, while the majority of coefficients or elements are zero or close to zero [56]. In fact, many real-world signals are sparse or compressible in either their original domain or some transform domain, such as Fourier or wavelet transforms [57]. An example is shown in **Figure 4**, where the raw CSI data is transformed into ADP to create a sparse representation. As may be observed in the ADP, the multipath components are concentrated into only a few clusters creating a sparse representation.

CS techniques have found many applications in wireless MIMO communication by exploiting the sparsity of channel model parameters [57, 58]. These applications include channel estimation, spectrum sensing, and localization. Channel estimation provides information on the AoA/AoD and ToA of the paths and thus the relative location of the UE with respect to the BS can be estimated.

In mmWave MIMO communication, channel estimation and localization are typically combined. The idea behind sparse channel estimation is that the system can make only a few random measurements which are then used to reconstruct channel model parameters using CS techniques. A commonly used CS technique in mmWave

MIMO channel estimation is Distributed Compressive Sensing - Simultaneous Orthogonal Matching Pursuit (DCS-SOMP). DCS-SOMP is typically used to estimate AoA/AoD and ToA [59–61]. Once the angle and delay channel parameters are recovered, the relative UE location can be estimated from the LOS path directly as shown in **Figure 2(a)**. When LOS is not available, the location can be estimated from the NLOS path by applying the virtual BS concept as shown in **Figure 2(b)**.

### 3.3 Fingerprinting-based localization

*Fingerprinting* is a data-driven localization technique that typically consists of two phases: offline phase and online phase. During the offline phase, the locations in the environment are mapped to a unique wireless measurement to create geo-tagged fingerprint database [62]. The unique wireless measurements are referred to as fingerprints. The measurements can be any wireless parameter such as RSSI, CSI/ADP, AoA/AoD or ToA. Then, during the online phase, the new measurement (fingerprint) is compared to the geo-tagged database to estimate the UE's location. The underlying principle behind fingerprinting is that the wireless channel between the UE and BS is uniquely determined by the scattering environment surrounding the UE's location [63]. Therefore, each location has a unique fingerprint. Matching a new wireless measurement to the measurements in the geo-tagged dataset typically involves a machine learning model. The training is performed during the offline phase. The most common fingerprinting models are based on Deep Learning (DL), Gaussian Progress Regression (GPR), or clustering and classification models.

RSSI-based fingerprinting is commonly used in wireless systems that have rich AP distributions such as Wireless Sensor Networks (WSNs) [64–66], Wi-Fi networks [67, 68], or Distributed Massive MIMO (DM-MIMO) systems [69, 70]. Since the RSSI provides a single measurement from the BS or AP, multiple APs are required to generate a unique fingerprint. On the other hand, single-site localization takes advantage of the multipath characteristics of the MIMO channel which are captured in CSI data or the angle and delay parameters that define the multipath. Furthermore, the CSI fingerprint can be used in its original form or it can be transformed into ADP.

#### *3.3.1 Application of deep learning techniques*

Deep Learning Neural Networks (DL NNs) require a large training dataset that covers the entire environment. The input to the NN is the wireless measurement and the output is the UE location. Several different NN architectures have been proposed in fingerprinting-based localization, including Multiple-Layer Perception (MLP) networks, [71, 72], Convolutional Neural Networks (CNNs), [51, 63, 73–75] and Recurrent Neural Networks (RNNs) [53].

Thus far, CNN models have demonstrated the highest localization accuracy performance. The CNN model treats the input fingerprint as a 2D image and performs series of convolutions over multiple layers to establish the spatial correlation in the 2D input. Typically, raw CSI or transformed ADP fingerprints are used for this application. The sparsity of ADP enhances the CNN model both from a computational complexity and a learning point-of-view [76]. RNN models are time series models that can track the changes of the input over time to predict the next UE location. RNN models can predict changes in the environment and account for these changes in the location estimation. RNN models are also used to predict the future location of the UE.

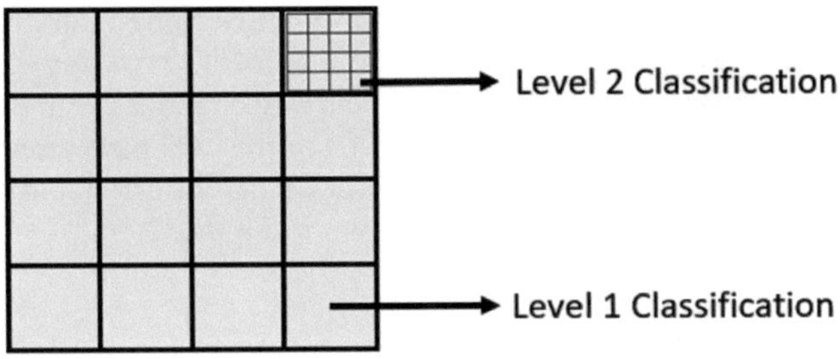


**Figure 7.**
*Fingerprinting based multi-level classification grid of the environment map.*

These networks can either be postulated as classification or regression models. In the classification models, the environment is usually divided into grids where each grid represents a class. If the area is larger, it is not uncommon to have multiple levels of classification, where each grid may be subdivided into smaller grids as shown in **Figure 7**. In general, the first level employs a CNN classification model (coarse search), whereas the second level utilizes a different machine learning algorithm to perform a fine search. In addition to increasing the complexity of the model, the multi-layer approach is more susceptible to errors. If at the first stage, the grid is classified incorrectly, then the error propagates into the second stage. Furthermore, the accuracy of the classification model is limited to the size of the grid. On the other hand, the goal of regression is to find a function or equation that best describes the relationship between the input and output variables. Therefore, regression models predict a continuous output variable and the accuracy is not limited to the grid as in classification.

### *3.3.2 Application of Gaussian process regression models*

A *Gaussian Process* (GP) is a collection of random variable functions indexed by time or space. The key property of a GP is that any finite subset of the random variables is jointly Gaussian distributed. That is, for any finite set of vector elements $\mathbf{x}_1, \dots, \mathbf{x}_n \in \mathcal{X}$, the associated set of random variables $f(\mathbf{x}_1), \dots, f(\mathbf{x}_n)$ follow a joint Gaussian distribution. The following notation is commonly used in literature to represent the GP

$$f(\mathbf{x}) \sim \mathcal{GP}(m(\mathbf{x}), k(\mathbf{x}, \mathbf{x}')), \tag{9}$$

where the mean and covariance functions are defined as

$$m(\mathbf{x}) = \mathbb{E}[f(\mathbf{x})], \tag{10}$$

$$k(\mathbf{x}, \mathbf{x}') = \mathbb{E}[(f(\mathbf{x}) - m(\mathbf{x}))(f(\mathbf{x}') - m(\mathbf{x}')] \tag{11}$$

for any $\mathbf{x}, \mathbf{x}' \in \mathcal{X}$ [77]. Therefore, the GP is entirely defined by its mean and covariance functions [78].

A Gaussian Process Regression (GPR) model is a non-parametric statistical model that uses a GP to model a continuous function and provides a probabilistic prediction with uncertainty estimates [79]. To define the GPR model, assume $\mathcal{D}_{train} \triangleq (\mathbf{X}, \mathbf{y}) \triangleq \{\mathbf{x}_i, y_i\}_{i=1}^{n}, \mathbf{x}_i \in \mathbb{R}^d, y_i \in \mathbb{R}$ to be an input–output pair training dataset. Furthermore, assume that a latent function $f(\cdot)$ is responsible for generating the observed output $y_i$ given the input vector $\mathbf{x}_i$. Then, GPR model can be defined as

$$y_i = f(\mathbf{x}_i) + \epsilon_i, \tag{12}$$

where $f(\mathbf{x}) \sim \mathcal{GP}(m(\mathbf{x}), k(\mathbf{x}, \mathbf{x}')), \epsilon \sim \mathcal{N}(0, \sigma^2\mathbf{I})$ is the noise of the system that has an independent, identically distributed (i.i.d.) Gaussian distribution with zero mean and variance $\sigma^2$, and $i$ refers to the $i^{th}$ observation.

GPR models often assume zero mean as default. The correlation between input points is defined by the covariance function (also known as the kernel). There is a variety of kernels, including exponential, matern, quadratic, and more, each with hyper-parameters that can be fine-tuned during training [80]. Given a new testing sample $\mathbf{x}_*$, the mean and variance (uncertainty) of the unknown output $\mathbf{y}_*$ are predicted as

$$\overline{\mathbf{y}}_* = \mathbf{K}_*^{\mathrm{T}} \left(K + \sigma_n^2 I\right)^{-1} \mathbf{y}, \tag{13}$$

$$\mathbb{V}\left[\overline{\mathbf{y}}_*\right] = \mathbf{K}_{**} - \mathbf{K}_*^{\mathrm{T}} \left(K + \sigma_n^2 I\right)^{-1} \mathbf{K}_*, \tag{14}$$

where $\mathbf{K} = K(\mathbf{X}, \mathbf{X}), \mathbf{K}_* = K(\mathbf{X}, \mathbf{X}_*)$, and $\mathbf{K}_{**} = K(\mathbf{X}_*, \mathbf{X}_*)$ [81]. $\mathbf{K}$ is the covariance matrix (also known as Gram matrix) whose entries are the kernel functions $k(\mathbf{x}_i, \mathbf{x}_j)$ [79].

In localization, the objective of the GPR model is to define the latent function $f(\cdot)$ in (12), where $\mathbf{x}_i$ is the channel parameter (fingerprint) and $y_i$ is the UE location. Given a new fingerprint $\mathbf{x}_*$, the UE location is predicted using (13), while the level of uncertainty in the prediction is estimated by (14). The fingerprint in distributed MIMO systems is usually the RSSI vector as proposed in [69, 70, 82]. On the other hand, the input in single-site MIMO systems can be AoA/AoD vector, CSI or ADP data as proposed in [83, 84].

The main advantage of GPR models over DL CNN models is that they can be trained on substantially smaller datasets. GPR models have shown the ability to train models with small-scale datasets due to the small number of hyper-parameters that define the model [84]. However, the GPR model does have its drawbacks. The main weakness of the GPR model lies in its training complexity, which is characterized by high computational and memory demands. Specifically, GPR training has a computational complexity of $O(n^3)$ and a memory complexity of $O(n^2)$, where $n$ represents the number of training points in the dataset [85].

### *3.3.3 Clustering and classification*

*Clustering* is a class machine learning algorithms used to group similar objects or data points together into clusters. The groupings are based on some similarity or distance measures. The goal of clustering is to identify patterns or structures in the data that may not be immediately apparent, and to group similar data points into clusters that can be easily analyzed or visualized. In localization, clustering techniques can be used to compare the test fingerprints to the fingerprints in the training database. K-means clustering and K-Nearest Neighbor (KNN) classification have widely been used in fingerprinting-based wireless localization and have shown to provide excellent accuracy given enough data point [86, 87]. The KNN location estimation is given by

$$\hat{p} = \frac{1}{K} \sum_{i=1}^{K} p_i, \tag{15}$$

| Technique | Parameters | Methods |
|---|---|---|
| Map-Assisted | CSI/ADP | MAP-CSI [88] |
| | AoA and ToA | MAP-AT [91, 92] |
| CS | AoA/AoD | DC-SOMP [59–61] |
| Fingerprinting DL | CSI/ADP | MLP [71, 72, 93, 94], CNN [51, 73–75, 95], RNN [53] |
| | AoA | MLP [71] |
| | AoA and ToA | MLP [71] |
| | RSSI | MLP [71] |
| Fingerprinting GPR | CSI/ADP | GPR [83], FC-AE-GPR [84], DCGPR [96] |
| | RSSI | DM-MIMO [69, 70, 82, 97] |
| Fingerprinting clustering | CSI/ADP | KNN [51, 63, 86, 87, 93, 98] |
| | AoA | WMSE [90], ASCW [89] |
| | RSSI | KNN [99] |

**Table 2.**
*Methods in MIMO localization, categorized according to the localization technique employed and the parameters utilized.*

where $K$ is the number of surrounding neighbors considered and $\boldsymbol{p}_i$ is the coordinate of the $i^{th}$ nearest reference point. Weighted KNN (WKNN) is an extension of KNN where the contribution of each neighbor is weighted. The WKNN is defined as

$$\hat{\boldsymbol{p}} = \sum_{i=1}^{K} w_i \boldsymbol{p}_i, \tag{16}$$

where $w_i$ is the weight of the $i^{th}$ reference point. Typically, the weight corresponds to the distance between the reference point and the input point. The closer the neighbor is to the input point, the more weight it carries in the final prediction. The weights can also be defined by some similarity criteria calculated between the input and the reference fingerprint. Various similarity criteria have been established in wireless MIMO communication, such as normalized correlation [53, 86, 88], Joint Angle Delay Similarity Coefficient (JADSC) [63], Angular Similarity Coefficient Weight (ASCW) [89], and Weighted Mean Square Error (WMSE) [90].

### 3.4 Summary of methods

**Table 2** provides a summary of the methods proposed in recent years that apply the localization techniques discussed in the previous subsection. The techniques are also grouped by the type of communication parameter used with the associated technique.

## 4. Challenges and opportunities

While MIMO systems offer many potential communication performance improvements and enable highly accurate localization models, several challenges still need to be addressed. This section aims to introduce some of the main challenges in MIMO localization.

### 4.1 Dynamic environments

The majority of the models presented above assume a *static environment*, where the objects within the environment of interest are not moving or changing. In real world scenarios, we observe *dynamic environments* where objects are constantly moving through the environment changing the scattering in the environment quickly and thoroughly [53]. The static environment can be altered by any of the following dynamic changes:

- LOS blockage: a new object blocks the LOS path between the UE and the BS.
- NLOS blockage: a new object blocks some NLOS paths between the UE and the BS.
- NLOS addition: scattering from surfaces of a new object adds some NLOS paths between the UE and the BS.

Some efforts have been undertaken to mitigate the impact of dynamic changes. For example, through the analysis of the time sequence of fingerprints, it becomes possible to identify the moment when a dynamic change anomaly occurred. Models can then be developed to identify and remove the effect of the dynamic change anomaly from the fingerprint sample. However, countering the effects of the dynamic environment still poses a challenge in many proposed approaches.

### 4.2 Dataset collection

Data-driven localization techniques, specifically DL techniques, thus far have shown the best performance when it comes to accuracy. However, there is a major challenge with real world deployment of these models. In particular, data-driven methods necessitate extensive datasets for training the models, which are obtained through costly measurement campaigns that can be difficult to perform. Furthermore, as the environment changes, the dataset becomes invalid and a new measurement campaign needs to be deployed.

### 4.3 Generalization

Generalization in massive MIMO refers to the ability of a system to maintain good performance in a wide range of scenarios, including different channel conditions and new environments. This is important for practical deployment of massive MIMO systems, as it ensures that the system will work well in real-world environments where the conditions may vary.

Transfer Learning (TL) has been suggested as a potential approach to improve generalization in machine learning [73]. This technique involves reusing a pre-trained model to enhance the learning and generalization of a new model. In TL, the pre-trained model is fine-tuned to the new environment using a small dataset representative of that environment. The goal is to leverage the knowledge gained from the prior environment to enhance the learning and generalization of the new environment.

Some studies have been exploring TL techniques to adapt their models to new environments [73, 100, 101]. However, TL does not solve the problem completely as it still requires some data collection in new environments. Generalization remains an open area of research in DL-based localization.

### 4.4 Adversarial attacks

An adversarial attack is a type of cyber-attack where an attacker modifies data to deceive or harm a machine learning system, causing it to produce incorrect or unexpected results. DL techniques are vulnerable to such attacks, and intentional CSI perturbations can significantly impact the accuracy of fingerprinting-based localization. While few studies have addressed adversarial attacks and defenses in the context of MIMO systems [102], it remains an open area of research.

## 5. Conclusions

This chapter offers a comprehensive overview of the localization techniques proposed in wireless MIMO communication systems in sub-6 GHz and mmWave frequency bands. Initially, the need for highly accurate positioning systems is introduced along with some applications in LBS. Subsequently, the wireless communication parameters that define the propagation within the MIMO channel model are introduced. This is followed by a discussion on several localization techniques in MIMO systems including map-assisted, CS based, and fingerprinting models. This chapter explains how each localization technique uses wireless communication parameters to localize the UE. Finally, the last section outlines the remaining challenges and possible opportunities for improvement on MIMO localization.

## Acknowledgements

This work is supported by the National Science Foundation under Grant No. CCF-1718195.

## Author details

Katarina Vuckovic*† and Nazanin Rahanvard†
University of Central Florida, Orlando, USA

*Address all correspondence to: kvuckovic@knights.ucf.edu

† These authors contributed equally.

## References

[1] Kenan M. Comparative analysis of localization techniques used in LBS. In: 2021 5th International Conference on Computing Methodologies and Communication (ICCMC). IEEE; 2021. pp. 300-304

[2] Djuknic GM, Richton RE. Geolocation and assisted GPS. Computer. 2001;**34**(2): 123-125

[3] Reed J, H, Krizman KJ, Woerner BD. An overview of the challenges and progress in meeting the e-911 requirement for location service. IEEE Communications Magazine. 1998;**36**:3037

[4] Majid Butt M, Rao A, Yoon D. RF fingerprinting and deep learning assisted UE positioning in 5G. In: 2020 IEEE 91st Vehicular Technology Conference (VTC2020-Spring). IEEE; 2020. pp. 1-7

[5] del Peral-Rosado JA, Raulefs R, López-Salcedo JA, Seco-Granados G. Survey of cellular mobile radio localization methods: From 1G to 5G. IEEE Communications Surveys Tutorials. 2018;**20**(2):1124-1148

[6] Kuutti S, Fallah S, Katsaros K, Dianati M, Mccullough F, Mouzakitis A. A survey of the state-of-the-art localization techniques and their potentials for autonomous vehicle applications. IEEE Internet of Things Journal. 2018;**5**(2):829-846

[7] Burghal D, Phadke G, Nair A, Wang R, Pan T, Algafis A, et al. Supervised learning approach for relative vehicle localization using V2V MIMO links. In: ICC 2022 - IEEE International Conference on Communications. IEEE; 2022. pp. 4528-4534

[8] Sarker MAL, Son W, Han DS. RIS-assisted hybrid beamforming and connected user vehicle localization for millimeter wave MIMO systems. Sensors. 2023;**23**(7)

[9] Wang H, Wan L, Dong M, Ota K, Wang X. Assistant vehicle localization based on three collaborative base stations via SBL-based robust DOA estimation. IEEE Internet of Things Journal. 2019;**6**(3):5766-5777

[10] Jia R, Kui X, Xia X, Xie W, Sha N, Guo W. Extrinsic information aided fingerprint localization of vehicles for cell-free massive MIMO-OFDM system. IEEE Open Journal of the Communications Society. 2022;**3**: 1810-1819

[11] Chen Y, Palacios J, González-Prelcic N, Shimizu T, Hongsheng L. Joint initial access and localization in millimeter wave vehicular networks: A hybrid model/data driven approach. In: 2022 IEEE 12th Sensor Array and Multichannel Signal Processing Workshop (SAM). Vol. 2022. IEEE. pp. 355-359

[12] Alexandropoulos GC, Vlachos E, Smida B. Joint localization and channel estimation for UAV-assisted millimeter wave communications. In: 2020 54th Asilomar Conference on Signals, Systems, and Computers. IEEE; 2020. pp. 1318-1322

[13] Rodriguez-Fernandez J, Gonzalez-Prelcic N, Heath RW. Position-aided compressive channel estimation and tracking for millimeter wave multi-user MIMO air-to-air communications. In: 2018 IEEE International Conference on Communications Workshops (ICC Workshops). Vol. 2018. IEEE; pp. 1-6

[14] Li X, Jie X. Positioning optimization for sum-rate maximization in

UAV-enabled interference channel. IEEE Signal Processing Letters. 2019;**26**(10): 1466-1470

[15] Bresson G, Alsayed Z, Li Y, Glaser S. Simultaneous localization and mapping: A survey of current trends in autonomous driving. IEEE Transactions on Intelligent Vehicles. 2017;**2**(3):194-220

[16] Placed JA, Strader J, Carrillo H, Atanasov N, Indelman V, Carlone L, et al. A survey on active simultaneous localization and mapping: State of the art and new frontiers. IEEE Transactions on Robotics. 2023;**39**:1-20

[17] Liu Y, Yujia F, Chen F, Goossens B, Tao W, Zhao H. Simultaneous localization and mapping related datasets: A comprehensive survey. arXiv preprint arXiv:2102.04036. 2021

[18] Zafari F, Gkelias A, Leung KK. A survey of indoor localization systems and technologies. IEEE Communications Surveys Tutorials. 2019;**21**(3):2568-2599

[19] Kim J, Hwangbo H, Kim SJ, Kim S. Location-based tracking data and customer movement pattern analysis for sustainable fashion business. Sustainability. 2019;**11**(22):6209

[20] Zhang D, Tan C. Application of indoor positioning technology in smart home management system. In: 2021 IEEE 2nd International Conference on Big Data, Artificial Intelligence and Internet of Things Engineering (ICBAIE). IEEE; 2021. pp. 627-631

[21] Snoonian D. Smart buildings. IEEE Spectrum. 2003;**40**(8):18-23

[22] Kehua S, Li J, Fu H. Smart city and the applications. In: 2011 International Conference on Electronics, Communications and Control (ICECC). IEEE; 2011. pp. 1028-1031

[23] Siano P. Demand response and smart grids—A survey. Renewable and Sustainable Energy Reviews. 2014;**30**: 461-478

[24] Chebrolu N, Lottes P, Läbe T, Stachniss C. Robot localization based on aerial images for precision agriculture tasks in crop fields. In: 2019 International Conference on Robotics and Automation (ICRA). IEEE; 2019. pp. 1787-1793

[25] Ding H, Zhang B, Zhou J, Yan Y, Tian G, Baoxing G. Recent developments and applications of simultaneous localization and mapping in agriculture. Journal of Field Robotics. 2022;**39**(6): 956-983

[26] Abouzar P, Michelson DG, Hamdi M. RSSI-based distributed self-localization for wireless sensor networks used in precision agriculture. IEEE Transactions on Wireless Communications. 2016;**15**(10): 6638-6650

[27] Mohanty MK, Thakurta PKG, Kar S. Efficient sensor node localization in precision agriculture: An ann based framework. Opsearch. 2023:1-24

[28] Mamehgol Yousefi DB, Mohd Rafie AS, Al-Haddad SAR, Azrad S. A systematic literature review on the use of deep learning in precision livestock detection and localization using unmanned aerial vehicles. IEEE Access. 2022;**10**:80071-80091

[29] Boccardi F, Heath RW, Lozano A, Marzetta TL, Popovski P. Five disruptive technology directions for 5G. IEEE Communications Magazine. 2014;**52**(2): 74-80

[30] Agiwal M, Roy A, Saxena N. Next generation 5G wireless networks: A comprehensive survey. IEEE

Communications Surveys & Tutorials. 2016;**18**(3):1617-1655

[31] Lu L, Li GY, Lee Swindlehurst A, Ashikhmin A, Zhang R. An overview of massive MIMO: Benefits and challenges. IEEE Journal of Selected Topics in Signal Processing. 2014;**8**(5):742-758

[32] Gao X, Edfors O, Rusek F, Tufvesson F. Massive MIMO performance evaluation based on measured propagation data. IEEE Transactions on Wireless Communications. 2015;**14**(7):3899-3911

[33] Payami S, Tufvesson F. Channel measurements and analysis for very large array systems at 2.6 GHz. In: 2012 6th European Conference on Antennas and Propagation (EUCAP). IEEE; 2012. pp. 433-437

[34] Ahmad T, Li XJ, Seet B-C. 3D localization using social network analysis for wireless sensor networks. In: 2018 IEEE 3rd International Conference on Communication and Information Systems (ICCIS). IEEE; 2018. pp. 88-92

[35] Bjornson E, Van der Perre L, Buzzi S, Larsson EG. Massive MIMO in sub-6 GHz and mmwave: Physical, practical, and use-case differences. IEEE Wireless Communications. 2019;**26**(2):100-108

[36] Chataut R, Akl R. Massive MIMO systems for 5g and beyond networks—Overview, recent trends, challenges, and future research direction. Sensors. 2020; **20**(10):2753

[37] Wang X, Kong L, Kong F, Qiu F, Xia M, Arnon S, et al. Millimeter wave communication: A comprehensive survey. IEEE Communications Surveys Tutorials. 2018;**20**(3):1616-1653

[38] Dan W, Wang J, Cai Y, Guizani M. Millimeter-wave multimedia communications: Challenges, methodology, and applications. IEEE Communications Magazine. 2015;**53**(1): 232-238

[39] mmMAGIC. Measurement Campaigns and Initial Channel Models for Preferred Suitable Frequency Ranges h2020-ict-671650-mmmagic/d2. 1 v1. 0. 2016

[40] Ali A, González-Prelcic N, Heath RW. Estimating millimeter wave channels using out-of-band measurements. In: 2016 Information Theory and Applications Workshop (ITA). IEEE; 2016. pp. 1-6

[41] Alrabeiah M, Alkhateeb A. Deep learning for mmwave beam and blockage prediction using sub-6 GHz channels. IEEE Transactions on Communications. 2020;**68**(9):5504-5518

[42] Sim MS, Lim Y-G, Park SH, Dai L, Chae C-B. Deep learning-based mmwave beam selection for 5G nr/6G with sub-6 GHz channel information: Algorithms and prototype validation. IEEE Access. 2020;**8**:51634-51646

[43] Shen Y, Win MZ. On the use of multipath geometry for wideband cooperative localization. In: GLOBECOM 2009 - 2009 IEEE Global Telecommunications Conference. IEEE; 2009. pp. 1-6

[44] Liu L, Oestges C, Poutanen J, Haneda K, Vainikainen P, Quitin F, et al. The COST 2100 MIMO channel model. IEEE Wireless Communications. 2012; **19**(6):92-99

[45] Ali A, González-Prelcic N, Heath RW. Millimeter wave beam-selection using out-of-band spatial information. IEEE Transactions on Wireless Communications. 2017;**17**(2): 1038-1052

[46] Alkhateeb A, Heath RW. Frequency selective hybrid precoding for limited feedback millimeter wave systems. IEEE Transactions on Communications. 2016; **64**(5):1801-1818

[47] Hassan K, Masarra M, Zwingelstein M, Dayoub I. Channel estimation techniques for millimeter-wave communication systems: Achievements and challenges. IEEE Open Journal of the Communications Society. 2020;**1**:1336-1363

[48] Dong P, Zhang H, Li GY, Gaspar IS, Alizadeh NN. Deep CNN-based channel estimation for mmWave massive MIMO systems. IEEE Journal of Selected Topics in Signal Processing. 2019;**13**(5): 989-1000

[49] Gao S, Dong P, Pan Z, Li GY. Deep learning based channel estimation for massive MIMO with mixed-resolution ADCs. IEEE Communications Letters. 2019;**23**(11):1989-1993

[50] Jin Y, Zhang J, Ai B, Zhang X. Channel estimation for mmWave massive MIMO with convolutional blind denoising network. IEEE Communications Letters. 2019;**24**(1):95-98

[51] Sun X, Chi W, Gao X, Li GY. Fingerprint-based localization for massive MIMO-OFDM system with deep convolutional neural networks. IEEE Transactions on Vehicular Technology. 2019;**68**(11):10846-10857

[52] Chen T, Zhang Z, Wang P, Balachandra S, Ma H, Wang Z, et al. Sparsity winning twice: better robust generalization from more efficient training. In: International Conference on Learning Representations. arXiv; 2022

[53] Hejazi F, Vuckovic K, Rahnavard N. DyLoc: Dynamic localization for massive MIMO using predictive recurrent neural networks. In: IEEE INFOCOM 2021 - IEEE Conference on Computer Communications. IEEE; 2021. pp. 1-9

[54] Shahrasbi B, Rahnavard N. Model-based nonuniform compressive sampling and recovery of natural images utilizing a wavelet-domain universal hidden Markov model. IEEE Transactions on Signal Processing. 1 Jan 2017;**65**(1):95-104. DOI: 10.1109/TSP.2016.2614654

[55] Tuan Nguyen M, Teague KA, Rahnavard N. CCS: Energy-efficient data collection in clustered wireless sensor networks utilizing block-wise compressive sensing. Computer Networks (IEEE). 2016;**106**:171-185

[56] Baraniuk RG. Compressive sensing [lecture notes]. IEEE Signal Processing Magazine. 2007;**24**(4):118-121

[57] Rani M, Dhok SB, Deshmukh RB. A systematic review of compressive sensing: Concepts, implementations and applications. IEEE Access. 2018;**6**: 4875-4894

[58] Gao Z, Dai L, Han S, Chih-Lin I, Wang Z, Hanzo L. Compressive sensing techniques for next-generation wireless communications. IEEE Wireless Communications. 2018;**25**(3):144-153

[59] Talvitie J, Valkama M, Destino G, Wymeersch H. Novel algorithms for high-accuracy joint position and orientation estimation in 5g mmwave systems. In: 2017 IEEE Globecom Workshops (GC Wkshps). IEEE; 2017. pp. 1-7

[60] Shahmansoori A, Garcia GE, Destino G, Seco-Granados G, Wymeersch H. Position and orientation estimation through millimeter-wave MIMO in 5G systems. IEEE Transactions on Wireless Communications. 2018;**17**(3):1822-1835

[61] Trivedi MA, van Wyk JH. Localization and tracking of high-speed trains using compressed sensing based 5G localization algorithms. In: 2021 IEEE 24th International Conference on Information Fusion (FUSION). IEEE; 2021. pp. 1-8

[62] Alamu O, Iyaomolere B, Abdulrahman A. An overview of massive MIMO localization techniques in wireless cellular networks: Recent advances and outlook. Ad Hoc Networks. 2021;**111**:102353

[63] Sun X, Gao X, Li GY, Han W. Single-site localization based on a new type of fingerprint for massive MIMO-OFDM systems. IEEE Transactions on Vehicular Technology. 2018;**67**(7):6134-6145

[64] Puckdeevongs A. Indoor Localization using RSSI and artificial neural network. In: 2021 9th International Electrical Engineering Congress (iEECON). IEEE; 2021. pp. 479-482

[65] Niu R, Vempaty A, Varshney PK. Received-signal-strength-based localization in wireless sensor networks. Proceedings of the IEEE. 2018;**106**(7): 1166-1182

[66] Csík D, Odry Á, Sarcevic P. Comparison of RSSI-based fingerprinting methods for indoor localization. In: 2022 IEEE 20th Jubilee International Symposium on Intelligent Systems and Informatics (SISY). IEEE; 2022. pp. 000273-000278

[67] Zhang G, Wang P, Chen H, Zhang L. Wireless indoor localization using convolutional neural network and gaussian process regression. Sensors. 2019;**19**(11):2508

[68] Lezama F, González GG, Larroca F, Capdehourat G. Indoor localization using graph neural networks. In: 2021 IEEE Urucon. IEEE; 2021. pp. 51-54

[69] Savic V, Larsson EG. Fingerprinting-based positioning in distributed massive MIMO systems. In: 2015 IEEE 82nd Vehicular Technology Conference (VTC2015-Fall). IEEE; 2015. pp. 1-5

[70] Moosavi SS, Fortier P. Fingerprinting localization method based on clustering and Gaussian process regression in distributed massive MIMO Systems. In: 2020 IEEE 31st annual international symposium on personal, Indoor and Mobile Radio Communications. Vol. 2020. IEEE. pp. 1-7

[71] Bhattacherjee U, Anjinappa CK, Smith LC, Ozturk E, Guvenc I. Localization with deep neural networks using mmwave ray tracing simulations. In: 2020 SoutheastCon. IEEE; 2020. pp. 1-8

[72] Decurninge A, Ordóñez LG, Ferrand P, Gaoning H, Bojie L, Wei Z, et al. CSI-based outdoor localization for massive MIMO: Experiments with a learning approach. In: 2018 15th International Symposium on Wireless Communication Systems (ISWCS). IEEE; 2018. pp. 1-6

[73] De Bast S, Guevara AP, Pollin S. CSI-based positioning in massive MIMO systems using convolutional neural networks. In: 2020 IEEE 91st Vehicular Technology Conference (VTC2020-Spring). IEEE; 2020. pp. 1-5

[74] Vieira J, Leitinger E, Sarajlic M, Li X, Tufvesson F. Deep convolutional neural networks for massive MIMO fingerprint-based positioning. In: 2017 IEEE 28th Annual International Symposium on Personal, Indoor, and Mobile Radio Communications (PIMRC). IEEE; 2017. pp. 1-6

[75] De Bast S, Guevara AP, Pollin S. CSI-based positioning in massive MIMO systems using convolutional neural networks. In: 2020 IEEE 91st Vehicular

Technology Conference (VTC2020-spring). IEEE; 2020. pp. 1-5

[76] LeCun Y, Bengio Y, Hinton G. Deep learning. Nature. 2015;**521**(7553):436-444

[77] Yiu S, Yang K. Gaussian process assisted fingerprinting localization. IEEE Internet of Things Journal. 2016;**3**(5): 683-690

[78] Görtler J, Kehlbeck R, Deussen O. A visual exploration of gaussian processes. Distill. 2019;**4**(4):e17

[79] Rasmussen CE, Williams CKI. Gaussian Processes for Machine Learning. Vol. 11. Cambridge, Massachusetts, USA: The MIT Press; 2005

[80] Duvenaud D. Automatic Model Construction with Gaussian Processes [Thesis]. Cambridge, UK: University of Cambridge; 2014

[81] Wang J. An Intuitive Tutorial to Gaussian Processes Regression. arXiv; 2021

[82] Prasad KNRSV, Hossain E, Bhargava VK. Machine learning methods for RSS-based user positioning in distributed massive mimo. IEEE Transactions on Wireless Communications. 2018;**17**(12): 8402-8417

[83] Moosavi SS, Fortier P. A fingerprint localization method in collocated massive MIMO-OFDM systems using clustering and Gaussian process regression. In: 2020 IEEE 92nd Vehicular Technology Conference (VTC2020-Fall). IEEE; 2020. pp. 1-5

[84] Vuckovic K, Hejazi F, Rahnavard N. CSI-Based Data-Driven Localization Frameworking Using Small-Scale Training Datasets in Single-Site MIMO Systems. arXiv; 2023

[85] Liu H, Ong Y-S, Shen X, Cai J. When Gaussian Process Meets Big Data: A Review of Scalable GPs. arXiv; 2018

[86] Qiu J, Kui X, Shen Z. Cooperative Fingerprint Positioning for Cell-free Massive MIMO Systems. Vol. 2020. 2020 International Conference on Wireless Communications and Signal Processing (WCSP). IEEE; pp. 382-387

[87] Wang X, Lin L, Lin Y, Chen X. A fast single-site fingerprint localization method in massive MIMO system. In: 2019 11th International Conference on Wireless Communications and Signal Processing (WCSP). IEEE; 2019. pp. 1-6

[88] Vuckovic K, Hejazi F, Rahnavard N. MAP-CSI: Single-site map-assisted localization using massive MIMO CSI. In: 2021 IEEE Global Communications Conference (GLOBECOM). IEEE; 2021. pp. 1-6

[89] Liao C, Xu K, Xia X, Xie W, Wang M. AOA-assisted fingerprint localization for cell-free massive MIMO system based on 3D multipath channel model. In: 2020 IEEE 6th International Conference on Computer and Communications (ICCC). IEEE; 2020. pp. 602-607

[90] Shen Z, Kui X, Xia X. 2D fingerprinting-based localization for mmwave cell-free massive MIMO systems. IEEE Communications Letters. 2021;**25**(11):3556-3560

[91] Kanhere O, Shihao J, Xing Y, Rappaport TS. Map-assisted millimeter wave localization for accurate position location. In: 2019 IEEE Global Communications Conference (GLOBECOM). IEEE; 2019. pp. 1-6

[92] Seow CK, Tan SY. Non-line-of-sight localization in multipath environments. IEEE Transactions on Mobile Computing. 2008;7(5):647-660

[93] Sobehy A, Renault É, Mühlethaler P. Generalization aspect of accurate machine learning models for CSI-based localization. Annals of Telecommunications. 2022;**77**(5-6):345-357

[94] Fan J, Chen S, Luo X, Zhang Y, Li GY. A machine learning approach for hierarchical localization based on multipath MIMO fingerprints. IEEE Communications Letters. 2019;**23**(10): 1765-1768

[95] Widmaier M, Arnold M, Dorner S, Cammerer S, ten Brink S. Towards practical indoor positioning based on massive MIMO systems. In: 2019 IEEE 90th Vehicular Technology Conference (VTC2019-Fall). 2019. pp. 1-6

[96] Wang X, Patil M, Yang C, Mao S, Patel PA. Deep convolutional Gaussian processes for mmwave outdoor localization. In: ICASSP 2021 - 2021 IEEE International Conference on Acoustics, Speech and Signal Processing (ICASSP). IEEE; 2021. pp. 8323-8327

[97] Walaa Y, Al-Rashdan, Tahat A. A comparative performance evaluation of machine learning algorithms for fingerprinting based localization in DM-MIMO wireless systems relying on big data techniques. IEEE Access (IEEE). 2020;**8**:109522-109534

[98] Sobehy A, Renault É, Mühlethaler P. CSI-MIMO: K-nearest neighbor applied to indoor localization. In: ICC 2020 - 2020 IEEE International Conference on Communications (ICC). IEEE; 2020. pp. 1-6

[99] René JE, Salazar A, Beltrán K, Caisaluisa O. Subscriber location in 5G mmwave networks - machine learning rf pattern matching. In: 2022 IEEE International Autumn Meeting on Power, Electronics and Computing (ROPEC). Vol. 6. IEEE; 2022. pp. 1-6

[100] Stahlke M, Feigl T, Castañeda García MH, Stirling-Gallacher RA, Seitz J, Mutschler C. Transfer learning to adapt 5G AI-based fingerprint localization across environments. In: 2022 IEEE 95th Vehicular Technology Conference: (VTC2022-Spring). IEEE; 2022. pp. 1-5

[101] Guo Z, Lin K, Chen X, Chit C-Y. Transfer learning for angle of arrivals estimation in massive MIMO system. In: 2022 IEEE/CIC International Conference on Communications in China (ICCC). IEEE; 2022. pp. 506-511

[102] Boora U, Wang X, Mao S. Robust massive MIMO localization using neural ODE in adversarial environments. In: ICC 2022 - IEEE International Conference on Communications. IEEE; 2022. pp. 4866-4871